MEAT CUTS AND MUSCLE FOODS - 2nd EDITION

D0881426

Meat Cuts and Muscle Foods - 2nd Edition

H.J. Swatland
University of Guelph
Ontario
Canada

NOTTINGHAM
University Press

AI/AU Tampa Library
4401 N. Himes Ave.
Tampa, FL 33614

Nottingham University Press
Manor Farm, Main Street, Thrumpton
Nottingham, NG11 0AX, United Kingdom

NOTTINGHAM

First published 2004
© Nottingham University Press 2004

All rights reserved. No part of this publication
may be reproduced in any material form
(including photocopying or storing in any
medium by electronic means and whether or not
transiently or incidentally to some other use of
this publication) without the written permission
of the copyright holder except in accordance with
the provisions of the Copyright, Designs and
Patents Act 1988. Applications for the copyright
holder's written permission to reproduce any part
of this publication should be addressed to the publishers.

British Library Cataloguing in Publication Data
Meat Cuts and Muscle Foods:
Swatland, H.J.

ISBN   1-904761-15-1

Disclaimer

Every reasonable effort has been made to ensure that the material in this book is true, correct, complete
and appropriate at the time of writing. Nevertheless, the publishers and authors do not accept responsibility
for any omission or error, or for any injury, damage, loss or financial consequences arising from the use
of the book.

Typeset by Nottingham University Press, Nottingham
Printed and bound by Hobbs the Printers, Hampshire, England

PREFACE

MEAT CUTS, MUSCLE FOODS AND FOOD MYOSYTEMS

Skeletal muscle in all its forms, from steak to scallop, is a vital component of the human diet and a commercial commodity of strategic importance. Although we have no single word in the English language for all the different types of muscle foods we consume, there are commercial and scientific reasons for combining them. In the English language, about the only suitable word food scientists have invented is "myosystem".

WHY INFORMATION IS NEEDED

Where once we had separate shops for meat, poultry and seafood, now we often have supermarkets offering a wide range of muscle foods under one roof, if not in the same refrigerated display case. Increasing amounts of product cutting and packaging are being done centrally. Many retailers no longer break meat carcasses into retail cuts, but now handle a much wider variety of products and need a more extensive knowledge of different types of meat cuts and muscle foods. Scientifically, many of the facts and theories relating to one type of muscle food may be applicable to another. This book provides access to the myriad of international meat cuts and muscle foods, using information collected from many sources - old and new, common and rare.

THREE TYPES OF INFORMATION

(1)   Meat cuts. Wholesale and retail meat cuts are described, reconciling two opposite trends. On one hand, in major markets such as those of Europe and North America, various industry-wide committees have standardized cutting and nomenclature. On the other hand, the maintenance of traditional or ethnic meat-cutting patterns is commercially advantageous for niche marketing. Meat cutting also has a rich linguistic history worthy of scholarly attention. I have followed the etymologies of a variety of different dictionaries listed in the references. In many cases, I have guessed how word roots and meat cuts might be related, and I look forward to seeing my guesses corrected by others at some future date. The objective is to define standardised patterns of meat cutting and ethnic variations in ways understandable to anyone who can read English and/or recognise a diagram of a skeleton.

(2)   Muscle foods. A wide diversity of sources is indexed, including beef, pork, lamb, veal, poultry, game and fish, plus a number of invertebrates, such as crabs, lobsters, shrimps, squid and scallop, producing edible myosystems dominated by striated muscle. Edible snails and the like are excluded because their myosystems make a minimal contribution to the properties of the whole food unit. Meat from a great variety of wild animals is consumed by hunters, and ranch production of game animals may help provide a sustainable meat supply in arid regions. Numerous examples of both categories are given, but a complete list would have turned the book into a zoology text. Dogs and monkeys are hunted and bartered as a source of meat in Africa (Anadu et al., 1988; Simoons, 1996), but their meat production has not yet been developed scientifically. Horse meat, on the other hand, has been studied scientifically and is traded internationally. Thus, for this book, dogs and monkeys are out, but horses are in. Meat I do not think we should be eating is excluded, such as meat from whales and endangered species not suitable for ranching. The objective is to

provide a ready reference to the main-stream muscle foods available commercially or being developed scientifically around the world.

(3) Food myosystems. The key scientific concepts required in understanding food myosystems are briefly outlined. The objective is to help explain the properties of different meat cuts and muscle foods in terms of meat quality.

DEALING WITH COMPLEXITY

Where there have been disagreements between sources, as in the gross anatomy of muscles or taxonomy of fishes, the simplest solution has been adopted. Although nomenclature and taxonomy are important for professional anatomists and zoologists, most food technologists and scientists simply require a reasonable frame of reference to define a particular food myosystem.

STANDARD FORMAT

Much of the valuable information gathered by my predecessors, such as Gerrard (1949) and Schön (1958), is difficult to standardise. I have attempted to transfer line drawings from various sources into a standard format with skeletal reference points, but the skeletal references are missing in many of the originals and some originals were for animals on the hoof, not for carcasses on the rail. Hindlimb suspension of a carcass causes major changes of muscle distributions, especially in the hindlimb. Thus, information drawn from sketches of carcasses of various shapes and sizes could not be standardised with complete reliability.

LEXICAL ORDER

Many languages have their own lexical order which determines where non-English letters of the alphabet or those with accents are placed when listed alphabetically. It is difficult to combine different lexical orders and, anyway, this book is structured primarily to lead from local names to English descriptions. Thus, the lexical order used here ignores accents and groups non-English letters with their nearest English equivalent. Transliteration was required for several languages. However, the book works equally well in the reverse direction - away from English. Start with the name of the country, then proceed to the non-English labels on the diagrams.

INDEX TO GEOGRAPHICAL REGIONS COVERED

Regions in bold are new, updated or expanded for this second edition.

→ **FURTHER READING**

The arrow → indicates other entries for further reading. But it was not needed to refer readers to individual mammalian muscles, all of which have their own entries.

ANATOMICAL TERMS

A number of anatomical terms are needed to describe the relative positions of structures within the body. Words such as front and back are of little use because of the drastic changes in posture and limb position occuring when a standing animal becomes a hanging carcass.

Anterior - towards the head
Posterior - towards the tail
Dorsal - towards the upper part or back of the standing animal
Ventral - towards the lower part or belly of the standing animal
Medial - towards the midline plane separating right and left sides of the body
Lateral - towards the sides of a standing animal
Proximal - towards the body in a limb of the animal
Distal - away from the body in a limb of the animal

These adjectives may be combined. For example, antero-dorsal means towards the head as well as towards the upper part or back of the standing animal. For an introduction to the bones of the carcass → skeleton.

MISTAKES AND OMISSIONS

The first edition of this book (published in 2000) was based on my collection of meat cutting patterns and muscle foods started in 1960 at Smithfield College, inspired by the fascinating collection gathered by Gerrard (1949). A second edition was justified by the first edition selling out and by new information gathered. The collection is still incomplete globally, with many countries and languages missing, although nearly all the major international traders are represented. Given the vast scope of the material, in over 30 different languages, there is no hope of complete multilingual cross-referencing of all meat cuts and fish. I apologise in advance for the omissions and errors, especially for those countries where I was unable to locate published information or a current local authority.

I have done my best to illustrate the book but I lack the skills of a professional illustrator. I quite agree with a reviewer of the first edition who found the diagrams lacked artistic merit.

Any errors becoming apparent after publications will be noted on my home page where I list my all too numerous errata in other publications, **http://www.aps.uoguelph.ca/~swatland/gasman.html** This also provides access to further information on the growth and structure of meat animals and to meat science in general.

THANKS TO THE PEOPLE WHO MADE THIS BOOK POSSIBLE

It is a great pleasure to acknowledge the personal communications and donations of technical information for the first edition I received from researchers and librarians around the world, in particular: S. Alexic, Ghaleb Alhadrami, Jan Andersen, Shai Barbut, S.A.S. Barnabas, Ole Braathen, the late Roberto Chizzolini, H.W. Cyril, Hector Delgado, Phil and Lauralee Edgell, S. Fadda, O. Fornias, Val Heeb, Oli Thor Hilmarsson, Louw Hoffman, Parichat Hongsprabhas, Karl Honikel, Masakazu Irie, Jennifer Janz, Seon Tea Joo, James Kuo, Facundo Laborde, R. Lahucky, Andrey Lissitsyn, Kerstin Lundström, Roberto de Oliveira Roça, M. Lourdes Pérez, Gabriel Monin, Edward Pospiech, Eero Puolanne,

Marku Raevuori, the late Derrick Rixson, Jorge Simões, M. Süth, D. Suwattana, Molly Swatland, Nelcindo Terra, Rose Toorop, K. Vareltzis, Eva Wiklund, Alfons Willam, and X. Yan. Thanks also to Ralston Lawrie for his early support of the idea for this book

The second edition was made possible largely by new contributions from S.S. Abiola, Basem Abdallah, Izaz Anjum, Noreddine Benkerroum, Ahmed Daoudi, Carmen Gallo, Louise Houston, and Ricardo Salazar.

For both editions, my special thanks to Bridget O'Brien for help with proof reading, to Sarah Keeling for production, and to editor Simon Robinson for advice and encouragement.

ABA DAS COSTELAS

A ladder of beef rib sections in → Portugal.

ABA DESCARREGADA

Abdominal muscles of beef in → Portugal.

A-BAND

The A-band is one of the → transverse striations along the → myofibre. It gets its name from being anisotropic - appearing bright with crossed polarisers under a light microscope. It is dominated by → myosin in thick → myofilaments.

ABASTERO

A wholesale butcher in Spanish, and a hindlimb beef cut in → Chile containing the *gastrocnemius*.

ABATS COMESTIBLES

French for edible offal.

ABBREVIATIONS

In North America, the following abbreviations are approved for use in retail labelling for meat cuts: barbecue, BAR B Q; bone in, BI; boneless, BNLS; double, DBLE; large, LGE; New York, NY or N.Y.; pork, PK; pot roast, POT-RST; round, RND; roast, RST; shoulder, SHLDR; square, SQ; steak, STK; and trimmed, TRMD.

ABDARI

The shoulder of a beef or pork carcass in → Korea.

ABDUCTOR POLLICIS LONGUS

A very small triangular muscle of the mammalian forelimb, inserted on the dorsal surface of the distal end of the radius. The muscle has pennate tendinous bands leading to a tendon which passes medially to insert on the third metacarpal in ruminants and on the second metacarpal in the pig.

ABONU

Yoruba for thin flank of a carcass in → Nigeria.

ACCLIMATISATION SOCIETIES

Societies to promote the introduction of "useful" species from one geographic locality to another. The definition of useful often includes an edible myosystem, but sometimes the useful species turns out to be a pest in its new environment, like the rabbit in Australia. The Société d'Acclimatation was formed in Paris in 1854 by the zoologist and embryologist, Isidore Geoffroy Saint-Hilaire.

A similar society was formed in London in 1860 by Frank Buckland (pioneer aquaculturist), with prompting from Robert Owen (comparative anatomist). Acclimatisation societies had a major impact on our present patterns of food production. With the conclusion of the colonial era, our current acclimatisation societies are mostly entrepreneurial and based on a single species, such as the introduction of meat-producing ratites to North America and Europe.

ACÉM

In → Brazil, a cut of beef located dorsally in the shoulder, medially to the paleta .

ACÉM COMPRIDO

Located in the shoulder, medial to the scapula in a beef carcass in→ Portugal.

ACÉM REDONDO

A postero-dorsal thoracic cut producing most of the rib roasts of a beef carcass in → Portugal.

ACHTERMUIS

A beef cut prepared from the semitendinosus in the → Netherlands, literally, after the main "ball" of hindlimb muscles.

ACHTERPOOT

The hind trotter of a pork carcass in the → Netherlands .

ACHTERSCHENKEL

The hindshank of a beef or veal carcass in the → Netherlands.

ACTIN

Actin is a globular protein (G-actin) which polymerizes to form a fibrous protein (F-actin). The thin → myofilament is formed from a double helix of two strands of F-actin, with the cross-over points about 35.5 nm apart (Figure 1). G-actin molecules have a diameter of about 5.5 nm. Each complete thin myofilament has a length of about 1 µm and contains 300 to 400 G-actins.

Figure 1. *A short part of the length of a thin myofilament formed from a double helix of actin molecules.*

The polymerization of G-actin to F-actin requires energy from ATP (→ adenosine triphosphate), and the resulting ADP is bound into the myofilament. Thin myofilament assembly by actin polymerization is controlled by proteins such as gelsolin, villin, and capZ binding either to the fast-growing or barbed end of the new myofilament, or to tropomodulin in association with tropomyosin binding at the slow-growing pointed end of the new myofilament. The overall length of the thin myofilament may be controlled by → nebulin.

G-actin has a site on its surface where it reacts with → myosin molecule heads during muscle contraction. When G-actin is polymerized to F-actin, these sites all face towards the H zone and away from the Z-line in which the thin myofilament is anchored. Thus, within an individual → sarcomere, the thin myofilaments on either side of the H zone are mirror images and are compatible with the arrangement of myosin molecules in the thick myofilament. If a thick myofilament penetrates the → Z-line, the immediate thin myofilaments in the next sarcomere are facing in the wrong direction to interact with the myosin molecule heads of the intruding thick myofilament.

(Sources: Weber *et al.*, 1994; Labeit and Kolmerer, 1995a)

α-ACTININ

A widespread cytoskeletal protein which attaches → actin filaments to cell membranes by binding with → vinculin. A large amount of α-actinin occurs in → Z-lines where it forms filaments that anchor the thin myofilaments.

(Source: McGregor *et al.*, 1994)

ADDUCTOR FEMORIS

A large muscle located medially in the mammalian hindlimb (Figure 2).

Figure 2. A lateral view of beef adductor muscle showing its ventral origin from the ischium and pubis (1) and insertion on the femur (2).

It is usually visible on a side of a commercial carcass, where it dominates the D-shaped area of muscle exposed just distal to the pubis. Thus, the *adductor femoris* is one of the few major hindlimb muscles from which data such as pH can be collected on an intact side. The main origins of the *adductor femoris* are on the ventral region of the ischium and pubis, and on the subpelvic ligament. It is inserted on the posterior face of the shaft of the femur. *Adductor femoris* is seen in round steaks and ham slices, located between the *pectineus* (which is anterior) and the *semimembranosus* (which is posterior) but, in lean animals, the separations between the *adductor femoris* and these other two muscles may be difficult to detect without probing.

ADENOSINE TRIPHOSPHATE

Adenosine triphosphate (ATP) stores energy. To add a third phosphate to a molecule which already has two phosphates (adenosine diphosphate, ADP) requires work to overcome the natural electrostatic repulsion between the phosphates. When the third phosphate is later released, the work can be recaptured to drive other events requiring energy, such as muscle → contraction. Thus, living muscle contains ATP and is capable of contraction. Whereas, when meat sets in → rigor mortis, the ATP is depleted, contraction is no longer possible, and thick and thin → myofilaments lock together in rigor mortis where they overlap.

(Source: Bendall, 1973)

ADIPOSE CELL

Adipose tissue (fat) is composed of globular cells tightly pressed together. Mature cells reach a diameter of 100 μm or more and are filled by a single large droplet of triglyceride. Thus, the nucleus and cytoplasm of an adipose cell are restricted to a thin layer under the plasma membrane, which accounts for the low water content of fat. Mature adipose cells have little cytoplasm and few organelles. The triglyceride filling most of the adipose cell is not directly bounded by a membrane, but is restrained by a cytoskeletal meshwork of 10-nm filaments, best seen in poultry. Adipose cells are kept in place by a meshwork of fine → reticular fibres. Large adipose depots are usually subdivided into layers or lobules by septa of fibrous connective tissue. Adipose tissue is well supplied with capillaries, which are normally emptied when animals are exsanguinated at slaughter. However, a residual trace of haemoglobin may give beef fat an amber tinge.

(Source: Allen *et al.*, 1976)

AFOR

Ibo for thin flank in → Nigeria.

AFTURPARTUR

The hindquarter of a carcass in → Iceland.

AFTURSTYKKI

The gluteal region of a carcass in → Iceland.

AGING

→ Conditioning

AGTERSKENKEL

The hindshank of a beef carcass in → South Africa.

AGUJA

A beef cut from the shoulder in → Spain and → Argentina. With aguja meaning a needle, the naming may relate to the narrowness of the anterior *longissimus thoracis* seen in entrecôte steaks?

AGUJAS

A thin section of beef flank in → Mexico (part of the pecho).

AIGUILLETTE

In beef cutting in → France, the aiguillette is centred around the *tensor fascia lata*. Its derivatives included the aiguillette de coeur and aiguillette baronne. The aiguillettes ferrée are rib cuts of beef, similar to the → plat-de-côtes.

AINE

Originating from the tranche grasse of a beef carcass in → Belgium.

AITCH BONE

In ancient times, two of the bones contributing to the pelvis, the pubis plus the ischium, were called the nache bone in Middle English. Thus, the nache or nage was the rump region. When used as two separate words, aitch bone probably refers to the bone but, when run together in England, the aitchbone is a well known cut of beef in → England, primarily in the Midlands and around London.

AKA

Ibo for the forequarter in → Nigeria.

ÅL

Danish for an eel, → Anguilliformes.

ALASKA KING CRAB

A long-legged deep-sea crab, *Paralithodes camtschatica*, → Crabs.

AL-ASSE

Brisket of beef in → Jordan.

ALCATRA

In → Brazil and → Portugal, a cut of beef, right across the carcass level with the sacrum.

AL DAHAR WA ADHLLA

In the → United Arab Emirates, the posterior thoracic region of camel.

AL-DAHER

A cut of veal containing the thoracic vertebrae in → Jordan.

AL-DELEA

Rib of veal in → Jordan.

AL-EKWEH

A cut of beef through the sacrum , ilium and pubis in→ Jordan.

ALETILLAS

A beef cut in → Chile containing the costal cartilages.

ALEWIFE

A bony fish, *Alosa*, of the herring family, → Clupeiformes.

AL FAKHED

Proximal hindlimb of camel or lamb in → Jordan.

AL-FAKHETH

A cut of veal from the anterior hindlimb in → Jordan.

AL GAIMAH

In the → United Arab Emirates, Al Gaimah Alkhalfia and Al Gaimah Alamamia are the distal extremities of hind and forelimb, respectively, of the camel .

AL GATAN

In the → United Arab Emirates, the lumbar region of camel.

AL-HABRA

A veal cut containing the femur in → Jordan.

ALIGÆS

Icelandic for a farm-reared → goose.

ALIÖND

Icelandic for a farm-reared → duck.

ALIPEGO

Beef *transversus abdominis* and *obliquus abdominis* in → Costa Rica.

AL-KATEF

Forelimb of a lamb or veal carcass in → Jordan.

AL-KHAD

Ischial cut of beef in → Jordan.

AL-KHASERAH

Flank of beef, veal or lamb in → Jordan.

ÁLL

Icelandic for → eel, → Anguilliformes.

ALLIGATOR

→ Crocodile

AL-LOHAH

A cut of beef through the humerus in → Jordan.

AL LUAH WA ALKATIF

Scapular region of camel in the → United Arab Emirates.

AL-MOZEH

Fore- or hindlimb shank of veal in → Jordan.

ALOYAU

In → Belgium , aloyau is a beef porterhouse steak. It may also be a pork cut from this region. A similar usage of the term occurs in Quebec. In → France, the aloyau is a large primal cut of beef including the following cuts: the romsteck (*gluteus* muscles), the aiguillette baronne (*tensor fascia lata*), the filet (*psoas major* and *minor*) plus the faux-filet (contre-filet or *longissimus dorsi*).

ALPACA

The alpaca, *Lama pacos*, is famous for its long, fine wool. The meat is consumed domestically when only a few animals are kept, but the surplus from larger ranches may be used for → charqui.

(Source: Calle Escobar, 1984)

AL-RAKABEH

Neck of lamb in → Jordan.

AL REGABA

Camel neck in the → United Arab Emirates.

AL-SADER

Sternal region of beef, veal or lamb in → Jordan.

AL SADR

Sternal region of camel in the → United Arab Emirates.

AL-SHAIKEH

Posterior hindlimb cut of beef in →Jordan.

AL SHAKELA

Posterior abdominal region of camel in the → United Arab Emirates.

AL-THAHER

Vertebral column of a lamb carcass in → Jordan.

AL-WERK

A veal cut from the posterior hindlimb in → Jordan.

AL-ZEND

Hindlimb cannon bone and foot of a lamb carcass in → Jordan.

AMORPHIN

An amorphous protein in the spaces between → α-actinin filaments of the → Z-line.

(Source: Chowrashi and Pepe, 1982)

ANCHOA

Anchovy in Spanish

ANCHOVY

A small herring-like bony fish, *Engraulis*, → Clupeiformes.

ANCONEUS

The *anconeus* (*anconaeus*) is a small muscle of the mammalian forelimb, originating on the distal, posterior surface of the humerus and inserted onto the olecranon process. It covers the olecranon fossa of the humerus and may be difficult to separate from the lateral head of the *triceps* in a lean carcass, especially in lamb. The *anconeus* is divided into two parts in the pork carcass.

ANDROSTENONE

→ Boar taint

ANGEL SHARK

Squatina, also called the monkfish, has a body shape intermediate between the sharks and rays (→ Chondrichthyes).

ANGRA

Neck of beef, lamb or camel in → Morocco.

ANGUILLIFORMES

The eel family has only one genus, *Anguilla* (Figure 3). The freshwater eels have small cycloid scales in their skin and their teeth are small. They are catadromous (living in freshwater and travelling out to sea to spawn). Both the American (*A. rostrata*) and the larger European eel (*A. anguilla*) breed near the Sargasso Sea and the leptocephalus larvae take several years to return to the rivers from which their parents originated. The immature eels returning to freshwater are thin, with a leaf-like shape, and are called elvers. In their freshwater habitat, eels have a greenish colour. Eels migrating to the Sargasso Sea to mate are silver coloured and their digestive system is atrophic. The myosystem of the adult eel is firm and white with a distinctive sweet flavour, and is often used for jellied or marinated products. Large silver eels with a high fat content produce a better smoked product than smaller yellow eels.

(Source: Jessop, 1993)

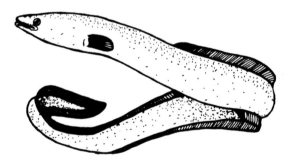

Figure 3. The eel, Anguilla.

ANITRA

Italian for → duck.

ANKA

Swedish for → duck.

ANOA

The anoa is a small bovine found in Sulawesi. The lowland and mountain anoas (*Bubalus depressicornis* and *B.*

quarlesi, respectively) are related to water buffalo but are not much larger than goats. They are hunted for their meat.

(Source: NRC, 1983)

ANSIM

In → Korea, the fillet of a beef or pork carcass.

ANTRYKOT

In → Poland, this is an → entrecôtes, a cut from the posterior thoracic region in beef or lamb.

APA

Yoruba for the forequarter in → Nigeria.

APATA UKWU

Ibo for the hindquarter in → Nigeria.

APODEME

In → Crustaceans, an apodeme is an internal projection of the exoskeleton to a muscle, like an ossified tendon. Removal of apodemes improves texture.

ARAIGNÉE

In beef cutting in → France, this is part of the tende de tranche in the medial part of the hindlimb.

ARCTIC CHARR

Regardless of spelling, char or charr, *Salvelinus alpinus*, is a freshwater or sea-running circumpolar salmonid, but now also farmed at lower latitudes, → Salmoniformes.

ARGENTINA - BEEF CUTS

Two of the beef cutting patterns used in Argentina are shown in Figures 4 and 5, the former more suited to a band saw and the latter to cutting by hand. Both take a box-like cuadril including the whole of the pelvis and a loin-ball or bola de lomo based on the *quadriceps femoris* muscles. The numerous infoldings of connective tissue in the *gastrocnemius* may have given rise to the name tortuguita (full of winding bends), but the other names are more difficult to solve linguistically. The terminology for narrow and wide beef-steaks (bifes angostos and anchos, respectively) makes sense anatomically but, relative to Spanish beef cutting, it is unusual to see the lomo restricted to the psoas muscles. The aguja may derive its name from the needle-like shape of the sections of anterior *longissimus thoracis* appearing as entrecôtes steaks.

The cut through the forelimb ventral to the scapula may be straight to give a rectilinear cut of meat destined for the cooking pot (puchero) or may be curved, following

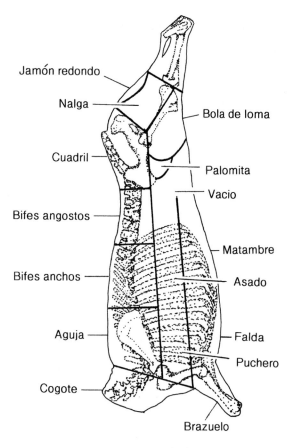

Figure 4. Argentina - beef cuts.

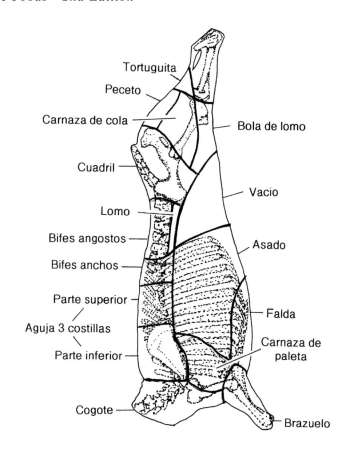

Figure 5. Argentina - beef cuts.

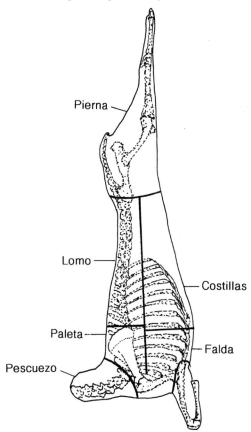

Figure 6. Argentina - lamb cuts.

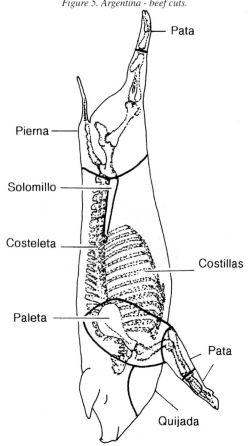

Figure 7. Argentina - pork cuts.

the *triceps* to produce shoulder meat (carnaza de paleta). The Argentinean falda is the brisket, mainly in the sternal region, unlike in → Spain where the falda (skirt) is more understandably abdominal in location. In Argentina, the abdominal muscle is simply the void or vacio. By name, the asado ribs clearly are destined for roasting. The neck and arm are the cogote and brazuelo, respectively.

(Sources: Fadda, 1999; Laborde, 1999)

ARGENTINA - LAMB CUTS

The paleta is a square-cut shoulder. The breast or falda includes the ventral rib region (Figure 6).

(Source: Fadda, 1999)

ARGENTINA - PORK CUTS

The main features of pork cutting in Argentina (Figure 7) are the pierna or ham (which becomes jamón when cured), the chops (costeleta), ribs (costillas) and shoulder (paleta). The solomillo is the *psoas* muscles plus *iliacus*. The jowl is called the quijada or jaw.

(Sources: Fadda, 1999)

ARMBONE CHUCK

In → North American beef cutting, an armbone chuck is the chuck plus the attached shank.

ARM CHUCK ROAST

In → North American beef cutting , this is an arm steak or pot roast (Figure 8).

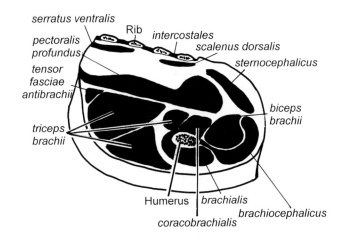

Figure 8. North American beef arm pot roast from the chuck.

ARM STEAKS OR POT ROAST OF BEEF

In → North American beef cutting , arm steaks (when cut thinly), arm pot roasts (when cut thickly) or cross cut ribs

(when subdivided) are cut from the primal chuck, perpendicularly to the humerus (Figure 8).

ARTICULARIS GENU

A small muscle in the mammalian hindlimb, originating proximally to the trochlear surface of the femur and inserted onto the patella and adjacent ligaments and cartilages.

ASADO

Roasting beef ribs in → Argentina. In → Chile the asado del carnico , literally meat roast, is the *subscapularis* while the asado de tira, literally strip roast, is taken through the ventral rib region.

ASIENTO

Gluteal muscles in → Chilean beef cutting.

ASTAXANTHIN

The colour of muscles from salmonid fish (→ Salmoniformes) is important for consumer acceptance, where a high degree of pigmentation is attractive. Like other vertebrates, fish are unable to synthesize carotenoids and must acquire them from their diet. The pinkness of skin and musculature in wild salmon originates from the ingestion of crustacean pigments, primarily astaxanthin, and sets a standard by which farmed salmonid products are judged. As a dietary supplement for farmed fish, however, carotenoid pigments are relatively expensive.

(Source: Storebakken *et al.*, 1987)

ATP

→ Adenosine triphosphate.

AUDE

In → France, this is an alternative name for the flanchet in the flank region of the beef carcass.

AUERHAHN

German for → grouse.

AUQUENIDOS

Auquenidos are South American camelids. The → llama and → alpaca are domesticated. The vicuña (*Vicugna vicugna*) is wild. It is captured, sheared for its fine wool, then released. The wild guanaco (*Lama guanico*) is the largest of the auquenidos.

AUSTRALIA - BEEF CUTS

Beef cutting in Australia (Figures 9 and 10) shares some similarities to that in → New Zealand, although forequarters are separated from hindquarters at a different position. Beef cutting in Australia follows an English pattern, and the hindquarter is separated from the forequarter between ribs 10 and 11 to leave three ribs on the hindquarter. As in → North America, however, the sirloin is taken level with the ilium, although the Australian sirloin butt is quite curved and may be divided into a top sirloin butt (dorsal, containing the ilium) and bottom sirloin butt (ventral). The gluteal muscle group may be removed as rostbiff, leaving the *tensor fascia lata* as bottom sirloin triangle. In Australia, sirloin butt (AMLC, 1996) also may be called a rump (AMLC, 1979). Also similar to North American terminology, the lumbar *longissimus dorsi* is called a striploin. The muscles around the femur are divided into (1) an anterior knuckle cut containing the *quadriceps femoris* group of muscles, (2) a medial inside cut (*semimembranosus, adductor* and *pectineus*), and (3) a lateral outside cut (*biceps femoris* and *semitendinosus*). The outside may be subdivided into eye round (*semitendinosus*) and outside flat (*biceps femoris*).

Obliquus internus abdominis and *obliquus externus abdominis* are used for flap meat and flank plate. From ribs 4 to 13, a cube roll contains *longissimus dorsi* and adjacent muscles (*spinalis, semispinalis* and *multifidus dorsi*). The chuck roll is anterior to the cube roll, and contains the same muscles, plus parts of *iliocostalis, serratus dorsalis cranialis, rhomboideus, subscapularis, teres major* and *serratus ventralis*. A chuck roll with overlying muscles (*trapezius, infraspinatus* and *latissimus dorsi*) is called a spencer roll. The rib blade meat contains *latissimus dorsi, trapezius* and *serratus ventralis* from ribs 6 to 8. The current Australian designation of the clod within the primal chuck meat is rather unusual because it is centred on the scapula (not the humerus or clod bone) and is dominated by muscles such as *infraspinatus, deltoideus, teres minor* and *triceps brachii*. It looks as if the Australian clod is a contraction of a larger shoulder clod containing both scapula and humerus, such as that found in → New Zealand. For the brisket point end deckle off, the wider definition of the → deckle as fat and connective tissues between the ribs and the pectoralis muscles of a beef brisket is used (not a *latissimus dorsi* or *trapezius* deckle). In Australia, the deckle is removed from the cut.

(Source: AMLC, 1996)

AUSTRALIA - LAMB CUTS

The eight-way cutting pattern for Australian lamb (Figure 11) is rectilinear and provides six cuts for fresh-meat air-freight to North America (excluding the cervical region and postero-ventral rib and flank). The chump (ilium) remains on the leg, but the leg may be broken into a sirloin or chump, and into inside, outside and knuckle cuts following the pattern used for Australian beef . The eye (*longissimus dorsi*) may be isolated from the shortloin. An already → frenched rack is available as an export cut.

(Source: AMLC, 1991).

AUSTRIA - BEEF CUTS

In Austria (Figure 12), the primal cuts of beef (Rindfleisch) are the Englischer (posterior thoracic and lumbar vertebrae), Gustostücke vom Knöpfel (proximal hindlimb), Gustostücke vom Vorderviertel (proximal forelimb), Hinteres (anterior ribcage), Gulaschfleisch (neck and distal limb) and Vorderes (ventral neck and abdominal muscles). The Gustostücke vom Knöpfel is divided into Weisses Scherzel (white cut or *semitendinosus*), Tafelstück (table cut or *biceps femoris*), Schale (mostly *semimembranosus, adductor* and *pectineus*), Nuss (*quadriceps femoris* muscles), Tafelspitz (table point around the sacrum), Hüferscherzel (gluteal muscles) and Hüferschwanzel (*tensor fascia lata*).

The loin is subdivided into the Lungenbraten (*psoas* muscles and *iliacus*), Beiried (lumbar) and Rostbraten (posterior thoracic). The shoulder is lifted from the ribcage and divided into Dicke Schulter (thick shoulder or *triceps brachii*), Schulterscherzel (shoulder cut or *infraspinatus*), Mageres Meisel (*supraspinatus*) and Kavalierspitz (*subscapularis*). The remaining rib cuts are the Hinteres Ausgelöstes (dorsal) and Dicker Spitz (rib midlength). The brisket is divided into the Brustkern (anterior) and Mittleres- und Dünnes Kügerl (posterior). The Platte is the flank, and the Hals is the neck. The tail is the Schlepp.

(Source: Willam, 1999)

AUSTRIA - LAMB CUTS

For lamb (Lammfleisch), the leg or Schlögel is divided into the anterior Nuss, medial Schale, lateral Fricandeau, distal Hintere Stelze and ilial Schlussbraten (Figure 13). After removal of the Rücken and Filet, the shoulder or Schulter is lifted from the ribcage to leave the neck (Hals), anterior breast (Brust) and belly meat (Bauchfleisch).

(Source: Willam, 1999)

AUSTRIA - PORK CUTS

The primal cuts of pork (Schweinefleisch; Figure 14) are the Karree (thoracic and lumbar vertebrae), Schlögel (hindlimb), Schulter (forelimb) and Bauch (sternal and abdominal muscles). The Karree contains the filet (*psoas* muscles and *iliacus*), kurz and lang Karree (short and long, lumbar and posterior thoracic vertebrae, respectively) and Shopfbraten (anterior thoracic vertebrae). The Schlögel is divided into the Kaiserteil or Schale (medial), the Fricandeau (lateral) and the Nuss (anterior). The shoulder is lifted from the ribcage and divided into Dünne and Dicke Schulter (thin and thick, dorsal and ventral to the spine of the scapula, respectively). The Brust (breast) is separated from the Bauchfleisch (belly meat) level with rib 4.

(Source: Willam, 1999)

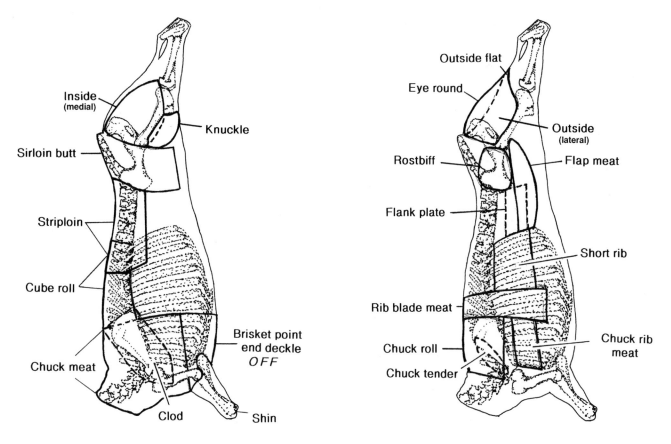

Figure 9. Labels: Inside (medial), Sirloin butt, Striploin, Cube roll, Chuck meat, Knuckle, Brisket point end deckle OFF, Clod, Shin

Figure 9. Australia - beef cuts.

Figure 10. Labels: Outside flat, Eye round, Rostbiff, Flank plate, Rib blade meat, Chuck roll, Chuck tender, Outside (lateral), Flap meat, Short rib, Chuck rib meat

Figure 10. Australia - beef cuts.

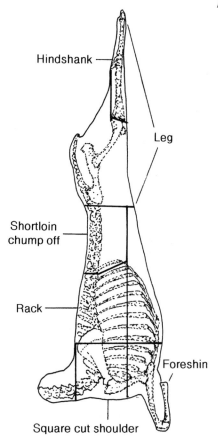

Labels: Hindshank, Leg, Shortloin chump off, Rack, Foreshin, Square cut shoulder

Figure 11. Australia - lamb cuts.

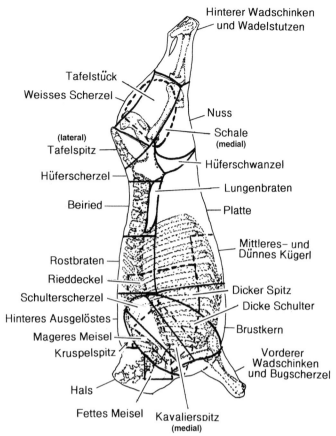

Hinterer Wadschinken und Wadelstutzen
Tafelstück
Weisses Scherzel
Nuss
Schale (medial)
(lateral)
Tafelspitz
Hüferschwanzel
Hüferscherzel
Lungenbraten
Beiried
Platte
Mittleres- und Dünnes Kügerl
Rostbraten
Rieddeckel
Schulterscherzel
Dicker Spitz
Dicke Schulter
Hinteres Ausgelöstes
Mageres Meisel
Brustkern
Kruspelspitz
Vorderer Wadschinken und Bugscherzel
Hals
Fettes Meisel
Kavalierspitz (medial)

Figure 12. Austria - beef cuts.

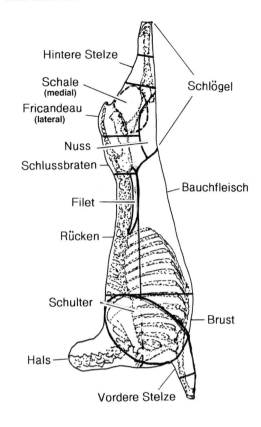

Hintere Stelze
Schale (medial)
Schlögel
Fricandeau (lateral)
Nuss
Schlussbraten
Bauchfleisch
Filet
Rücken
Schulter
Brust
Hals
Vordere Stelze

Figure 13. Austria - lamb cuts.

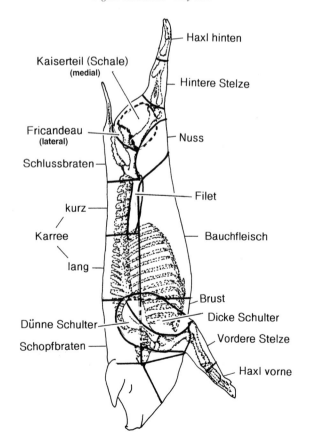

Haxl hinten
Kaiserteil (Schale) (medial)
Hintere Stelze
Nuss
Fricandeau (lateral)
Schlussbraten
Filet
kurz
Karree
Bauchfleisch
lang
Brust
Dünne Schulter
Dicke Schulter
Schopfbraten
Vordere Stelze
Haxl vorne

Figure 14. Austria - pork cuts.

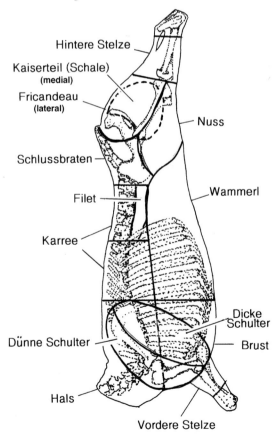

Hintere Stelze
Kaiserteil (Schale) (medial)
Fricandeau (lateral)
Nuss
Schlussbraten
Filet
Wammerl
Karree
Dicke Schulter
Dünne Schulter
Brust
Hals
Vordere Stelze

Figure 15. Austria - veal cuts.

AUSTRIA - VEAL CUTS

Veal cutting (Figure 15) is a simplification of beef cutting, but using a mixture of pork and lamb names. The only name not mentioned above is the Wammerl instead of Platte or Bauchfleisch.

(Source: Willam, 1999)

AVANT COEUR

Another name for the côte au spirling from the anterior rib region of a pork carcass in → Belgium.

BABILLA

In →Spain, a cut anterior to the femur in a beef carcass.

BABIRUSA

The babirusa (*Babyrussa babyrussa*) would be the ideal animal to slip into a pig slaughter line on April fool's day. It looks like a tusked pig but has a four-chambered stomach like a sheep. In Sulawesi it is hunted for its meat.

(Source: NRC, 1983)

BACKEN

In → Germany, the Bake or Backen is the pork jowl

BACK OF BEEF

In → North American beef cutting, a back is the chuck plus rib.

BACK RIBS OF BEEF

In beef cutting in → England, anterior to a 4-rib forerib is a 4-rib middle ribs cut. The back ribs cut is the dorsal part of the middle ribs cut, and contains some of the scapula.

BACKRIBS OF PORK

In → North America, backribs of pork originate from the primal loin, and are composed of ribs plus intercostal muscles from the widest part of the rib cage.

BACK STRAP

Parts of the → ligamentum nuchae.

BADARD

In → France, the beef badard is composed of the tende de tranche plus tranche grasse.

BAGSKANK

In → Denmark, a bagskank is a hindshank of beef or pork.

BAGTÅ

Rear trotter of a pork carcass in → Denmark.

BAIXOS

Breast of lamb in → Brazil.

BAJOUE

The jowl of a pork carcass in Quebec or Belgium.

BAK-KNOKE

In → Norway, the bak-knoke may be the hindshank (tibia) of a beef or veal carcass or part of the hindshank (proximal tibia and fibula) or a pork carcass.

BAKLÄGG

The hindshank of a beef carcass in → Sweden.

BALI CATTLE

→ Banteng

BALLERON

A large emulsion → sausage flavoured with pistachio nuts and smoked beef tongue.

BANGERS

The slang name for full-sized (about 15 x 4 cm) sausages in England. English sausages traditionally contain filler such as damp rusk (bread crumbs) and generate a fair amount of steam when cooked. When packed in a natural → casing, if the cook forgets to "fork" them, the bangers honestly do go bang and turn themselves inside out.

BANKEKJØTT

The lateral part of the beef round in → Norway.

BANTENG

The banteng is *Bos javanicus*, a bovine of Southeast Asia. Bantengs are sometimes called bali cattle when completely domesticated, but they are not really cattle if that term is reserved for *Bos taurus* or *B. indicus*. Banteng-cattle hybrids have elongated muscles in the hindlimb, and heavy muscling over the back, loin and shoulder. Banteng bulls may grow to 1.5 m at the shoulder and reach a live weight of 500 kg. They have tight skin with short hair, a short neck, an inconspicuous dewlap and a high dressing percentage. Males have a well developed thoracic crest, but it is not a hump. Horns are more strongly developed in bulls, where they grow outward and upward from a horny boss over the frontal bones. Banteng meat is like high quality beef. It is tender and lean, with little trace of intramuscular fat. Two thirds of body fat is mesenteric, with the remainder subcutaneous. Banteng meat is exported from Indonesia to Hong Kong and Japan.

(Source: NRC, 1983)

BÁRÁNY

Bárány is Hungarian for lamb, and is used as the prefix for a range of lamb cuts in → Hunagary, such as báránycomb (leg), báránygerinc (saddle), báránylapocka (shoulder) and bárány eleje (short forequarter with breast).

BARBECUE

From the Spanish word barbacoa, a barbecue may be a frame on which to sleep or cook, or an area on which to dry cocoa beans. Thus, any meat that is tender enough to be cooked without immersion in fluid can be barbecued. In North America, barbecue ribs typically are backribs, that is, ribs plus intercostal muscles from the region between rib chops and spareribs.

BARDIÈRE

In → France, pork bardière is the back-fat removed as a sheet for wrapping (barde) other rolled joints of meat.

BARON

A baron of beef is generally a very large roasting joint suitable for a carvery or at a banquet. To create a baron of beef in England, the beef carcass is not split left from right sides so that a full baron of beef is composed of left and right pairs of rumps, loins and a couple of ribs. Smaller versions include a short baron (pairs of rumps and loins) and double sirloins. In Canada, a baron (baron de ronde) is a round of beef (*biceps femoris, semitendinosus, semimembranosus, adductor* and *pectineus*), and is boned and rolled as a large carvery joint. In → France, both sides of lamb selle and côtelettes filet are called a baron, like a miniature baron of beef.

BARRACUDA

There are many types of barracuda (order Sphyraeniformes) around the world. In the Caribbean, the great barracuda, is *Sphyraena barracuda*. I have no idea why it has been reported as poisonous because the ones I have eaten were very good.

(Sources: Pivnička and Černý, 1987; Randall, 1996)

BARRAMUNDI

Lates calcarifer is a large fresh-water fish reaching 50 or 60 kg, a member of the → Perciformes. Although widely dispersed from the Persian Gulf to China, it is particularly important in Australia where it is both farmed and caught wild. Like other perch-like fish its first dorsal fin has a number of strong spines.

BARRIGA

In → Brazil, a belly of pork or veal.

BARROW

A male pig castrated at a relatively early age.

BARTAVELLE

French for the red-legged → partridge, *Alectoris rufa*.

BAS DE CROUPE

Bottom round rump in Quebec.

BAS DE SURLONGE DÉSOSSÉ

Bottom sirloin butt in Quebec.

BASEMENT MEMBRANE

The basement membrane is a viscous layer around → myofibres, immediately on the surface of the plasma membrane. It contains fibronectin, a glycoprotein with a high molecular weight and an affinity for → collagen. Although basement membranes may appear amorphous, many are now thought to be composed of a network of irregular cords. The cords contain an axial filament of Type IV collagen, plus ribbons of heparin sulphate proteoglycan and fluffy material such as laminin, entactin, and fibronectin.

(Source: Li *et al.*, 1997)

BASS

A perch of the family Serranidae, → Perciformes.

BASSE-CÔTE

Equivalent to the côte découverte of a beef carcass in → Belgium.

BASSE CUISSE

Equivalent to the tranche grasse of a beef carcass in → Belgium.

BASTERMA

Salted meat in Egypt.

(Source: Elkhateib, 1997)

BATTI

Dorsal abdominal muscles of beef, veal or camel in → Pakistan.

BAUCH

Belly of pork in → Germany.

BAUCHFLEISCHE

Belly meat on a lamb or pork carcass in → Austria.

BAVETTE

Equivalent to the flanc of a beef carcass in → Belgium.

BAVETTE D'ALOYAU

In → France, the beef bavette d'aloyau is a cut composed of abdominal muscle. It is part of the primal panneau and includes *obliquus internus abdominis* and *obliquus externus abdominis*. The same term may be used in → Northern Ireland.

BEAR

A hundred years ago, black bear (*Ursus americanus*) was

available as seasonal fare in Canadian butchers' shops but is now only available to hunters. The meat is coarse and dark, and carcasses taken prior to hibernation are extremely fat. Polar bear, *Ursus maritimus*, is a staple food of Inuit hunters and, as everyone knows, polar bear liver is poisonous because of its high vitamin A content.

BECCACCIA

Italian for → woodcock.

BECCACCINO

Italian for → snipe.

BÉCASSE

French for → woodcock.

BÉCASSINE

French for → snipe.

BED

In the Midlands of → England, the bed is a cut of beef anterior to the femur.

BEEF

Species

In the developed countries, beef is mostly produced from cattle of the genus *Bos*. However, there are several other genera which will interbreed with *Bos* to produce fertile, beefy offspring: *Bison*; *Poephagus*, the yak; *Bibos*, the gaurs; *Bubalus*, the Indian buffalo; *Sunda*; and *Syncerus*, the African buffalo. Several different species of *Bos* are recognized such as *B. nomadicus* and *B. indicus*. *B. indicus* may have been derived from *B. nomadicus* and is now recognized by the following features: a prominent shoulder hump of muscle supported by dorsal spines of the vertebrae, a long face with drooping ears, upright horns, small brow ridges, a prominent dewlap, slender legs, and uniform colouration (white, grey or black).

In North America, cattle were introduced by Spanish settlers to the Southwest USA, giving rise to breeds such as the Texas Longhorn. But in the north, the cattle were mainly derived from primitive breeds brought by French and British settlers. *Bos primigenius* genes were probably carried by the Spanish cattle while the primitive northern breeds were probably a mixture of *B. primigenius* and *B. longifrons*. In early settlements, the primary importance of cattle was their ability to pull a plough or a cart, and they were not normally slaughtered until the end of their working life. Improved British breeds of cattle developed in the late 1700s and 1800s were imported into North America to form the basic Shorthorn, Angus, Hereford stock. Between 1905 and 1920, *Bos indicus* (Brahman cattle from India) were introduced into the southern USA for their heat tolerance. The most recent phase of beef

breed development in North America has been a re-introduction of Continental European beef breeds with rapid early growth and a large mature frame size - animals preserved as draft animals where steam engines were scarce. The large size of some of these breeds suggests that they may contain genes derived from *Bos primigenius*. There is renewed interest in early maturing breeds, such as Angus, because of their ability to produce fine-grained, tender beef. Whether or not these features will survive the current selection practices in favour of rapid early growth and a large mature frame remains to be seen.

(Sources: Zeuner, 1963; Ucko and Dimbleby, 1969)

BEEFALO

A hybrid of 3/8 American → bison and 5/8 domestic cattle. Beefalo are ranched in the USA, where they forage with grain supplementation. Beefalo meat is retailed in local markets, but reaches a larger customer base by mail order. The meat is very lean and similar to grass-fed beef in taste.

BEEF JERKY

Salted, dried beef.

BEERWURST

A German salami → sausage in a large diameter (around 7 to 8 cm) natural → casing, typically with some venation (real or artificial).

BEIKON

Icelandic for bacon.

BEINFLEISCH

In → German beef cutting, the foreshank may be called the legmeat or Beinfleisch.

BEIRIED

The lumbar region of a beef carcass in → Austria.

BELGIUM - BEEF CUTS

At the crossroads of Europe, with Dutch-speaking Flemings in the north and French-speaking Walloons in the south, beef cutting and nomenclature in Belgium are complex. Figure 16 provides some ideas, but is far from complete or reliable. The main hindquarter cut is centred around a topside, labelled the pièce levee (also called or containing the haie, tache noire, crosse cuisse, noix plannure and pièce à l'abattu). Located antero-distally is the tranche grasse (containing grosse tête, aine, tulipa, basse cuisse, or pièce ronde, or boule d'aloyau). The filet d'Anvers or plein filet is located posteriorly. The hindshank or jarret is quite short. It looks as if the rognon or graisse de rognon (fat) is removed dorsally from the flanc (containing flanchet, flanchis, hampe, hap, paillasse or bavette). The last two lumbar vertebrae start the aloyau

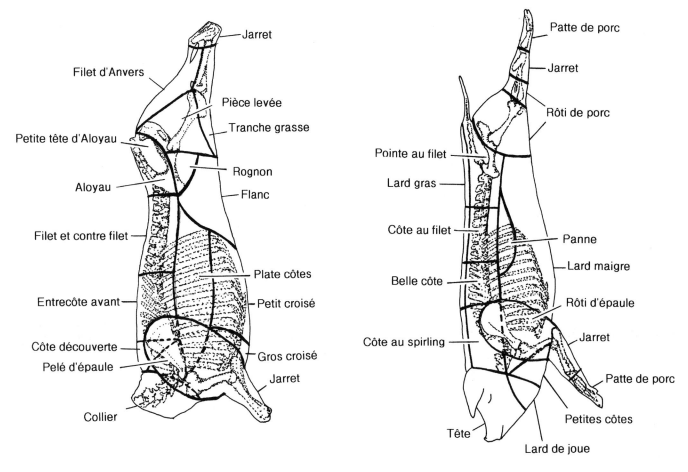

Figure 16. *Belgium - beef cuts.*

Figure 17. *Belgium - pork cuts.*

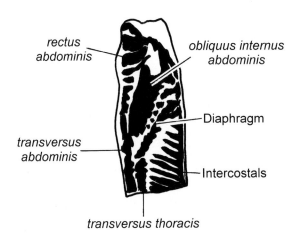

Figure 18. *North American belly of pork.*

(phalan) which is completed posteriorly by the petite tête d'aloyau (containing tête de phalan, tête de noix, pointe d'aloyau or câwi).

In the forequarter, the steaks from the anterior thoracic through to the lumbar region are, respectively, côte découverte (basse-côte, côte d'échine, spirling, langue de boeuf or trois côtes), entrecôte avant (tranche à la côte, côte couverte, premières côtes or belles côtes), and filet et contre filet. What happens to the shoulder is unclear from the available documentation. The names pelé d'épaule or pelé de platine suggest some muscles at least are peeled from the scapula. Medial to the scapula is a ladder of rib sections called the plate côtes or raccourci de côtes. From anterior to posterior, respectively, the sternal region is separated into the gros croisé (brosse, grosse croisure or bouilli d'épaule) and petit croisé (petite croisure or pointe de poitrine). The foreshank is the jarret. The neck may be called collier, collet or charbonnet.

(Source: OEEC, 1961)

BELGIUM - PORK CUTS

The hindlimb is split into a two-part leg roast (Figure 17), which may be called rôti de porc, rôti jambon or rôti à la tache noire, plus a shank (jarret) and trotter (patte de porc). The forelimb is lifted from the ribcage as a shoulder roast which may be called rôti d'épaule, rouelle de porc or roulé de porc. Various levels of fat cuts are stripped from the carcass: back-fat (lard gras or lard de dos), kidney fat (panne, paine, or graisse de rognon) and jowl (lard de joue or bajoue). The sequence of chops through the vertebral column is, from anterior to posterior: (1) the côte au spirling, spirling, côte à l'oriette, côte au juif or avant coeur; (2) the belle côte, moyenne côte or côte; (3) the côte au filet, filet or côte à la noix; and (4) the pointe au filet, aloyau, noix de porc, chely de porc or fricandeau. The belly of pork may be called lard maigre, petit salé, entrelardé, lard de poitrine or lard molé. The petites côtes is a cut from the anterior part of the sternum.

(Source: OEEC, 1961)

BELLES CÔTES

Rib of beef, entrecôte avant, in → Belgium. The same term is used in Belgium for a pork cut at the thoraco-lumbar junction.

BELLY OF PORK

In → North American pork cutting, the belly is separated from the loin with a cut following the curvature of the vertebral column (Figure 18). One end of the curve is just ventral to the ilium, the other end is just ventral to the blade of the scapula. The ribs and their immediately adjacent muscles are separated from the belly as spareribs. The remaining muscles of the abdomen, together with those that insert on the ribcage, constitute the side of pork, which may be cured and smoked to make slab bacon. However, belly of pork is a generic term common to many English-speaking countries, and its outline and uses vary considerably.

BELMONT STEAK OR ROAST

In → North America, retail cuts of beef taken through the scapula of the primal chuck.

BELONIFORMES

In this order of teleost fish, the most important food fish is in the family Scomberesocidae, the Atlantic saury, *Scomberesox saurus* (Figure 19). Large schools of fish may occur, often with tunny or mackerels in pursuit. The Atlantic saury has limited commercial importance, although small quantities are canned on a seasonal basis. The body is green-blue dorsally, and the belly is silver. The overall shape of the saury is elongated with a long beak. The length reaches about 40 cm. The flesh is oily, but with a pleasant taste.

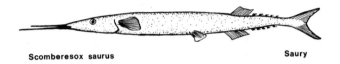

Scomberesox saurus　　　　　　　　　**Saury**

Figure 19. The saury, Scomberesox saurus.

(Source: Scott and Messieh, 1976)

BÉLSZÍN

A posterior thoracic rib cut of beef in → Hungary.

BEQUET

A French term for the thick muscular part of the diaphragm.

BERGIE FILET

In South Africa, Bergie filet is the *iliotibialis* muscle of the → ostrich.

BEST END NECK

In → English lamb cutting, the best end neck contains posterior thoracic vertebrae and the dorsal parts of their ribs. Best end neck also may be used for a cut of → New Zealand lamb, close to the rack. Best neck is an English cut of veal from the posterior thoracic region.

BIBRINGE

The postero-ventral region of the beef ribcage in → Norway.

BICEPS BRACHII

A medium-sized muscle of the mammalian forelimb with an appreciable connective tissue content. It occupies a position similar to that of the *biceps* in the human arm but, in meat animals, some of the neck muscles (such as *brachiocephalicus*) overlap the top of the forelimb so that the *biceps brachii* is covered and not exposed , as in the human arm. The origin of the *biceps brachii* is on the tuber scapulae (next to the socket joint on the scapula) and most of its insertion is on the proximal part of the radius and ulna.

BICEPS FEMORIS

A very large muscle (> 7% total muscle mass) on the lateral surface of the mammalian hindlimb. Its origins are extensive and include the dorsal spine of the sacrum, the sacroiliac and sacrosciatic ligaments, and the lateral surface of the ischium, near the acetabulum. Its insertions include the fascia lata, lateral patellar ligament, and anterior tibial crest. This massive muscle is seen all through the beef round and pork ham. It has a deep fissure so that the *biceps femoris* may look like two separate muscles when it is seen in transverse section. In pork, the main muscle and its accessory lobe often differ in degree of myoglobin redness. The *biceps femoris* in meat animals includes part of the equivalent muscle to the human *gluteus superficialis.*

BIFES

Beefsteaks in → Argentina, with narrow steaks (bifes angostos) in the lumbar region and wide steaks (bifes anchos) in the posterior rib region.

BIFTEK

French for beef steak in general, but it may refer specifically to the *obliquus internus abdominis* of the fausse hampe (false diaphragm).

BIFTEK D'ALOYAU

In Quebec, this is a beef T-bone steak. Biftek de coquille d'aloyau is bone-in strip-loin steak (transverse processes of lumbar vertebrae plus *longissimus dorsi* and associated muscles).

BIFTEK DE CONTRE FILET

Steaks of beef *longissimus dorsi* and adjacent muscles from the lumbar region. Contre filet is a generic term now widely used in many countries.

BIFTEK DE FAUX-FILET

Steaks of beef *longissimus dorsi* and adjacent muscles from the posterior thoracic region.

BINNEBOUD

A medial cut from the round of a → South African beef carcass containing *semimembranosus, adductor* and *pectineus.*

BIODRÓWKA

Posterior lumbar chops from a pork carcass in → Poland.

BISON

The bison (*Bison bison*), known as the buffalo in American folklore, was nearly hunted to extinction but now is protected and ranched. Ranch bison is sold commercially. The traditional prime parts were the tongue and the hump, but steaks and burgers are now the main commodity. Bulls can reach a massive size, around 1,000 kg live weight, but commercial meat is from smaller animals. After being on a finishing ration to a slaughter age of 14 to 15 months, carcasses weigh around 276 and 248 kg for bulls and cows, respectively. Dressing percentages are about 60%, but the carcasses are light in the hindquarter. Bison meat is dark, but this does not show after cooking. Treating bison carcasses like those of beef animals, cooler shrink losses are slightly higher than those for beef, probably because of the lighter fat cover and larger area of exposed muscles. However, shrink losses from bison carcasses may be improved by blast cooling. The saleable yield of meat trimmed to retail standards is around 78% of cold carcass weight. Shear values of the meat tend to have a wider range than those of beef but may be generally improved by electrical stimulation. The ultimate pH of bison meat is within a typical range for red meat (say pH 5.4 to 5.7). The main marketing feature of bison meat is that it tends to have a lower fat content than beef.

(Sources: Hawley, 1989; Janz, 1999)

BISTECA

Beef steaks in → Brazil.

BISTECCHE DI COSTA

The prime rib region of a beef carcass in → Italy.

BISTECCHE DI LOMBO

The lumbar steaks of an→ Italian beef carcass.

BIVENTER CERVICIS

→ *Semispinalis capitis.*

BLACK PUDDING

Blood sausage, typically in a large diameter (about 10 cm) natural → casing.

BLAD

A shoulder cut from a → South African beef carcass containing the scapula and humerus plus associated muscles.

BLADE BONE

The scapula

BLADE LOIN ROAST OF PORK

In → North American pork cutting, a whole pork loin includes the vertebral column and axial bones and muscles from the region of the scapula back to the pelvis (Figure 20). The blade loin roast is cut from the anterior part of the loin, but does not include much of the scapula, most of which remains in the shoulder.

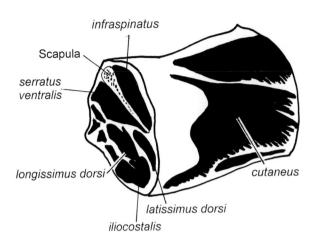

Figure 20. *North American blade loin roast.*

BLADE ROAST OR STEAKS OF BEEF

In → North America , the chuck may be sliced in planes parallel to rib 5 to make blade steaks or blade pot roasts.

BLANC DE VOLAILLE

French for white poultry meat.

BLANQUETTE

Proximal wing of poultry in → Morocco.

BLATWURST

A German black pudding or blood → sausage.

BLEIKJA

Icelandic for → charr.

BLESBOK

Blesbok, *Damaliscus dorcas*, is an African antelope reaching 59 to 100 kg.

(Source: Dorst and Dandelot, 1970)

BLOC

Beef chuck in Quebec.

BLOCK

Chuck of beef in Quebec.

BLOEMSTUK

A beef cut from the *triceps brachii* of a → Netherlands beef carcass.

BLOOMSTUK

Part of the dorsal thoracic region in a → Netherlands pork carcass.

BLUE CRAB

Western Atlantic shore crab, *Callinectes sapidus*, → Crabs.

BLUE DISCOLOURATION

In processed crab meat, blue discolouration ranges from light blue, through grey, to black. It may develop during or after heat treatment, and may be caused by a constituent of the blood because it is reduced by exsanguination before processing. The problem may originate from iron or copper compounds, phenolic compounds such as tyrosine forming melanin, copper proteins or biuret complexes, or haemocyanin. Whatever the cause, blue discolouration is different from the black discolouration caused by an interaction between the product and the metal of the can to form ferrous sulphide.

(Source: Boon, 1975)

BLUME

In → German beef cutting, the Blume refers to the body part near the tail, but the Blume also may be a cut straight across the carcass similar to a North American sirloin.

BOAR

A relatively large entire male pig.

BOAR TAINT

Boar taint is a bad smell which can occur when pork fat from boar meat hits a hot frying pan. Few people can detect boar taint in fat from young males, but the risk causes male pigs to be castrated in some countries. A major cause of boar taint is the concentration of sex steroids in the fat, particularly 5α-androst-16-ene-3-one, commonly called androstenone. Androstenone smells strongly of animal urine, but testicular steroids in the 16-androstene family also have a musk-like odour. Androstenone carried by the blood from the testes may accumulate in adipose tissue and *parotid* and *submaxillary* salivary glands. From the salivary glands, androstenone is normally transmitted as a pheromone in the boar's breath or saliva to the sow during mating. Boar taint is a heritable trait but can be suppressed immunologically. Other causes of boar taint include skatole and indole, with a faecal odour, produced from the amino acid tryptophan in the gut.
(Source: Mågård *et al.*, 1995; Lundström *et al.*, 1994; Babol *et al.*, 1996)

BOCKWURST

A pale emulsion → sausage (pork and/or veal), traditionally sold raw and cooked by the consumer.

BOCZEK

Polish for bacon, the carcass cut of this name includes muscle and subcutaneous fat lateral to the ribs and from the anterior abdominal region in a → Polish pork carcass.

BOEGLAPJE

A cut prepared from the *biceps femoris* of a → Netherlands beef carcass.

BØF

Beefsteak in → Denmark.

BOG

The bog is a shoulder cut from a → Norwegian or → Swedish beef, pork, lamb or veal carcass. In beef, the bog may be subdivided to give thick marrow-bone slices either with or without the scapulo-humeral ball joint called bog m/kuleben and u/kuleben, respectively, in Norway, and märgipan in Sweden. Bog also is used for a shoulder of reindeer in → Sweden.

BOGBLADSSTEK

The scapular portion of the bog from a pork carcass in → Sweden.

BOG MED BEN, KLACKAD

In a → Swedish veal carcass, the bone-in heel of the shoulder is called the bog med ben, klackad.

BOG MED BRÖST

Shoulder with breast, a cut of lamb in → Sweden.

BÓGSNEIÐAR

Shoulder steaks in → Iceland.

BÓGUR

The shoulder of a carcass in → Iceland.

BOK

Abdominal and posterior rib cuts of a carcass in → Slovakia. In beef, the abdominal muscles (bok-slabina) are separated from the posterior ribs (bok s kost'ou).

BOLA

In Cuban beef cutting, the *quadriceps femoris* group of muscles is called the bola or ball. This ball-shaped muscle group also gives rise to the name for Mexican escalops of veal from this region (escalopas de bola), as well as to beef cuts from this region in → Argentina.

BOLICHE BLANCO

The *semitendinosus* in → Cuban beef cutting.

BOLICHE FRANCÉS

The *supraspinatus* in → Cuban beef cutting.

BOLO

In → South Africa, the *triceps brachii* of a beef carcass.

BOLOGNA

An emulsion → sausage.

BONES

→ Skeleton

BONG

Fore- or hindshank of beef, veal or camel in → Pakistan.

BOOK STEAK

In → North American beef cutting, a book steak is a slice across the *infraspinatus* muscle of the shoulder.

BORSSTUK

The brisket of a → South African beef carcass.

BORST

The brisket of a beef carcass or breast of a veal carcass in the → Netherlands.

BOSTON BLUEFISH

A pollack, *Pollachius virens*, → Gadiformes.

BOSTON BUTT

A North American cut of pork initially including all the cervical vertebrae, the first few thoracic vertebrae and most of the scapula. Boston butts may be sold whole or cut in half as roasts, or may be boned out and sliced (Figure 21).

BOSTON CUT, ROAST OR BROILING STEAK

In the eastern style of → North American beef cutting, this may be a roast taken at the edge of the *triceps brachii*, ventral to the scapula in the cross rib region of the chuck. A Boston cut roast (also known as an English cut roast) may be located in the chuck over the ventral part of rib 5.

(Source: Bull, 1951)

BOTTOM ROUND

In→ North American beef cutting, the bottom or outside round is taken from the lateral face of the hindlimb and contains *semitendinosus* (eye of the round) and *biceps femoris*.

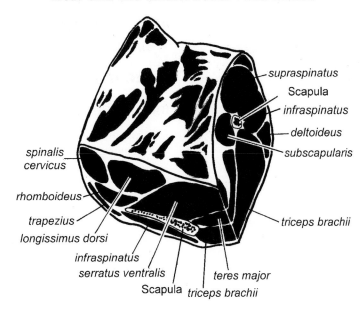

Figure 21. North American Boston Butt.

BOUCAN

French for preserving meat by drying and barbecuing, hence the English word, buccaneer. Pirates used to make vegetarians walk the plank.

BOUCANÉ

Traditional French product made by salting, drying and hot smoking belly of pork.

(Source: Poligné et al., 2001)

BOUDIN

French blood sausage, spread via the displaced Acadians of Nova Scotia to Louisiana, to become Cajun-style boudin. It also contains pork and onion, and sometimes fruit such as plum or apple.

BOUILLI D'ÉPAULE

Equivalent to the gros croisé, or anterior part of the brisket on a → Belgian beef carcass.

BOULE D'ALOYAU

A → French beef cutting term, the ball of the sirloin is the *quadriceps* group of muscles, otherwise known as the tranche grasse. The same term is also used in → Belgium for the tranche grasse.

BOURGUIGNON

In → French beef cutting terminology, this Burgundian cut is taken from above the ribs (dessus de côtes) in the train-de-côtes.

BOUSSUITE

Flank of lamb or mutton in → Morocco.

BOUT

In the → Netherlands, a bout is a whole hindlimb as, for example, in a veal carcass.

BOUT DE CÔTES

Short ribs of beef in Quebec.

BOUT GRAS

In French beef cutting terminology, the fat tip refers to the anterior part of the beef brisket.

BOUUITE

Flank of beef in → Morocco.

BOV

In → Denmark, the shoulder of either a beef or pork carcass, but used specifically for a shoulder cut of the pork carcass.

BOVENBIL

The bovenbil (literally above the buttock) is a medial cut from the hindquarter of beef in the → Netherlands, and contains *semimembranosus, adductor* and *pectineus*.

BOVENFRICANDEAN

A medial cut from the ham of a → Netherlands pork carcass.

BRACHIALIS

A medium sized muscle of the mammalian forelimb, slightly smaller than the nearby *biceps brachii*. It wraps around the humerus in a spiral from its origin in the proximal part of the muscular-spiral groove to its insertion on a tuberosity on the medial surface of the radius (Figure 22). In the pig, the *brachialis* is quite a large muscle with a divided tendon at its insertion.

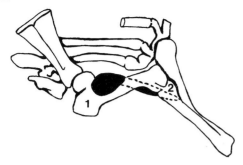

Figure 22. *Lateral view of beef* brachialis *(solid black and broken line) showing its origin on the humerus (1) and insertion on the radius (2).*

BRACHIOCEPHALICUS

A broad, flat muscle extending along the side of the neck from the head to the forelimb in mammals (Figure 23). It has proved difficult for anatomists to decide whether this muscle is split into two or three parts, or whether it is formed by the fusion of two or three muscles with separate origins, all inserting on the crest of the humerus, close to *pectoralis superficialis*. The simplest subdivision of the muscle is into a dorsal part (*cleidocervicalis*) originating occipitally and a ventral part (*cleidomastoideus*) originating from the mandible and mastoid. However, meat cuts taken from the neck region are poorly defined commercially and of no great value, so that subtleties of muscle anatomy are unimportant with respect to food myosystems.

Figure 23. *Lateral view of beef* brachiocephalicus *(solid black) showing its insertion on the humerus (1).*

BRACIOLE DI COSTA

The prime rib region of an → Italian beef carcass, also named bistecche di costa.

BRACIOLE DI LOMBO

The lumbar region of an → Italian beef carcass, also named bistecche di lombo.

BRACIOLI

In → North America, a retail cut of beef made by thinly slicing through the *adductor* and *semimembranosus* of the top round.

BRATWURST

A pork and/or veal emulsion → sausage with a very pale appearance. Traditionally, bratwurst was coarsely ground, packed in a natural → casing and cooked by the consumer.

BRAUNSCHWEIGER

A creamy emulsion → sausage, dominated by liver but usually containing at least some skeletal muscle.

BRAWN

Meat fragments set in a thick gelatin so that the product can be thinly sliced.

BRAWNER

A castrated male pig.

BRAZIL - BEEF CUTS

There are similarities between beef cutting in → Portugal and Brazil (Figure 24), but they are more linguistic than anatomical. In Brazil, a side of beef is split into the forequarter (quarto dianteiro) and hindquarter (quarto traseiro) quite far anteriorly, between ribs 5 and 6, tightly on the scapular cartilage. From the forequarter, the limb bones and associated muscles are removed as a primal cut called the paleta. From the paleta is separated the pá, which contains the scapula and humerus together with their surrounding muscles. Further separation produces the raquete (*infraspinatus*), peixinho (*supraspinatus*), and the coração da paleta (the heart of the paleta, which is the *triceps brachii*). The extensor and flexor muscles of the foreshank are removed as the músculo do dianteiro. The remainder of the forequarter after removal of the paleta (dianteiro-sem-paleta) is divided into the pescoço (neck), acém (*trapezius, rhomboideus* and *serratus ventralis*), costela-do-dianteiro (a ladder of midlength rib sections with associated muscles), and peito (brisket). As in other countries with humped cattle (like → South Africa), the hump of *trapezius* and *rhomboideus* may be removed separately as the cupim. Cupim has a taste similar to beef brisket in other countries. Although cupim has a high collagen content, prolonged cooking and a high fat content greatly reduce its toughness.

From the hindquarter is obtained the traseiro-serrote, which is the hindlimb plus the vertebral column. The characteristic shape so produced, instead of being called a pistol cut as in many other countries, is called the serrote or hand-saw. From the traseiro-serrote is cut the lombo or loin which includes *longissimus dorsi* all the way from rib 6 to its posterior attachment to the anterior face of the ilium. The lombo is separated to obtain the contrafilé (including *longissimus dorsi* and all associated muscles dorsal to the ribs and transverse processes of the lumbar vertebrae), which is subdivided longitudinally into the filé-de-costela (from rib 6 to 10) and the filé-de-lombo (rib 11 to the ilium). The superficial muscles of the filé-de-costela (*trapezius* and *rhomboideus*) are removed as the capa-de-filé. *Psoas major* and *minor* plus *iliacus* are

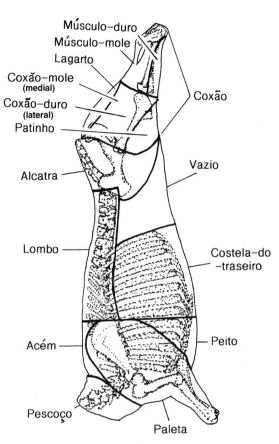

Músculo-duro
Músculo-mole
Lagarto
Coxão-mole (medial)
Coxão-duro (lateral)
Patinho
Coxão
Alcatra
Vazio
Lombo
Costela-do-traseiro
Acém
Peito
Pescoço
Paleta

Figure 24. Brazil - beef cuts.

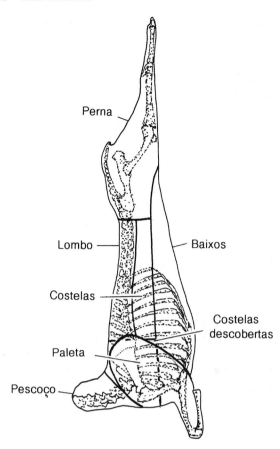

Perna
Lombo
Baixos
Costelas
Costelas descobertas
Paleta
Pescoço

Figure 25. Brazil - lamb cuts.

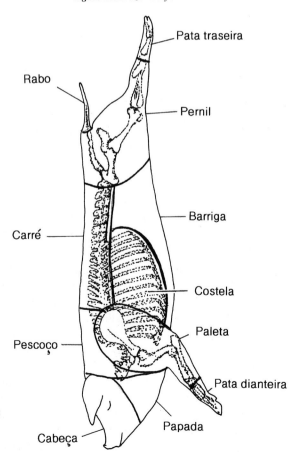

Pata traseira
Rabo
Pernil
Barriga
Carré
Costela
Paleta
Pescoço
Pata dianteira
Cabeça
Papada

Figure 26. Brazil - pork cuts.

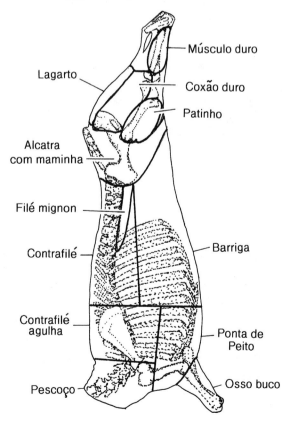

Músculo duro
Lagarto
Coxão duro
Patinho
Alcatra com maminha
Filé mignon
Contrafilé
Barriga
Contrafilé agulha
Ponta de Peito
Pescoço
Osso buco

Figure 27. Brazil - veal cuts.

the filé-mignon. Bisteca are steaks from anywhere along the lombo, while tibone are steaks from its lumbar region.

The hindlimb is separated into the alcatra, which includes the gluteal muscles plus *tensor fascia lata*. The alcatra commands a premium price in Brazil, considerably more than rib or T-bone steaks. The reason seems to be that none of the cuts are particularly tender, and the gluteal muscles are superior in taste and succulence. The distal remainder of the hindlimb is the coxão (thigh) which is split into the coxão-duro (the tough thigh or *biceps femoris*), the lagarto (*semitendinosus*), patinho (*quadriceps femoris* group), músculo-mole (*gastrocnemius* and associated muscles), coxão-mole (mostly *pectineus, adductor, semimembranosus, gracilis* and *sartorius*) and músculo-duro (extensors and flexors around the tibia). Likening the *semitendinosus* to a lizard (lagarto) is, I suppose, the cultural equivalent of likening this muscle to a salmon (as in Cumbria and some parts of Scotland). The ponta-de-agulha comprises the remaining flank and postero-ventral rib region, which is divided into vazio (flanco) and costela-do-traseiro, respectively. The *obliquus abdominis internus* is called the fralda (shirt-tail).

(Sources: ABIF; de Oliveira Roça, 1999; Terra, 1999; Pedrão *et al.*, 2003)

BRAZIL - LAMB CUTS

The paleta or shoulder is lifted from the ribcage to "discover" the underlying rib cut (costelas descobertas) from which is removed the neck (pescoço) containing all but the last cervical vertebrae (Figure 25). After removal of the perna or leg, which includes the last lumbar vertebra, the remainder of the carcass is split longitudinally into the ribs (costelas) and the breast (baixos).

(Sources: de Oliveira Roça, 1999)

BRAZIL - PORK CUTS

As shown in Figure 26, the pernil or leg of pork includes the whole hindlimb and sacral region, from which is removed the tail (rabo) and trotter (pés or pata traseira). The carré extends from the ilium anteriorly to end somewhere from the third to fifth thoracic vertebra, with the remaining thoracic and all the cervical vertebrae in the pescoço. The *psoas* and *iliacus* muscles may be removed as a filé. The paleta containing the bones and muscles of the forelimb is lifted from the ribcage, and the trotter is removed (pés or pata dianteira). This leaves the head (cabeça) from which are removed the ear (orelha) and jowl (papada).

(Sources: de Oliveira Roça, 1999; Terra, 1999)

BRAZIL - VEAL CUTS

The pattern for veal cutting in Brazil (Figure 27) is similar to that for beef, but with a major difference in the shoulder region. Instead of lifting a paleta from the ribcage, the scapula remains with the underlying ribcage and is the cut which produces "needle-thin" contrafilé agulha.

(Source: Terra, 1999).

BRAZO

A foreleg cut from a beef carcass in → Spain.

BRAZUELO

The foreshank of an → Argentinean beef carcass.

BREAD AND BUTTER CUT

→ North American retail cuts of beef through the *triceps brachii*, ventral to the scapula in the chuck.

BREAM

Various species of *Abramis*, → Cypriniformes, some with a high oil content.

BREAST

In → North American lamb cutting, the breast and foreshank may be removed with a single cut from the anterior of the sternum to the ventral part of rib 12. Alternatively, the dominant cut may be made between ribs 5 and 6, to separate the rib from the shoulder, and to divide the breast into anterior and posterior sections. In → English lamb cutting, the anterior of the breast is usually restricted to the sternum and associated muscles, and does not include the humerus, radius and ulna, which remain on the shoulder. The muscle groups of the foreshank and breast are shown in Figure 28.

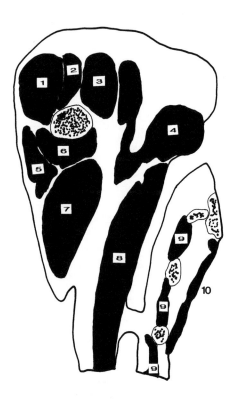

Figure 28. *North American foreshank (muscles 1 to 8) and breast (muscles 9 to 10) of lamb with some landmark muscles;* extensor carpi radialis *(1),* brachialis *(2),* biceps brachii *(3),* pectoralis superficialis *(4), lateral (5), medial (6) and long (7) heads of* triceps brachii, pectoralis profundus *(8), intercostals (9) and* transversus thoracis *(10).*

BRILL

An oval flatfish with smooth scales, *Scophthalmus rhombus*, → Pleuronectiformes.

BRINGA

The brisket of a → Swedish reindeer or beef carcass. On a Swedish veal carcass, after the shoulder or bog is lifted from the ribcage, the underlying brisket with bone is the bringa med ben.

BRISKET

In → North America, the brisket is a primal cut of beef removed with the shank. It has a high bone content (the sternum), but the pectoral muscles are ideal for curing and smoking. Even in a lean animal, the brisket usually contains a coarse textured fat contributing to the dewlap of the live animal. The anterior end may be called the point of the brisket, with the more posterior part being called the navel end, as in → New Zealand beef cutting. The brisket is widely distributed in many → English , → Scottish and → Irish patterns of beef cutting.

BRISKET POINT END DECKLE OFF

In → Australian beef cutting, the brisket point end deckle off is a cut involving the anterior sternum. Here the wider definition of the deckle as tissues between the ribs and the *pectoralis* muscles of a beef brisket is used. Thus, the Australian deckle is fat and connective tissue which is removed from the cut of meat, not muscle (like *latissimus dorsi* or *trapezius*) which remains in the cut.

BRIZOLES

In → Greece, lumbar or posterior thoracic cuts of beef or pork.

BROEK

→ Broekvet

BROEKVET

The ventral region of the belly in a → Netherlands pork carcass.

BROSSE

Equivalent to the gros croisé of a → Belgian beef carcass.

BRUST

In cutting → German beef and → Austrian pork, lamb and veal, the breast or Brust underlies the Schulter. In German lamb cutting, the Brust underlies the Bug or shoulder.

BRUSTBEIN

A → German beef cut including the anterior part of the sternum.

BRUST FLANKEN RIBS

In → North American beef cutting, brust flanken ribs (brust is German for breast) are cut across the ribs, posterior to the humerus and dorsal to the scapula. The dominant lateral muscle is *pectoralis profundus*, while medially may be found *rectus thoracis*, *serratus ventralis cervicis* and intercostal muscles.

BRUSTKERN

A centre cut from the brisket of a beef carcass in → Switzerland. It is the anterior part of the beef brisket in → Austria.

BRUSTSPECK

In → Swiss pork cutting, the breast bacon includes most of the ventral thorax.

BRUSTSPITZ

In → Swiss beef cutting, the breast-point is a cut from the anterior sternal region.

BRUSTSPITZE

In → German pork cutting, the Brustspitze is a cut from the anterior sternal region.

BRYST

The brisket or sternal region of a→ Norwegian beef, lamb or veal carcass.

BRYSTFLÆSK

Belly of pork (breast pork) in → Denmark.

BUCKLAND

Francis Trevelyan Buckland (1826 to 1880) was the son of a famous geologist and palaeontologist (William Buckland). After service as a surgeon in the British army, he devoted himself to aquaculture and became Inspector of Fisheries for the British government in 1867. He was founder of the → Acclimatisation Society. His enthusiasm for introducing new muscle foods was motivated by a desire to improve the nutrition of ordinary people and balanced by his instinct for conservation.

(Source: Bompas, 1886)

BUFFALO

African buffalo (*Syncerus caffer*) come in several sizes, ranging from the dwarf forest buffalo of West and Central Africa to the heavy Cape buffalo of South Africa. For bulls and cows, respectively, carcass weights of 380 and 326 kg have been recorded, with dressing percentages up to about 50%. For North American buffalo, → bison.

(Sources: Eltringham, 1984)

BUG

→ German for shoulder, as in a shoulder of lamb or thick shoulder-piece of beef, Dickes Bugstück.

BUIK

The belly of a → Netherlands pork carcass.

BUKLIST

Belly of pork in → Norway.

BURBOT

Burbots are in the cod family (→ Gadiformes), grouped with the → cusk and → ling. Several genera are marine, but *Lota lota* is freshwater.

BURK

Gluteal muscles of beef, veal or camel in → Pakistan.

BUT

Hindlimb of beef, pork and lamb in → Serbia & Montenegro.

BUTIKSSKINKA

A boneless shop-ham from the skinka of a → Swedish pork carcass.

BUTLER STEAK

In → North American beef cutting, this is a slice across the *infraspinatus* muscle of the shoulder.

BUTT

A large primal cut of an → Irish beef carcass containing the ilium, pubis and associated muscles.

BUTT CHOPS OF PORK

In Canada, chops through the Boston butt containing sections of scapula.

BUTTERFLY

Incomplete slicing of a steak, cutlet or chop to leave a hinge that allows the product to be opened like butterfly wings to increase the display and cooking areas.

BUTTOCK

In → English beef cutting, the buttock is the silverside plus the topside.

CABEÇA

The head of a pork carcass in →Brazil or → Portugal.

CABEZA

The head of a → Spanish pork carcass. In → Mexico, the term is used to denote the head end of the loin, as cabeza de lomo.

CABOB

→ Kabob.

CABRITO

Meat from a young → goat, also called chevon.

CACHAÇO

The neck of a → Portuguese beef, pork, lamb or veal carcass.

CACHO DE PALETA

Beef *supraspinatus* in → Costa Rica.

CADERA

A cut of beef through the ischium in → Spain, the whole of the pelvis in → Mexico, or the anterior pelvis in → Ecuador.

CAILLE

French for → quail.

CAIMAN

→ Crocodile

CALIFORNIA ROAST

Although it may be unreliable (see fanciful names), in → North American beef cutting California roasts and steaks seem to originate from the chuck but with the *subscapularis* and other lateral muscles around the scapula removed.

CALLIE

A fanciful name, sometimes used in North America for smoked picnic of pork.

CALPAIN

Calpains are calcium-activated cysteine proteases involved in → conditioning of meat. The calcium ion concentration required for activation differs between μ-calpain (requiring micromolar levels) and m-calpain (requiring millimolar levels). Many calpains have been discovered. Calpains 1, 2 and 3 have attracted most attention in meat conditioning. Calpain 10 is the latest.

(Sources: Goll *et al.*, 2003; Ilian *et al.*, 2004a, 2004b)

CALSEQUESTRIN

Calsequestrin is a protein that binds and stores calcium ions within the → sarcoplasmic reticulum. Calsequestrin concentration is higher in fast-contracting than in slow-contracting myofibres.

(Sources: Tume, 1979; Maier *et al.*, 1986)

CAMELS

Camels are important meat producing animals. *Camelus dromedarius*, the Arabian camel or dromedary, has one hump, slender limbs and runs swiftly. *Camelus bactrianus*, the Bactrian camel of the cold deserts of Asia, has two humps and thick limbs. Camel meat is rather like beef and has a similar nutrient value. Prime meat from younger animals can be cooked quickly with dry heat, while meat from the extremities of younger animals and all the meat from older animals is best cooked with moist heat. Zoologically, the camels are somewhere between pigs (which have upper incisors) and cattle and sheep (which lack upper incisors and, instead, have a horny pad). Camels have vestigial upper incisors, plus well developed canine teeth that may be tusk-like (canines are present in pigs but not in cattle and sheep). The shape of a camel carcass (→ United Arab Emirates) is quite different from that of cattle, sheep and pigs. Apart from the obvious shape of the hump dorsally, note how the insertion of hindlimb muscles is restricted dorsally to the pelvis: they do not blend in with the abdominal muscles. Thus, when the hindlimb is stretched back it forms an indentation ventral to the ilium (i.e., the ventral angle between the Al Shakela and the Al Fakhed in the → United Arab Emirates). This is because the camel has long limbs capable of greater rotation relative to the vertebral axis. When a camel is sitting, the distal end of the femur projects downwards towards the ground. Whereas, in sitting cattle and sheep, the distal end of the femur projects upwards. The camel also has a broad cutaneous pad on each foot instead of two hooves. Camels are ruminants with four stomach compartments, but the omasum is indistinct, not hard and round as in cattle.

(Sources: Clutton-Brock, 1987; Elgasim and Alkanhal, 1992)

CAMPANELLO

An → Italian cut of beef containing the *gastrocnemius*.

CANADA - MEAT CUTS

→ North America.

CAÑADA

In Cuban beef cutting, the *adductor, pectineus* and *semimembranosus* muscles are removed as a boneless cut medial to the femur. The name is from the cañada (pronounced canyada), a conspicuous groove between muscles.

CANADIAN BACON

In the USA, bacon is traditionally made from pork belly, not from pork loins. Thus, in the USA, English-style back bacon containing *longissimus dorsi* is called Canadian bacon.

CANADIAN POT ROAST

In → North America, a boneless pot roast of beef cut from the plate.

CANARD

French for → duck. Canard sauvage is wild duck.

CANDLE FISH

An oily Pacific salmonid, *Thaleichthyes*, → Salmoniformes.

CANE RAT

The cane rat or grasscutter, *Thryonomys swinderianus*, is an important source of meat in West Africa. It is a relatively large, herbivorous rodent reaching 7 kg live weight. Its meat is rated highly, above other sources of bushmeat, so that there is hunting pressure on the natural population. Scientific husbandry of the cane rat might produce a dispersed, local meat supply like that provided by poultry in other areas.

(Sources: Dorst and Dandelot, 1970; Anadu et al., 1988)

CANET

A French term for the thick muscular part of the diaphragm.

CAPARAÇON

In → French beef cutting, this is a primal cut that includes the bavette d'aloyau, plat-de-côtes, bavette, flanchet, tendron and poitrine.

CAPELIN

A small smelt-like bony fish from Arctic or sub-Arctic seas, → Salmoniformes.

CAPERCAILLIE

The largest grouse, *Tetrao urogallus*, → Galliformes.

(Source: Vesey-Fitzgerald, 1949; Bertin, 1967)

CAPILLARY

The smallest branch of the blood vascular system bringing oxygen to the → myofibre surface and removing metabolic waste products. The capillary supply to myofibres specialized for repetitive contraction using blood-borne energy sources is particularly well developed.

(Source: Honig et al., 1992)

CAPON

A fat, castrated → chicken.

CAPPICOLA

Cured, smoked and dried pork (typically proximal hindlimb).

CAPRETTO

Meat from a young → goat in Italy.

CAPRIN

In → French beef cutting terms, the caprin (goat) may refer to beef from above the ribs (dessus de côtes) in the train-de-côtes.

CAPRIOLO

Italian for → venison from roe deer.

CAPYBARA

The capybara (*Hydrochoerus hydrochoeris*) is the largest living rodent. It reaches 55 kg. It is semiaquatic, herbivorous and occurs east of the Andes, from Panama down to Paraguay. It is well muscled and the meat is excellent. The taste is difficult to describe – something like chicken leg meat eaten together with a small amount of cucumber? However, the fatty acid composition varies considerably between different muscles and one might expect the taste to exhibit comparable variability. Capybara meat can be quite tender and succulent.

(Source: Roca *et al.*, 1999; Bressan *et al.*, 2003)

CAP-Z

A protein of the → Z-line involved in → actin polymerisation.

(Source: Casella *et al.*, 1987)

CARA

A medial cut of beef from the pierna of the hindlimb in → Mexico.

CARBONADES

A French term for neck meat.

CARIBOU

→ Reindeer.

CARNAZA DE COLA

A hindlimb cut of beef in → Argentina.

CARNAZA DE PALETA

The *triceps brachii* muscle of an → Argentinean beef carcass.

CARNOSINE

Carnosine is a → myofibre dipeptide with acid-base buffering capacity. It acts as an antioxidant and may be involved in regulating calcium-ion release channels of the sarcoplasmic reticulum.

(Source: Rao and Gault, 1989; Chan *et al.*, 1993; Batrukova *et al.*, 1993)

CAROTENE

The yellow colouration of fat by ß-carotene from dietary sources is common in ruminants, and may be a cause for down-grading in countries dominated by grain-fed animals normally producing white fat. Fat with a high ß-carotene content has a lower overall reflectance than white fat, particularly from 440 to 500 nm where strong absorbance by ß-carotene is separable from the Soret absorbance band of any residual haemoglobin.

(Source: Simonne *et al.*, 1996)

CAROTTE

The *tensor fascia lata* muscle of a beef carcass in some parts of France.

CARP

A freshwater pond fish, → Cypriniformes.

CARRE

A primal rib cut of an → Italian pork carcass. In → Brazil, the carré is much longer and extends from the scapula to the ilium.

CARRÉ DÉCOUVERTES

In → French lamb cutting, removing of the scapular portion of the épaule uncovers the deeper vertebral muscles of the carré découvertes.

CARRÉ DE FILET

In → French pork cutting, carré de filet is a cut including the rib chops.

CASINGS

Various parts of the alimentary canal are used as natural casings to enclose various types of → sausages. The cleaning and preparation of natural casings is highly effective, leaving no detectable trace of the intestinal contents. When initially obtained in the abattoir and emptied, natural casings have five layers. From lumen (inner cavity) to exterior these are: (1) the mucosa, composed of epithelial, glandular and vascular components; (2) the submucosa composed of connective tissues that strengthen the gut wall; (3) a circular layer of smooth muscle cells; (4) a longitudinal layer of smooth muscle cells; and (5) an irregular layer of visceral fat covering the outside. The fat is removed manually and by machine brushing. The intestinal contents are squeezed out and washed away in a process called stripping. Finally, the layers of muscle and mucosa are removed as the intestine passes between a pair of rollers in a process called sliming. This leaves the strong connective tissues of the submucosa as the sausage casing. Thus, the dominant material of a natural casing is → collagen fibres.

The commercial properties of casings originate from the high collagen content of the submucosa, together with smaller amounts of elastin. Casings often are turned inside-out to facilitate processing. Clean casings are preserved with dry sodium chloride prior to sale. With approximate metric lengths given in parentheses, the commonly used beef casings are: the weasand from the oesophagus (0.6 m), rounds from the small intestine (32 m), the bung from the caecum (1.8 m), and middles from the large intestine (8 m). Beef casings are tough and strong, and are usually removed before consumption of the product. The commonly used hog casings are rounds or small casings from the small intestine (18 m), the cap from the caecum (0.4 m), middles from the middle part of the large intestine (1 m), and the bung from the terminal end of the large intestine (1 m). Being intermediate in tenderness, hog casings either may be eaten or removed before consumption of the product. The small intestine from sheep provides 27 m of tender casing that gives smaller sausages their "snap".

CASQUE

A French term for the forequarter, typically of lamb or mutton.

CATFISH

A freshwater bony fish, → Cypriniformes.

CAUCARA O FALDA DORADA

Ventral abdominal muscles of a beef carcass in → Ecuador.

CÂWI

Equivalent to the petite tête d'aloyau of a → Belgian beef carcass.

CEABET

Neck of → reindeer in the Saami language.

CECINA

Cured, smoked, dried meat in Spain and South America.

(Source: Calle Escobar, 1984; Garcia *et al.*, 1997)

CENTER ROUND STEAK

A large retail cut of beef in → North America. It is composed of a round steak, but it only contains the muscles level with or posterior to the femur.

CERDO

Spanish for a pig. Carne de cerdo is pork.

CERF

French for → venison from red deer.

CERVELA

A short, thick emulsion → sausage, strongly spiced with garlic.

CERVELAT

A semidry → sausage.

CERVICOHYOIDEUS

A very small, thin, ribbon of muscle that crosses the mammalian trachea near the thyroid gland. Its remains, if any, are generally desiccated in the neck of a dressed carcass.

CERVO

Italian for → venison from red deer.

CHAANP

The thoracic vertebral region of a carcass in → Pakistan.

CHÃ DE FORA

In → Portuguese beef cutting, the chã de fora contains the muscles lateral and anterior to the femur.

CHAEKEUT

The lumbar region of a → Korean beef carcass.

CHAINETTE

In French beef cutting, these are small steaks taken across the anterior thin end of *longissimus dorsi*, medially to the scapula.

CHAMBÃO

The foreshank (chambão anterior) or hindshank (chambão posterior) of a → Portuguese beef carcass. The foreshank of a veal carcass is called the chambão da mão (of the hand).

CHAMBARETE

The foreshank of a beef carcass in → Mexico.

CHAMORRO

Distal portion of a leg of pork in → Mexico.

CHAP

An alternative English name for the jowl of a pork carcass.

CHAPELET

In → French beef cutting, chapelet or chaplet is the plat-de-côtes, where the ribs appear in a series like beads in a rosary.

CHARBONNET

Collier or neck of beef in → Belgium.

CHAROLAISE

In → French beef cutting, the charolaise is a cut made transversely across the forelimb, just proximal to the olecranon process of the ulna.

CHARQUI

Intermediate moisture (45%) meat products with a high (15%) sodium chloride content in tropical countries. Typically with some protein denaturation (A-band and M-line) and fluid channels created during dehydration. For → alpaca charqui, the meat is cut into slices 0.5 to 1 cm thick, treated with salt, and then soaked in brine for two or three days. Two to three weeks open-air drying in the Andes essentially is freeze-drying. Final drying is done under a roof.

(Source: Calle Escobar, 1984; Biscontini *et al.*, 1996)

CHARR

A salmonid bony fish, → Salmoniformes.

CHATEAUBRIAND

This name is frequently used for the thick part of the → filet of beef, where the *psoas major* is joined by the *iliacus* to become the *iliopsoas*.

CHELY DE PORC

In → Belgium, this is equivalent to the pointe au filet, a cut of pork containing sacral and posterior lumbar vertebrae.

CHEVON

Meat from a young → goat, also called cabrito.

CHEVRETTE

Meat from a young → goat in France.

CHEVREUIL

French for → venison from roe deer.

CHICAGO ROUND

A primal cut of beef in → North America, essentially the same as a hip of beef, without the anterior part of the sirloin tip which remains on the primal sirloin.

CHICKEN

The domestic chicken is descended from the Red Jungle Fowl, *Gallus gallus*. Chicken is used as a generic term in some countries whereas, in others, chickens are categorised by age and type. In France, for example, the age range is from young poussin to the poule, an old fowl. Similarly, in North America, although different breeds usually become anonymous after slaughter, a series of carcass types is defined by age and size. Rock Cornish are 4 to 5 weeks of age, their carcasses weigh less than about 0.8 kg, and they may be males or females. Broilers or fryers are about 5 to 8 weeks of age, have carcass weights from 0.8 to 1.8 kg, and may be males or females. Roasters are males or females older than 9 weeks with carcasses over 1.8 kg. Capons are castrated males over 9 weeks of age and with carcasses over 1.8 kg. Surgical castration is difficult (because the male gonads are inside the body cavity near the kidneys) and has been replaced by hormonal castration in most countries.

Chicken carcasses with tender meat may be identified by their soft, pliable and smooth-textured skin, and by their flexible sternal cartilage. Chickens such as roosters and mature hens producing relatively tough meat are identified by their greater age, coarse skin, and stiff sternal cartilage. Several features may be used as a guide to the age of a chicken. Young birds have unwrinkled combs with sharp points. In older birds, the comb becomes wrinkled with blunt points. The plumage becomes worn and faded in older birds, unless the birds have just moulted. With age, the subcutaneous fat becomes darker and lumped under the main feather tracts, and the pelvic bones become less pliable.

Old chickens have large scales which are rough and slightly raised and their oil sac becomes enlarged and hardened. Older male chickens develop long spurs.

The downward stroke of the wing during flight originates from the large *pectoralis* muscle over the sternum (Figure 29). The wing is elevated by the *supracoracoideus* muscle, which is located between the *pectoralis* and the sternum. Although the *supracoracoideus* is adjacent and parallel to the *pectoralis*, it causes an opposite movement (raising instead of lowering the wing) because its tendon inserts onto the opposite side of the humerus to the *pectoralis* tendon. In poultry, the *pectoralis* muscles are the largest body muscles, comprising about 8% of the total body weight. Small muscles located within the wing are concerned with controlling the shape and degree of rotation of the wing during flight.

Look carefully at the insertion of the flight muscles in the breast region. It might appear at first sight that most of the very large sternum is used for the attachment of the *pectoralis*. However, careful dissection will reveal that the only part of the sternum directly in contact with the *pectoralis* is a zone around the *supracoracoideus* muscle. Most of the *pectoralis* originates from the wishbone or furcula, and from a membrane stretched between the furcula and the coracoid. The main tendon of the muscle is continued intramuscularly as a layer of connective tissue that divides the muscle into two heads which probably have different functions during flight.

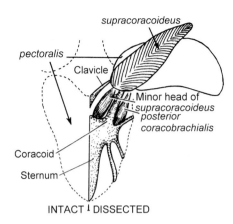

Figure 29. *Dissection of a chicken breast on its left side.*

The muscles in a chicken's leg are adapted for bipedal locomotion. The *sartorius* or *iliotibialis cranialis*, instead of being a medial strap-like muscle, is triangular in cross section and is the most anterior muscle in the leg. Posterior to the *sartorius* are muscles in similar general positions to those of mammals (Figures 30 and 31). They are not, however, exactly equivalent muscles. Following Halvorson (1972), mammalian muscle names are used here for convenience. Comparative anatomists have radically different names for these muscles. Muscle anatomy is essential for understanding meat cuts in large animals - but the most that normally happens to a chicken leg is to be cut into thigh and drumstick.

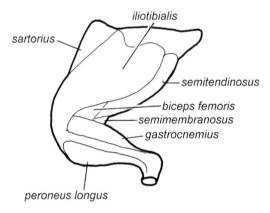

Figure 30. *Simplified plan of lateral muscles in a chicken leg.*

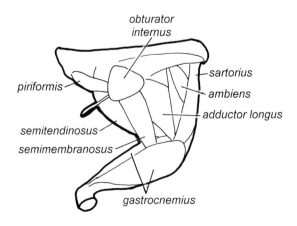

Figure 31. *Simplified plan of medial muscles in a chicken leg.*

Chicken carcasses may be subdivided so that customers may purchase half carcasses (right or left sides), white or red meat (breast versus leg), and premium or economy parts of the carcass. If right and left sides are quartered, the cut follows the posterior edge of the last rib and is continued through the vertebral column. The wings may be removed with a cut through the shoulder joint (glenoid). Sometimes the wing is subdivided by a cut immediately distal to the radius and ulna. The breast is composed of the sternum and its associated muscles. The sternum may be sold intact (whole breast) or split into left and right halves. There is some variation with regard to the inclusion of the clavicles and ventral parts of the ribs. Sometimes the whole breast is split into three parts with the wishbone (pulley bones) as the central unit. Neck skin is normally excluded from all the breast cuts.

The chicken leg may be removed at the acetabulum so that the pelvic muscles, but not the pelvic bones, accompany the leg. The proximal part of the leg, the thigh, may be separated from the distal part, the drumstick, at the joint between the femur and the tibia. The way in which the back is cut is highly variable, depending mainly on how the other cuts have been defined. Whole back includes the pelvic bones, scapulae, dorsal parts of the ribs, and the vertebrae from the posterior part of the neck to the tail.

CHIKI

Hausa for the thin flank in → Nigeria.

CHILE – BEEF CUTS

The primary separation into forequarter (paleta) and hindquarter (pierna) is made between ribs 9 and 10. The forelimb is removed from the ribcage and used for choclillo (*supraspinatus*), punta de paleta (*infraspinatus*), posta de paleta (*triceps brachii, teres minor, tensor fasciae antebrachii, anconeus* and other nearby parts of muscles), lagarto (*biceps brachii*) and osobuco de mano (forelimb extensors and flexors). Medially the forelimb yields the sobrecostilla (*serratus ventralis, scalenus dorsalis,* and nearby parts of *rhomboideus* and *splenius*) and asado del carnicero (*subscapularis* and *teres major* with adjacent parts of *latissimus dorsi* and *rhomboideus*). The axial frame beneath the forelimb is subdivided into cogote (neck muscles), huachalomo (muscles along the thoracic vertebrae), lomo vetado (muscles overlying thoracic vertebrae 4 to 9), tapapecho (anterior muscles over the sternum including *pectoralis profundus* and *superficialis*), aletillas (abdominal muscles attached to the ventral ribs and costal cartilages), plateada (mainly *latissimus dorsi* and part of *trapezius*), and asado de tira (muscles between and immediately lateral to the ribs). Not shown in Figure 32 are the malaya (*cutaneus* muscle) and entraña (diaphragm).

The loin or lomo liso is long because of the four ribs on the hindquarter. Then follows the asiento containing the gluteal muscles, the punta de picano (*tensor fascia lata*), and the posta rosado with the *quadriceps* group. The *biceps femoris* forms punta de ganso proximally and ganso distally. The *semitendinosus* is removed separately as the pollo ganso, leaving the medial muscle mass of *semimembranosus, adductor* and *pectineus* as the posta negra. The abastero contains *gastrocnemius*. The hindshank is called osobuco de pierna. The filete is composed primarily of *psoas major* and *minor,* and *iliacus*. The flanks is divided into the dorsal tapabarriga, the posterior palanca, and the anterior coludas. Parts of the diaphragm in the hindquarter are called pollo barriga. The muscles to the tail are removed as the cola.

(Source: Gallo, 2003).

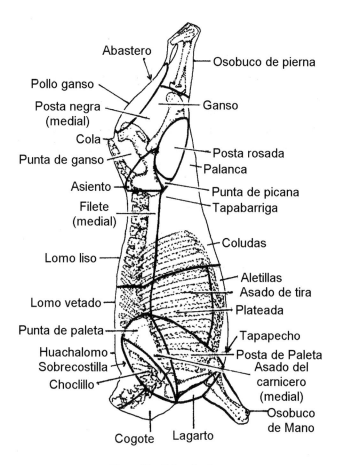

Figure 32. *Chilean beef cuts.*

CHILVER

A female lamb.

CHINA - BEEF CUTS

With the vast geographical diversity within the People's Republic of China, not to mention the problem of transliteration, meat cutting in China is a difficult topic. Figure 33 gives only a synopsis of the main features. The primal cuts follow a familiar pattern, although all the ribs are maintained in the forequarter.

(Source: Yan, 1999).

CHINA - LAMB CUTS

A lamb cutting pattern for the People's Republic of China is given in Figure 34, with the shoulder being lifted from the rib cage and breast.

(Source: Yan, 1999).

CHINA - PORK CUTS

A pork cutting pattern for the People's Republic of China is given in Figure 35. The shoulder is lifted from the rib cage.

(Source: Yan, 1999).

CHINE

To butchers in English-speaking countries, the chine or chine-bone is the vertebral column, but which of the two linguistic roots we have inherited is uncertain. On one hand is the Old French, eschine for spine, while on the other hand is the Old English, cinu for a deep cleft or splitting. Both make sense to us in the meat trade, as we chine our way down a chine. In the Midlands of → England, the chine is a cut of beef in the posterior thoracic region. In → Ireland, the chine may be the dorsal part of a shoulder of pork.

CHINE AND END

In → English lamb cutting of an unsplit carcass, a chine and end is left and rights sides of loin, plus best end neck, plus middle neck.

CHINYA

Hausa for the hindquarter in → Nigeria.

CHIROMIRI

→ Greek for ham, following the outline of the bouti on a pork carcass.

CHISPE

In a → Portuguese pork carcass, the hindlimb chispe is distal to the patella, while the forelimb chispe (da mão, of the hand) is distal to the elbow.

CHOCLILLO

A cut of beef in → Chile containing the *supraspinatus* and having the shape of a corn cob.

CHONDRICHTHYES

Description

The Chondrichthyes are the cartilaginous fish. Their skeletons are composed of stiff white cartilage, and the hardest parts of their bodies are the sharp, plate-like denticles protecting the skin. Dermal denticles are greatly enlarged in the mouth where an intucking of the epidermis provides a continuous conveyer-belt of teeth derived from enlarged denticles. Shagreen is a form of leather produced from shark skin. The sharp denticles are ground down and dyed to form a remarkably durable and attractive surface.

In the Chondrichthyes, the gills open separately onto the surface of the body to form a series of slits (usually five) plus a small round accessory hole called the spiracle, just behind the head. The familiar sharks and rays are in the order Euselachii, and are distinguished by the position of their gill slits. The Pleurotremata have lateral gill slits, like sharks, while the Hypotremata are dorso-ventrally flattened with ventral gill slits, as in the rays. Typical examples are the dogfish for the Pleurotremata, and the skates and thornback ray for the Hypotremata. In the flattened skates and rays of the Hypotremata, the spiracle is located on top of the body behind the eye, from where water is drawn inwards through the gill chambers and vented to the exterior through the ventral gill slits. The wings of the skate (*Raja radiata*, the thorny skate, and *R. senta*, the smooth skate) produce a myosystem with a distinctive and delicate flavour, rather like scallop. After cooking, long strips of muscle can be removed from the cartilaginous rays quite easily. *Squatina squatina*, called either the angel shark or monkfish, has a body shape intermediate between the sharks and rays.

Myosytem

The skeletal musculature of the Chondrichthyes provides an excellent food myosystem with both a desirable taste and texture, but it has one unfortunate feature with regard to food technology. Marine vertebrates above the level of the jawless fish or Agnatha maintain their body fluids using a concentration of inorganic ions at about 40% of the level found in sea water. This requires a physiological mechanism for the maintenance of an osmotic equilibrium between the body fluids and the external aquatic environment. Thus, the Chondrichthyes retain a high concentration of urea in the blood. The gills and other exposed body surfaces are relatively impermeable to urea and so the body tends to take up water by osmosis. An osmotic equilibrium is maintained by the fish continuously excreting hypotonic urine. After the fish has been caught and processed, its muscles still retain a high concentration of urea. Because urea readily breaks down to form ammonia post-mortem, there may be a rapid deterioration of the stored product, with consequences best imagined rather than directly experienced. When cooking the fresh myosystem the problem of ammonia formation may be avoided by marinating the fillets with vinegar or lemon juice. Dogfish (*Squalus*) are metamorphosed to rock salmon by a magical battering in British fish and chip shops. Dogfish belly flaps are used for a smoked delicacy for German beer gardens, while the fins find a use in Japanese and Chinese cooking.

Both slow-red and fast-white myofibres occur in dogfish myomeres (→ histochemical types of myofibre). The red myofibres are lateral to the white myofibres, separated by a zone of intermediate myofibres. In frozen sections of a 50-

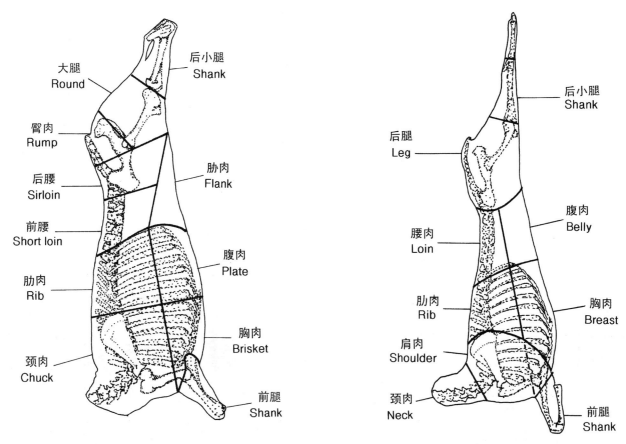

Figure 33. *China – beef cuts.*

Figure 34. *China – lamb cuts.*

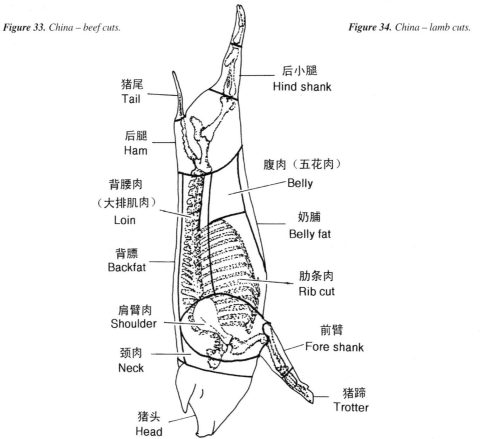

Figure 35. *China – pork cuts.*

cm dogfish, white myofibres might have a diameter of 150 to 170 μm, while red myofibres might have diameters of 18 to 35 μm. Red myofibres are innervated by widely distributed neuromuscular junctions, while white myofibres are innervated near the myocomma.

(Sources: Jordan, 1923; Hardy, 1959; Bone, 1966; Jones, 1967; Royce, 1972; Simms and Quin, 1973; Browning, 1974; Scott and Messieh, 1976; Lagler, 1977; McKone and LeGrow, 1983; Walsh, 1993)

CHOP

Pork, lamb and veal chops may be produced in many ways. If a whole pork loin includes the vertebral column and muscles from the scapula back to the pelvis, the chops may have a single eye of meat (*longissimus dorsi*) above a rib, or may have two eyes of meat, one above and one below a transverse process of a lumbar vertebra, as shown in Figure 36.

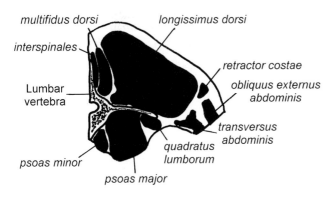

Figure 36. Pork loin chop.

CHORIZO

A highly spiced, dry → sausage containing pork. A high vinegar content and a covering of lard enhance its keeping properties in a hot climate.

(Sources: Ansorena et al., 1997)

CHRBÁT

The loin of a lamb or veal carcass in → Slovakia.

CHRBTOVÁ SLANINA

Backfat on a → Slovakian pork carcass.

CHUB

The chub is fresh-water fish, *Leuciscus cephalus*, found in many rivers of Europe (→ Cypriniformes). Its myosystem has a very large number of fine bones and is difficult to eat. The chub has a blocky shape. Also, in the meat industry, the word chub is widely used to indicate anything with a blocky shape, such as a large sausage or meat tightly packed into a bag to give it a bulbous shape.

CHUCK

In → North American beef cutting, the chuck is a large primal cut and is separated from the rib with a perpendicular cut through the vertebral column and the intercostal muscles between ribs 5 and 6 (Figure 37). Many of the smaller retail chuck steaks or roasts obtained on cutting the primal contain parts of the scapula and its surrounding muscles. In → England, the chuck is typically a medium-sized retail cut, although in → Ireland it may be a relatively large primal cut.

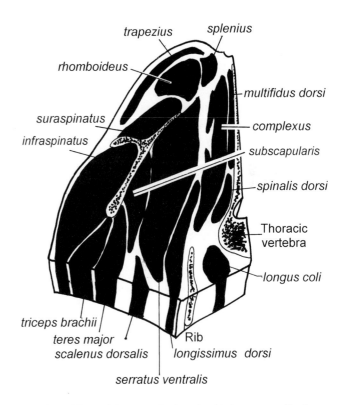

Figure 37. North American chuck steak or blade pot roast of beef.

In North American lamb cutting, a chuck of lamb is a primal cut composed of the neck, shoulder, foreshank plus breast of lamb.

CHUCK ARM ROAST

In → North American beef cutting, this is an arm steak or pot roast.

CHUCK RIB MEAT

In → Australian beef cutting, chuck rib meat is taken from the first five ribs.

CHUCK ROLL

In → Australian beef cutting, the chuck roll originates medially to the scapula and contains *longissimus dorsi, spinalis, semispinalis, multifidus dorsi, iliocostalis, serratus dorsalis cranialis, rhomboideus, subscapularis, teres major* and *serratus ventralis.*

CHUCK TENDER

In → North America, → Australia, → Ireland and → Northern Ireland, a chuck tender is a cut of beef prepared from the *supraspinatus* muscle.

CHUCK-WAGON CUT

→ Fanciful names.

CHULETA

Spanish for a chop.

CHULETAS

In Mexico, a general name for steaks or chops. In beef, they may be cut from *longissimus dorsi* of the anterior rib region. In pork, chuletas may be cut all along the vertebral column, giving chuletas de lomo (loin chops) and chuletas de cabeza de lomo (blade chops). In lamb, chuletas del centro are from the costilla, chuletas de pierna are cut through the ilium and gluteal muscles, and chuletas de aguja are cut from the ribcage after lifting off the scapula. In veal, chuletas del frente (front chops) are from the anterior rib region, chuletas del centro (centre chops) are from the posterior rib region, and chuletas de riñonada (kidney chops) are from the lumbar region.

CHUMP

A lamb chop from the posterior end of the loin. In → North America, the pin bone leg is removed by cutting perpendicularly through the vertebral column at a point level with the anterior face of the ilium. However, if the anterior part of the ilium remains in the posterior loin chump, it produces a chump chop with a very high bone content, even though both anterior and posterior faces look quite meaty. A chump of veal is cut to contain most of the pelvis . Chump is not often used as a term in relation to beef cutting, but sometimes it is, as in → Northern Ireland.

CICHLID

Freshwater tropical fish, → Perciformes.

CIMIER

In → French beef cutting, the haunch or cimier is the equivalent of the culotte, which is the posterior part of the romsteck. In Italian, the cimiero refers to the ischium.

CISCO

A member of the → Salmoniformes, *Stenodus leucichthyes,* caught in Russia and Alaska (Arctic Ocean, Caspian Sea, Ural and Volga rivers). Various other species of the genus *Coregonus* with a similar source extending to Canada also share this name.

CITY BUTT

Another name for a Boston butt of pork in the USA.

CIVET

Jugged → hare or stew in France.

CLEAR PLATE

A cut of pork in → North America, mostly subcutaneous fat with a little muscle. It originates over the → Boston butt and is used in canned pork and beans.

CLINQUETTE

An alternative name for plate de côtes in → French beef cutting.

CLOD

In many parts of → England, a clod is a cut of beef taken near to, or involving the clod bone, or humerus. Often the clod is part of a clod and sticking joint. The → Australian and → Irish usage differs from this in that the clod is centred on the scapula (not the humerus) and is dominated by muscles such as *infraspinatus, deltoideus, teres minor* and *triceps brachii.*

CLOD BONE

Humerus.

CLOD ROASTS AND STEAKS

In → North American beef cutting, clod roasts and steaks are cut across the ribs, posterior to the humerus in the chuck, with *triceps brachii* as the major muscle group. In → England, the term clod is also applied to numerous cuts of beef, but they usually exclude the ribs and are restricted to the humerus and its associated muscles.

CLUB STEAK

In → North American beef cutting and elsewhere, club steaks are cut from the anterior part of the loin, before the T-bone steaks. In → English beef cutting, club steaks are from the sirloin wing rib.

CLUPEIFORMES

This order of bony fishes contains some of the most valuable food fishes. In the family Clupeidae are the herring, sardine, sprat and shad, all with a fairly primitive type of body structure. The fins are supported by rays that are mostly branched and without any spines. The paired pectoral fins are located at the ventral edge of the gill openings while the paired pelvic fins are located quite far posteriorly, usually half way or more along the overall length of the fish. The tail is often deeply forked. Herrings occur in vast shoals feeding on plankton around the world, except in the Arctic and Antarctic. Gill-rakers (stiff combs on the inner edges of the gill arches) filter plankton from water streaming outwards over the gills. The myosystems generally have a conspicuous pattern of differentiation into red (slow) and white (fast) muscle (Figure 38).

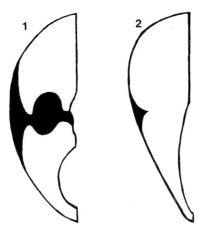

Figure 38. *Distribution of red muscle (black) in transverse steaks through tuna (1) and shad (2).*

The herrings of the family Clupeidae provide a high-quality food myosystem containing a high oil content, but with many intramuscular bones. Although the total tonnage of anchovies caught each year exceeds that of herrings, most herrings are used directly as human food whereas many of the anchovies are dried to make lower-priced products. Hence, the economic value of the herring catch is greater than that of anchovies. Most herring are caught by the centuries-old method of drift-net fishing, using a net suspended vertically in the water from floats, and weighted by a heavy cable running along the bottom edge of the net. Numerous nets may be connected to reach several kilometres in length, with the fishing vessel or drifter dragging lightly at one end. During the darkness of the night, shoals of herring ascend from the depths and are trapped by their gills. Individual fish become lodged with their heads through the meshes of the net, unable to retreat backwards because their gill covers are caught. When the nets are hauled aboard in the morning, the meshes are stretched open and the fish are released onto the deck and into the waiting holds below.

The northern hemisphere herring *Clupea harengus* (Figure 39) is of great historical importance in European trading and still has a vast present-day market. Although fresh herring is very tasty, most herrings are preserved in some way: (1), after removal of the head, fish may be salted in barrels on board the fishing vessel; (2), refrigerated herring may be gutted and smoked to produce red herring; (3), salted herring may be smoked to produce a bloater; or (4), herrings may be preserved in brine and then smoked to produce a kipper. The small yellow-orange eggs of herring may be used to make a type of caviar.

Sardinops is the pilchard or large sardine, while ordinary sardines are usually *Sardinia*, canned in oil. *Sprattus*, the sprat, is usually canned or smoked, while *Alosa*, the shad or alewife (Figure 39), is oven-cooked, although once it was valued for its ease of salting in barrels. The American shad, *Alosa sapidissima*, is the largest of the herring family and has razor-sharp scutes or modified scales along the ventral abdomen. It is prized for its taste, although it is very bony. However, many alewives such as *A. pseudoharengus* end up as canned pet food. The alewife *A. pseudoharengus* often coexists with a similar fish, *A. aestivalis*, the blueback herring. The former has a grey or pink-white body cavity lining, while the latter is sooty black. Least valuable of all, *Brevoortia*, the menhaden, has very bony muscles and is used for fish meal and oil (Figure 39).

The family Engraulidae contains the anchovies. These small (< 20 cm) fish are similar in appearance to small herrings but have an elongated snout, a larger mouth, and a rounded belly. They are widely distributed from tropical to temperate waters, with well developed fisheries off the coast of Chile, and smaller fisheries in the Northwest Pacific, in the Northeast Atlantic, in the Mediterranean, and off Australia. Like the herrings, the anchovies are planktonic filter feeders. Much of the catch is converted to oil, fish-meal or powdered products, but anchovies may also be salted and preserved whole for human consumption. *Engraulis*, the anchovy, may be eaten fresh in coastal areas (like a sardine) but it is usually salted or smoked, or used to make paste.

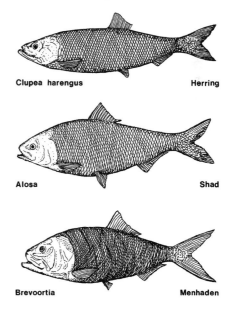

Figure 39. *Fishes of the herring family.*

(Sources: Hardy, 1959; Royce, 1972; Jessop, 1990; Dadswell, 1990)

COAL FISH

A cod-like bony fish or pollock, *Pollachius virens*, → Gadiformes.

COAMOHAS

Shoulder of → reindeer in the Saami language.

COARBBEALII

Leg of → reindeer in the Saami language.

COASTING

In → English beef cutting, this denotes forequarter flank plus brisket.

COCCYGEUS

A small triangular muscle originating from the deep surface of the sacrosciatic ligament and extending into the mammalian tail.

COD

A bony fish, *Gadus,* → Gadiformes. In beef cutting, the cod fat is subcutaneous fat ventral to the pubis.

CODILLO

The knee, or a tibial part of the ham on a → Spanish pork carcass.

CODORNIZ

Spanish for a → quail.

COGOTE

A boneless block of dorsal neck muscles in → Cuban beef cutting, or the neck of an → Argentinean beef carcass. In → Chile the cogote contains the ventral neck muscles.

COGOTE-NUCA

Neck of a beef carcass in → Ecuador.

COHETE

The *flexor digitorum superficialis* of the hindlimb in Cuban beef cutting. Figure 40 does not show very well how the cut gets its name. The muscle is composed of rocket-shaped subunits.

Figure 40. *The cohete* (flexor digitorum superficialis *of the hindlimb*) *in Cuban beef cutting.*

COLA DE BUEY

Spanish for oxtail, and a caudal cut from the beef carcass in → Chile.

COLD-SHORTENING

Rapid refrigeration of carcasses after slaughter reduces spoilage by bacteria but, if refrigeration is too rapid, it may cause the meat to become tough. In the live animal, muscle → contraction is turned on by calcium ions. Rapid refrigeration allows calcium ions to linger in the → sarcoplasm, where they cause muscle contraction. When contracted muscles develop → rigor mortis, their short → sarcomere lengths become fixed, and the greater overlap of thick and thin → myofilaments strengthens the meat. The main mechanism is that a low temperature reduces the ability of the → sarcoplasmic reticulum to sequester calcium ions. Lean animals lacking carcass insulation from fat and small animals with a large surface to volume ratio are particularly vulnerable to cold shortening.

COLITA DE CADERA

Tensor fascia lata of a beef carcass in → Ecuador.

COLLAGEN FIBRES

Microstructure

Collagen fibres are microscopic fibres composed of laterally linked tropocollagen molecules. Collagen fibres are extracellular and are formed by the activity of fibroblast cells, although the initial stages of collagen fibril formation occur within the cell. Collagen fibres in meat range in diameter from 1 to 12 µm. Unlike reticular and elastic fibres, collagen fibres seldom branch and, if branches are found, they usually diverge at an acute angle. Collagen fibres in meat are white, but they may be stained histologically with acid dyes such as eosin (pink). Unstained collagen fibres are visible with polarized light because they are birefringent. Collagen fibres have a wavy, crimped appearance which is lost when they are placed under tension.

Collagen fibres are composed of bundles of smaller collagen fibrils, themselves formed from elongated tropocollagen molecules with a staggered arrangement and lateral covalent bonding. For electron microscopy, negative staining with heavy metals causes the spaces between the ends of tropocollagen molecules to become electron dense, thus giving collagen fibrils a transversely striated appearance. Striation periodicity is 67 nm, but may shrink to 64 nm during processing for examination.

The distribution of collagen fibres in food myostems is very important. Thus, the price differentials between expensive and cheap cuts of beef are largely due to the distribution of collagen (expensive muscles have the least collagen). In fish muscle, collagen distribution is related to the mode of swimming, so that highly flexible species and parts of the body have high collagen levels.

Biochemistry

Tropocollagen molecules have three polypeptide strands linked by stable intramolecular bonds from the non-helical ends of the molecule. Stable disulphide bonds between cystine molecules in the triple helix also occur. Pyridinoline is a non-reducible, trifunctional cross-link between three tropocollagen molecules.

The great strength of collagen fibres originates from covalent bonding between adjacent tropocollagen molecules. During meat animal growth and development, the number of covalent cross-links increases and collagen fibres become progressively stronger. Meat from older animals, therefore, tends to be tougher than meat from the same region of younger animals. However, this is complicated in young animals by the rapid synthesis of large amounts of new collagen. New collagen has fewer cross-links so that, if there is a high proportion of new collagen, the mean degree of cross-linking may be low, even though all existing molecules are developing new cross-links. Thus, paradoxically, as the formation of new collagen slows down, the mean degree of cross-linking may increase.

Many of the intermolecular cross-links in young animals are reducible so that, although the collagen is strong, it is also fairly soluble. In older animals, reducible cross-links may be converted to non-reducible cross-links giving strong collagen which is far less soluble and more resistant to moist heat cooking. Changes in collagen solubility appear to have a greater impact on meat tenderness in beef from older animals than in beef from younger animals.

Differences in collagen cross-linking may occur between different muscles of the same carcass, as well as between the same muscle in different species. For example, collagen from beef *longissimus dorsi* is less cross-linked than collagen from beef *semimembranosus*, while collagen from pork *longissimus dorsi* tends to be less cross-linked than collagen from beef *longissimus dorsi*. Nutritional factors such as high-carbohydrate diet, fructose instead of glucose in the diet, low protein, and pre-slaughter feed restriction may reduce the proportion of stable cross-links. Non-enzymatic glycosylation (between lysine and reducing sugars) may be responsible for interactions between diet and collagen strength. In general, the turnover rate of collagen (typically 10% per day) is accelerated in cattle fed a high energy diet.

(Sources: Hall and Hunt, 1982; Yoshinaka *et al.*, 1988; Smith and Judge, 1991; Young and Braggins, 1993)

COLLET

In → French veal and lamb cutting, the collet is the neck. In → Luxembourg it may be the neck of beef. Collet may be used instead of beef collier in → Belgium.

COLLICLE

In North America, this is the kosher name for a cut of beef prepared from the *supraspinatus* muscle.

COLLIER

In → Belgian and → French beef cutting, the collier is a primal cut containing the neck. It may be subdivided into the veine maigre and veine grasse.

COLLO

The neck of a carcass in → Italy.

COLUDAS

Beef from over the posterior four ribs and false ribs in → Chile.

COMBER

A loin or saddle of lamb in → Poland.

COMPLEXUS

→ *Semispinalis capitis*.

CONDITIONING

Some types of meat, such as beef, are intrinsically tough and tasteless if eaten soon after slaughter. If conditioned (stored in a meat cooler at about 4°C either as a whole carcass or as a vacuum-packed primal) for a week or so after slaughter, the meat becomes more tender and tasty. It takes this long because living muscles protect themselves against autolysis (self-digestion) which might otherwise be initiated by the extreme changes in shape which accompany strong muscle contractions, and because meat cooler temperatures are far below the living body temperatures of mammals and birds. Thus, the reverse problem exists with tissues such as liver and with muscle foods from cold-blooded animals - they undergo autolysis too rapidly. With these latter commodities, where post-harvest storage causes problems, the term conditioning is not used (→ pyloric caecum). Thus, conditioning is a beneficial process applied to otherwise tough types of meat. It is expensive because of the cost of storing the meat and because of the fluid losses from the meat (drip and evaporation).

A variety of complex microstructural and biochemical changes occur as meat is conditioned. The strong rigor bonds between → myosin and → actin may be weakened, or myosin molecule heads may become detached from the thick → myofilaments. Calpains (calcium-activated proteases) may attack titin, nebulin, vinculin, desmin and dystrophin so that → sarcomere structure is greatly weakened. Collagenous connective tissues are weakened. Increasing ionic strength causes myofibrillar proteins to dissolve.

Gourmet conditioning of beef and other intrinsically tough meats is best done with aerobic exposure of primal cuts. It may take up to a month from the time of slaughter, during which time exposed muscle surfaces should become dry, black and lightly dusted with white mould. Thus, it is important to separate the cuts of meat, both from each other and from meat that is not being conditioned. In an age when the general public in most developed countries is truly paranoid about meat hygiene, gourmet conditioning must look like a dangerous procedure. But, of course, it is not. The inside of the meat is sterile (assuming the carcass has been properly inspected and no surface bacteria have been pushed deep into the meat on a skewer or by knife work). Free competition among bacteria for surface spoilage makes proliferation of pathogenic bacteria unlikely and, anyway, the surface receives the full impact of cooking. High temperature roasting is required to bring out the full flavour of the meat.

(Sources: Wu and Smith, 1987; Stanton and Light, 1990; Goll *et al.*, 1997)

CONEJO

Spanish for → rabbit.

CONIGLIO DA CARNE

Meat → rabbit in Italy.

CONNECTIN

→ Titin (α-connectin = titin 1).

(Source: Kimura and Maruyama, 1989).

CONTRA

A lateral cut of beef from the hindlimb or pierna in → Mexico.

CONTRA BRIZOLES

In → Greece, the lumbar region or *longissimus dorsi*.

CONTRACTION

The calcium ion concentration of the cytosol (the aqueous component of the → sarcoplasm) is regulated by the → sarcoplasmic reticulum. In a resting myofibre, the calcium ion concentration is very low (about 5×10^{-8} M). When prompted by a chain of events initiated by the nervous system, the sarcoplasmic reticulum releases calcium ions (to reach about 5×10^{-6} M). Thus, calcium ions provide the trigger for muscle contraction, as outlined below.

(1) The signal to start a contraction is initiated somewhere in the central nervous system, either as voluntary activity from the brain or as reflex activity from the spinal cord.

(2) A motor neuron in the ventral horn of the spinal cord is activated, either locally or by signals coming down the spinal cord, and an action potential passes outwards in a ventral root of the spinal cord.

(3) The axon carrying the action potential branches to supply a number of myofibres in the same muscle (a motor unit), and the action potential is conveyed to a motor end plate on each myofibre.

(4) At the motor end plate, the action potential causes the release of packets or quanta of acetylcholine into the synaptic clefts on the surface of the myofibre.

(5) Acetylcholine causes the electrical resting potential under the motor end plate to change, and this then initiates an action potential that passes in both directions along the surface of the myofibre.

(6) At the opening of each transverse tubule onto the myofibre surface, the action potential spreads deep inside the myofibre.

(7) Prompted by the transverse tubules, the sarcoplasmic reticulum releases calcium ions.

(8) The calcium ions are detected by → troponin molecules which cause → tropomyosin molecules to roll into the grooves on the thin myofilaments, thus allowing → myosin molecule heads to attach to → actin molecules and to move the thin → myofilament.

Relaxation occurs when the following sequence happens.

(9) Acetylcholine at the neuromuscular junction is broken down by acetylcholinesterase, thus terminating the stream of action potentials along the myofibre surface membrane.

(10) The sarcoplasmic reticulum ceases to release calcium ions, and immediately starts to resequester or obtain back all the calcium ions that have been released.

(11) In the absence of calcium ions, changes in the configuration of troponin and the location tropomyosin then block the action of the myosin molecule heads, and contraction ceases.

(Sources: Hasselbach and Oetliker, 1983)

CONTRAFILET

This is a very widespread name or concept in meat cutting. In general, if the psoas muscles are the filet (or fillet or filé), then the *longissimus dorsi* may be named the contrafilet because it is opposite or contra to the filet. A clear-cut example is seen in veal cutting in → Brazil, where there is a filé mignon (*psoas*) opposite a contrafilé (*longissimus dorsi*). However, this logic fails if the filet is a muscle other than the *psoas*. Thus, the meat cutting context must be known before the contrafilet can be defined.

CONTRATAPA

A cut from the posterior part of the round in a → Spanish beef carcass.

CONTREFILET

→ Contrafilet

COPERTINA

The *infraspinatus* of an → Italian beef carcass.

COPERTINA DI SOTTO

Muscles medial to the scapula (*subscapularis* and *teres major*) in an → Italian beef carcass.

COQ DE BRUYÈRE

French for → grouse.

CORAÇÃO DA PALETA

In → Brazil, the coração da paleta (the heart of the paleta) is the *triceps brachii*.

CORACOBRACHIALIS

A small muscle on the medial surface of the mammalian shoulder joint (Figure 41). In beef and lamb it may be divisible into two parts, but not in pork. It originates from the scapula, on the coracoid process and around the glenoid, and is attached medially on the humerus.

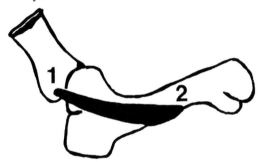

Figure 41. *Medial view of beef* coracobrachialis *(solid black) showing its origin from the scapula (1) and insertion on the humerus (2).*

CORDERO

Spanish for lamb.

CORONET

A cut from the scapular region in → Proten beef in England.

COSCIOTTO O QUARTO

A haunch or quarter of lamb in → Italy.

COSTAMERE

A rib-like band of → vinculin around a → myofibre.

(Source: Pardo *et al.*, 1983)

COSTA RICA - BEEF CUTS

In Costa Rica, the source of the beef is mostly *Bos indicus*, forage fed to a live weight of 450 to 500 kg at 3 to 3.5 years. With not many tender muscles, most beef is retailed as boneless cuts, including one or several muscles, as tabulated below. The terms may be useful for other Spanish-speaking countries in the Americas where documentation could not be obtained.

Alipego - *transversus abdominis* and *obliquus abdominis* (*internus* and *externus*)

Cacho de paleta - *supraspinatus*

Costilla - diaphragm, intercostals, *latissimus dorsi* and *serratus ventralis*

Falda - same as alipego

Lomito - *iliopsoas*

Lomo - *longissimus dorsi*

Lomo de paleta - *teres major* and *triceps brachii*

Lomo rayado - *spinalis dorsi*

Pecho - *pectoralis profundus* and *superficialis*

Pescuezo - *atlantis, brachiocephalicus, splenius* and *trapezius*

Quititeña - *serratus ventralis* and *subscapularis*

Ratón delantero - *biceps brachii, brachialis, coracobrachialis,* distal extensors and flexors of the forelimb, and *gastrocnemius*.

Ratón trasero - *gastrocnemius, peroneus, popliteus, tibialis,* and distal extensors and flexors of the hindlimb

(Source: Delgado, 1988)

COSTATE

An → Italian beef cut containing the anterior thoracic vertebrae after the removal of the shoulder.

COSTATO

Italian for the ribs of an animal.

COSTELA

A rib cut of pork in → Brazil (→ Costeletas).

COSTELA DO TRASEIRO

The postero-ventral region of a → Brazilian beef carcass.

COSTELAS

In → Brazil, a long cut of lamb through the midlength of the ribs.

COSTELETAS

In → Portugal, a costela is a rib. Costeleta is a chop. In pork, lamb and veal, the chops of the lumbar region are called costeletas do lombo, while those containing ribs are costeletas com pé (chops with shoulder). In Portuguese pork, chops from the cervical region are costeletas do fundo (deep chops), while in lamb the deep chops are from the anterior thoracic region. Costeleta is also used for chops on an →Argentinean pork carcass.

COSTILLA

Costilla is Spanish for a rib or cutlet. In → Costa Rica, the costilla is beef diaphragm, intercostals, *latissimus dorsi* plus *serratus ventralis*. In → Mexico, the beef costilla is a dorso-ventral cut through the scapula from the espaldilla y costillar while in lamb it is restricted to the thoracic vertebral region. The costilla of a Mexican lamb carcass may be a whole roast. In the Mexican veal carcass, the costillar de centro y lomo is separated into anterior chuletas del centro (centre chops) and posterior chuletas de riñonada (kidney chops). In → Argentina, the postero-ventral rib region of both pork and lamb carcasses is called the costillas.

COSTILLAS

A large cut containing the mid and ventral rib regions of a → Spanish beef or pork carcass. The "arched rib cut", costillas arqueados, contains the portions of rib close to the thoracic vertebrae in → Chile.

COSTILLITAS

Chops taken from the ventral rib region in a Mexican pork carcass.

COSTOLETTE

The rib region of an → Italian lamb carcass.

CÔTE

Rib of beef in Quebec.

CÔTE À LA NOIX

In → Belgium, this is a centre cut from a pork loin.

CÔTE À L'ECHINE

A cut from the anterior part of the sternum in a → Belgian pork carcass.

CÔTE A L'ORIETTE

The côte au spirling from the anterior rib region of a → Belgian pork carcass.

CÔTE AU FILET

In → Belgium, a centre cut from a pork loin.

CÔTE AU JUIF

Equivalent to the côte au spirling from the anterior rib region of a → Belgian pork carcass.

CÔTE AU SPIRLING

Anterior thoracic ribs from a pork carcass in → Belgium.

CÔTE COUVERTE

Rib of beef, entrecôte avant, in → Belgium.

CÔTE D'ÉCHINE

Equivalent to the côte découverte of a → Belgian beef carcass.

CÔTE DÉCOUVERTE

An anterior thoracic cut from a → Belgian beef carcass .

CÔTELETTES

French for a chop. In → French lamb cutting, the côtelettes of the thoracic region are subdivided into posterior premières and anterior secondes, and the loin chops are called the côtelletes filet. In → Luxembourg, the pork côtelettes extend as far forward as the neck.

CÔTES BASSES

Deep ribs of beef medial to the scapula in →Luxembourg.

CÔTES LEVÉES DE FLANC

In Quebec, side spareribs of pork.

CÔTES PREMIÈRES

Prime rib roast of beef in → Luxembourg.

COTTAGE BUTT

In → North America, cottage butt may be a smoked roll of pork shoulder, often tightly bound.

COTTO SALAMI

A smoked, emulsion → sausage.

COUDE

The distal half of an épaule in → French lamb cutting.

COUNTRY BACK BONES

In → North America, country back bones of pork do not include the vertebral column, but are the ribs plus intercostal muscles taken from between rib chops and spareribs.

COUNTRY STYLE SPARERIBS

In →North American pork cutting, although spareribs originate at the ventral or sternal region of the ribs, sometimes the dorsal part of the rib, complete with *longissimus dorsi*, is called country style sparerib.

COUTURIER

Couturier is the sartorius muscle. It gets its name from human anatomy, where tailors (French, couturier; English, sartor) used to sit cross-legged, working on their medial thigh muscles.

COXÃO

The hindlimb of a beef or veal carcass in → Brazil. The coxão-duro is the *biceps femoris* while the coxão-mole is mostly *pectineus, adductor, semimembranosus, gracilis*, and *sartorius*.

C-PROTEIN

About 40 molecules of C-protein occur in each thick → myofilament. C-protein has no ATPase activity, although it may inhibit actomyosin independently of calcium ion concentration. C-protein has a flexible rod-like shape, 32 nm by 3 nm, and may be involved in holding together the thick myofilament during muscle contraction or in regulating the length of the thick myofilament, although it might also bind in series to connectin. C-protein exists in different isoforms in fast and slow muscles and is especially abundant in fast myofibres.

(Source: Fürst *et al.*, 1992)

CRAB

Sources

Crabs are decapod crustaceans of the suborder Macrura-Reptantia (→ Crustacea). Shore crabs originate from coastal fisheries, while deep-sea fisheries produce long-legged crabs such as the Alaska king crab. The common edible crab in England (*Cancer pagurus*) was known to the Romans and, during the middle ages, local seacoast industries developed for the systematic harvesting of crabs in baited wicker traps or pots. Starting in the nineteenth century, the traditional industries were shaped by: (1), the development of transport systems, from railways to air freight, that enabled the rapid distribution of fresh produce to inland cities; (2), the development of microbiologically safe methods of food preservation; and (3), the statutory regulation of the types and total amounts of crustaceans that can be harvested.

Although king crabs (*Paralithodes camtschatica*) were first canned in Japan at the end of the nineteenth century, the deep-sea Alaska king crab fishery in North America was essentially a new industry created in the 1950s. By 1964, it peaked at an annual catch of about 40 million kilograms. New equipment was developed, primarily large steel-framed crab pots and lifting gear for deep-sea fishing, but also electronic navigation equipment. Fishing vessels were built to meet the exacting demands imposed by an on-board storage tank requiring a complete water change every 20 minutes. When moving to new fishing grounds, heavy crab pots stacked on deck give the boats a high center of gravity. This makes sailing difficult in the face of the hurricane-force winds and heavy icing that occur during the peak fishing months in the Winter.

The two main types of crabs are: (1) typical shore crabs with short legs and a carapace flattened from above (Figures 42 and 43), and (2) deep-water crabs with small bodies, long legs and a spider-like appearance (Figure 44). The anomuran spider-like crabs such as *P. camtschatica* may be readily distinguished because they have a total of only four pairs of large legs. All the other crabs have five pairs of large legs.

Deep-sea crabs

P. camtschatica, the Alaska king crab, is caught in heavy steel-framed rectangular pots in the deep, cold waters of the North Pacific. The commercial fishing grounds extend from Japan, past the Aleutians and the Bering Sea, and down as far as northern British Columbia. Alaska king crabs exhibit a range in colouration, from brownish-red or purple-red to greenish-white. The males grow larger than the females and may reach a carapace width of 28 cm and a weight of 11 kg. Crabs with a carapace width less than 18 cm are returned to the sea. Typical meat yields for the Alaska king crab range from 20 to 25% of the live weight, with the leg meat being by far the most valuable.

Chionoecetes opilio (Figure 44) is a true brachyuran spider crab caught in baited pots on muddy or sandy sea floors at a depth of from 75 to 450 m. It is caught over a wide geographical range in the cold waters around North America and is one of the most important species of crab caught off Eastern Canada. Together with other similar species (*C. bairdi*, *C. tanneri*, and *C. angulatus*) this important type of spider crab is usually sold as snow crab. Only the males are harvested, and these may grow to a carapace width of about 13 cm and a weight of about 0.7 kg (15 to 20% of which is meat). The market for snow crab (formerly called queen crab) was developed by the Alaskan crab fishery when king crabs became scarce due to initial overfishing. Sodium hexametaphosphate at 0.15 to 0.25% of total volume may be used to prevent struvite crystals forming in the canned product. Citric or comparable acids at 0.1% inhibit blue discolouration, aiming for a final pH of 6.85. Variation in the raw product is important: moulting crabs with paper shells giving a high pH (>7.6) after processing unless corrected. A variety of methods have been developed to avoid using acid or bisulphite to prevent blue discolouration.

Maia squinado (Figure 44) is another brachyuran spider crab and is the species that dominates the European market for spiny crabs or spider crabs, particularly in France as araignée de mer. Like other spider crabs, *Maia squinado* uses its spines to support a camouflage network of algae and detritus. The species occurs in the Atlantic and Mediterranean. In some areas it is purposefully fished, usually over sandy or detritus-littered sea bottoms, but in other areas it is caught accidentally in lobster pots. It is usually reddish in colour and may grow to a considerable size (to a carapace width of about 20 cm).

Shore crabs

The market for typical shore crabs is rather unbalanced in the number of species involved. One species, *Cancer pagurus*, (Figure 42) dominates almost the whole domestic market in England and the rest of Europe while, on the other side of the Atlantic, the East coast of North America provides a profusion of different species for the domestic market. There are major fisheries for *C. pagurus* off Northeast and Southwest England, Norway and France, with secondary centres off Ireland, Portugal and Spain. The European edible crab is usually caught in baited pots, but individual animals are often captured on the seashore at low tide. The pinkish-brown carapace has nine blunt projections on each side of the eyes and may reach a width of about 24 cm, although most of the catch is composed of smaller individuals. *C. pagurus* has a relatively high meat yield reaching 30% of live weight in males, depending on the relative size of the claws. A traditional method for preparing dressed crab is to extract the muscles of the legs and to serve them mixed with fragments of digestive gland (hepatopancreas) in the bowl-like inverted carapace.

Callinectes sapidus (Figure 43) is probably the most well known traditional crab of the East coast of the USA. It is called the blue crab, although the upper part of its carapace has a distinctly greenish appearance. The carapace grows to a maximum width of about 20 cm, but this includes the conspicuous lateral horns that are the distinctive feature of the species. Another diagnostic feature is that the hindmost large leg on each side forms a paddle for active swimming. Unlike many species of crab, the meat from the relatively slender claws is less highly valued than the backfin lump - the equivalent of a liver paté formed from the hepatopancreas. This explains why the species is often sold as soft-shelled crab immediately after moulting. After moulting, striated muscles swell to their new size by absorbing water and their characteristic taste and texture is lost for a while. This poses a serious loss in species that are valued primarily for their myosystems, but not in the meat of the dressed blue crab which is dominated by hepatopancreas.

Menippe mercenaria, the southern stone crab (Figure 43) is caught off South Carolina and Texas and provides a striking contrast to the blue crab since it normally has massive black-tipped claws that are valued for the myosystem they contain. The other legs have pointed tips used by the crab to wedge itself into mud holes or rock crevices that it then defends fiercely with its large claws. The grey-coloured carapace grows to a width of about 12 cm. Between these two extreme types (the blue crab versus the southern stone crab), are species with an intermediate form such as the rock crab, *Cancer irroratus* (Figure 43). Rock crabs may be caught with nets, dredges or traps, depending on whether the sea bed is rocky or sandy. Their geographical range is from Labrador down to Florida. The yellowish carapace is dotted with brown, it has nine blunt teeth on each side, and may reach a width of about 11 cm. About 45 crabs are needed for a kilogram of meat.

The Jonah crab, *Cancer borealis* (Figure 43), is a similar species to the rock crab but has a more northerly distribution (from Long Island to Nova Scotia). The Jonah crab may be distinguished from the rock crab by having sharper teeth on the sides of its carapace. The Jonah crab also is a more massive animal and may reach a weight of 1 kg, although the crabs caught near to shore are usually half this weight. Initially an uninvited visitor in lobster pots, there is a developing market for Jonah crabs, although their more massive exoskeleton makes meat extraction rather difficult. Another developing catch is for the deep-sea red crab, *Geryon quinquedens* (Figure 42). These crabs inhabit sandy or muddy bottoms on the continental shelf from Nova Scotia down to Brazil at a depth of from 300 to 900 m. Because of its rather long legs, *G. quinquedens* is a fragile crab that is better preserved by being enticed into a baited pot rather than by being trawled, although trawling is the easier method of fishing. Males with a carapace width of 15 cm may weigh about 1 kg. Females only reach half this weight.

On the West coast of North America from the Aleutian Islands down to California, is the dungeness crab, *Cancer magister* (Figure 42). There is an established fishery with a range of fresh and processed products. It is distinguished from the red rock crab, *C. productus* (Figure 42) by the relatively slender, light-coloured fingers to its claws. In India, *Scylla serrata* is available in large quantities throughout the year.

Land crabs

Land crabs with large claws provide a local source of skeletal muscle, such as *Geocarcinus lagostoma* in Brazil. However, be warned, some of them feed on toxic plant matter and require several days of clearance feeding on leftover salad!

(Sources: Wilder, 1966; Gangal and Magar, 1967; Motohiro and Inoue, 1970; Varga *et al.*, 1971; Gillies, 1971; Dewar *et al.*, 1972; Holmsen and McAllister, 1974; Dassow and Learson, 1976; Bigford, 1979; Edwards, 1979; FAO, 1981; Ke *et al.*, 1981; Butler, 1988; Bailey and Jamieson, 1990; Johnson *et al.*, 1998)

CRAMP BONE

Patella

CRAWFISH

A crustacean of the genus *Palinurus*. → Lobster.

CREATINE PHOSPHATE

Creatine phosphate may be used by creatine phosphokinase (CPK) to regenerate ATP from ADP (→ adenosine triphosphate) and, thus, creatine phosphate functions as a

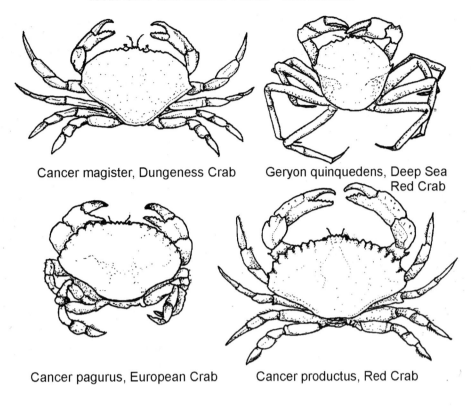

Cancer magister, Dungeness Crab

Geryon quinquedens, Deep Sea Red Crab

Cancer pagurus, European Crab

Cancer productus, Red Crab

Figure 42. *European and North American shore crabs.*

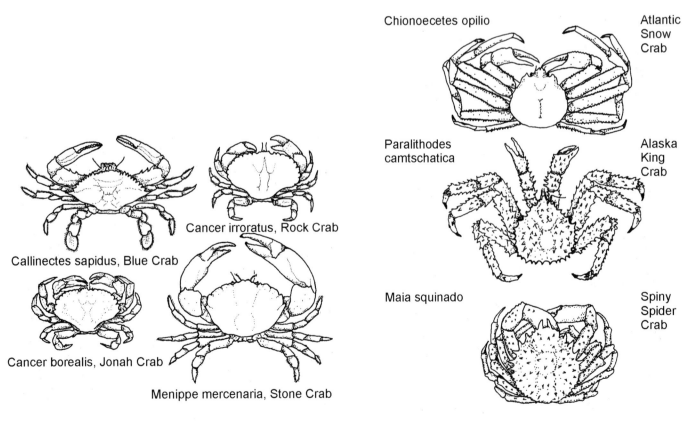

Cancer irroratus, Rock Crab

Callinectes sapidus, Blue Crab

Cancer borealis, Jonah Crab

Menippe mercenaria, Stone Crab

Chionoecetes opilio

Atlantic Snow Crab

Paralithodes camtschatica

Alaska King Crab

Maia squinado

Spiny Spider Crab

Figure 44. *Deep-sea crabs.*

Figure 43. *North American shore crabs.*

short-term, readily exhausted store of energy. CPK is a large enzyme. If it leaks from the → sarcoplasm into the blood, it becomes an indicator of muscle plasma membrane damage, as in pigs with → porcine stress syndrome or nutritional myodegeneration.

CROCODILE

Alligators, caimans and crocodiles are all in the Order Crocodylia. Hunting for their skins has pushed several species near to extinction. Illegal hunting remains a problem. But meat from alligators, caimans and crocodiles is now produced as a byproduct of ranching crocodylians for their skins and is commercially available as a gourmet item in the USA, Australia and elsewhere. In captivity, crocodiles may be fed with diets based on red meat or fish. Off-odours sometimes associated with a grassy or fishy taste originate from hexanal and heptenal.

(Sources: Broad and Luxmore, 1989; Piedra *et al.*, 1997; Baek and Cadwallader, 1997b)

CROP

In →New Zealand beef cutting, the crop is a cut including the mid-length portion of the ribs,.

CROSS-BRIDGES

In electron micrographs of skeletal muscle, the → myosin molecule heads protruding from thick → myofilaments often connect across to the thin myofilament and are called cross-bridges. In living muscle, cross-bridges are involved in thick and thin myofilaments sliding past each other during muscle → contraction, although exactly how they do it is still uncertain. In meat, immobile cross-bridges between overlapping thick and thin myofilaments are responsible for → rigor mortis.

The energy for muscle contraction is provided by the hydrolysis of phosphate from → adenosine triphosphate (ATP). Transduction from chemical to mechanical energy may be delayed until ADP and inorganic phosphate are released by myosin when it recombines with actin. The conformational change that causes cross-bridge movement may not be a simple angular change of the cross-bridge as pictured in many older or introductory textbooks. Instead, the movement may occur elsewhere in the molecule, perhaps in the long α-helix of the myosin molecule head snapping back to its starting position after being electrically charged then discharged. Alternatively, axial rotation may only occur in a minority of myosin heads at any one time, thus making it difficult to detect.

Whatever the details of cross-bridge movement, myofilament sliding and muscle contraction originate from very large numbers of myosin molecules. Each individual stroke by a myosin molecule head takes about 1 millisecond and produces a 12-nm movement. Although this is a very small distance, there are many thousands of sarcomeres in series and, over a short time, the sum of all these small distances may be measured in centimetres. The myosin head only releases its grip on an actin, and swings back for another power stroke with another actin, if it is recharged by another ATP molecule, which is why depletion of ATP causes rigor mortis and marks the conversion of muscle to meat.

(Sources: Barden and Mason, 1978; Rayment *et al.*, 1993; Elliott and Worthington, 1994)

CROSS-CUT CHUCK

In → North American beef cutting, a cross-cut chuck is the chuck plus the brisket, plus the shank.

CROSSE

A transverse cut across the distal part of the ulna and radius in a → French beef carcass.

CROSS RIB ROAST

A → North American beef roast composed mainly of *triceps brachii*, ventral to the scapula in the chuck.

CROWN ROAST OF LAMB

This method of preparing an attractive cut of meat, usually lamb but sometimes pork as well, appears in many countries. (Figure 45). Preparation starts with a long, straight length of vertebral column, plus dorsal parts of the ribs and associated muscles, such as a rack of lamb. The vertebral column is trimmed away so that the rack can be twisted into a crown and secured with string. Now the subcutaneous fat is in the center of the crown, while the medial (pleural) surfaces of the ribs face outwards. Intercostal muscles may be trimmed away to leave each rib projecting outwards like the point of a crown. The crown is carved by separating the tissue between the ribs to produce a series of rib chops which can be held by the rib when eaten.

Figure 45. *A crown roast of lamb.*

CRUSTACEA

Occurrence

Crustaceans such as → shrimps, → crabs and → lobsters originate from traditional fisheries supplying local markets in non-industrialized communities, as well as from technically sophisticated industries in developed countries producing a wide range of fresh-frozen and processed luxury foods. The most uniformly distributed crustacean fishery is for shrimps and, on some coastlines, such as around India and northern Australia, shrimps are the dominant crustacean catch. There are well developed fisheries for crabs from the South China Sea, through the Sea of Japan and round to Alaska and British Columbia, and encircling the Atlantic from the Gulf of Mexico, along the east coast of the United States and Canada, as well as Europe. The dominant lobster fisheries are along the Atlantic coasts of Europe, off Australia and Brazil, along the east coast of the USA, and north of the Cape of Good Hope.

Segmentation

Shrimps, lobsters, crabs, crayfish and krill are all crustaceans, one class of a familiar phylum of animals called Arthropods which includes the insects and spiders. The name arthropod means joint-footed, although the designation joint-legged might have been more appropriate. In practical terms, this means that food myosystems derived from crustaceans are obtained from the internal muscles that operate parts of the exoskeleton (Figure 46). The exoskeleton is composed of calcium carbonate and N-acetylglucosamines, frequently coloured by carotenoid pigments. Some parts of the exoskeleton are repeated serially along the length of the animal as, for example, in the segments of a lobster tail. The serial repetition of body segments, each with its own set of vital organs, is called metameric segmentation. In the Crustacea, however, the body segments are modified into groups that share a common function. Each group is called a tagma (plural, tagmata).

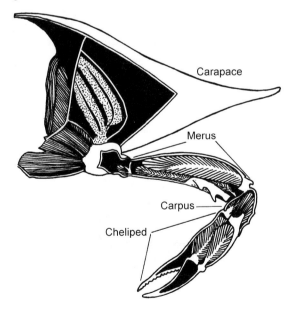

Figure 46. A transverse section through one side of the thorax of the blue crab, Callinectes sapidus, *showing the origin of muscles (striated) and gill (stippled)*

Classification

The relatively large crustaceans used as human food are in the subclass Malacostraca. Most adult Malacostracans have a head composed of five segments with antennae and mouth parts, a thorax of eight segments with accessory mouthparts (3 pairs) and legs (5 pairs), and an abdomen of six segments with small legs followed by a tail-fan or telson. Food myosystems are derived primarily from the thoracic legs used for walking or as grasping claws, and from the strong abdominal muscles that flap the telson downwards when the animal shoots backwards when startled. The animals that provide these myosystems are mostly members of a division of crustaceans called the Eucarida, all of which have a complete carapace or shell fused to the thoracic segments. There are five main orders in the Eucarida.

1. Order Euphausiacea. This contains the planktonic krill. These small (2 to 5 cm) shrimp-like crustaceans are harvested by Russian ships in the Antarctic.

2. Order Decapoda, suborder Macrura-Natantia. This contains shrimps and prawns with laterally compressed bodies and with legs adapted for swimming.

3. Order Decapoda, suborder Macrura-Reptantia. This contains the lobsters and crayfishes with strong abdominal segments and five thoracic legs adapted for grasping and/or walking.

4. Order Decapoda, suborder Anomura. This contains the Alaska king crab and similar species with reduced abdominal segments held under the carapace. The fifth pair of legs is greatly reduced in size so there are only four large pairs of legs adapted for walking and grasping.

5. Order Decapoda, suborder Brachyura. This includes the typical edible crabs with a vertically compressed body, a vestigial abdomen held tightly under the carapace, and large claws on the first of the five pairs of walking legs.

All the animals listed above are Malacostracans (with a 5:8:6 pattern of segments) and Eucaridans (with a complete carapace fused to all 8 thoracic segments). However, one other type of Malacostracan also yields a food myosystem, although of minor importance. The Hoplocarida have an incomplete carapace that only covers the front of the thorax. On the second pair of limbs there are large claws resembling those of a praying mantis, hence the common name for these animals - the mantis shrimps, *Squilla empusa*, of the east coast of the USA and *S. mantis* in Europe.

Processing

There are probably as many ways to process crustacean myosystems as there are methods to catch crustaceans, but certain operations and the sequences in which they are performed are fairly ubiquitous. First the catch must be sorted so that a mass of the same or similar species can be assembled. All too often the rejected animals are dumped overboard despite the fact that much of their protein content could be salvaged for animal feed or some other low-grade type of product. With crustaceans such as crabs and lobsters it is essential that dead or moribund animals are rejected. After the catch is sorted, some type of butchering is usually required for species that are to be completely processed and packaged for a period of storage. With crabs, for example, species which provide only a myosystem without any hepatopancreas are normally split open in the dorsal midline so that the shoulder meat is separated in a skeletal unit together with the leg meat.

The details of cooking procedures depend largely on the nature of the species involved. After being boiled, crabs may be chilled in cold water. This causes the meat to shrink, making its subsequent separation easier. There are a number of techniques for isolating the meat from crustacean exoskeletons, but the basic mechanical operations consist of: (1), compression rollers that force the meat out of tubular parts of the exoskeleton; (2), blowing with compressed air; and (3), shaking. The separation of shell debris, gill tissue and fat is often facilitated by a few minutes of floatation in brine with a high (> 90%) salinity. Ultraviolet light may be used to detect pieces of the carapace or shell remaining with the separated meat.

Before methods were developed for the preservation of lobster meat in top-quality condition by rapid freezing, the preparation of lobster paste was one of the few ways for preservation and distribution. Traditional lobster and crab pastes are still commercially important. The yield of boiled lobster is approximately 25% claws, 30% tail, 43% body, and 2% juice. The hepatopancreas is easily reduced to a smooth oily paste by cooking. Although the taste is very palatable, the colour may remain an unattractive green. The roe (immature eggs) of the female lobster may be waxy and crumble into separate eggs with a bright red colour and a good flavour. The fat content is approximately 12%. When extracted with rollers, the eight pairs of walking legs yield long thin strips of white meat covered with red skin. The yield of leg meat from boiled lobster may only be 3% by weight. The body meat is composed of rings of muscle attached to the body from where the tail is removed, and it is covered by the paper shell with an intense red colouration. The yield of body meat from boiled lobster is about 2.5%. The muscles that operated the legs and claws from within the body cavity are not easily extractable because they are enclosed in the rigid internal compartments of the exoskeleton. The yield is around 2.5%.

The preservation of top-quality crustacean myosystems by freezing was started in the late 1950s. The problems encountered with early methods included toughening of the meat, the development of off-flavours, and the difficulty of separating the meat from the shell after thawing and cooking. Lobsters immersed in water at 90°C for 70 seconds may be easily shucked, before or after freezing. After being packed and frozen in a 2 to 3% brine, the flavour may be acceptable for several months. The acceptable storage time of claw meat is much less than that of tail meat. Because of the small size of crustaceans such as shrimps and the relatively inaccessible location of many of the smaller muscles in the larger crustaceans, the automated recovery of crustacean myosystems has demanded considerable ingenuity by the designers of seafood processing equipment.

Myosystem structure

Crustacean → myofibres show a great range in diameter, up to 5 mm in the giant barnacle. By vertebrate standards, the range in sarcomere length also is vast, from 2 μm to 14 μm. The logarithm of twitch duration may be correlated with → sarcomere length, and sarcomere length may follow gradients along the limbs. Long sarcomeres may be near the body in crabs, while short sarcomeres may be near the body in lobsters. The plasma membranes of crustacean myofibres have a large number of clefts, into which open the transverse tubules for excitation-contraction coupling. Thus, gross electrical capacitance of crustacean myostems may be very high (30 μF/cm^2 membrane) relative to vertebrate muscle.

Slow (tonic) myofibres tend to have wide, irregular → Z-lines, numerous intermyofibrillar → glycogen granules and subsarcolemmal mitochondria, a high ratio of thin to thick → myofilaments (up to 7:1), low amounts of → sarcoplasmic reticulum, and H bands and M-lines which may be indistinct or absent. T-tubules may traverse the sarcomere at its midlength, level with the H zone, or at the A-I junction, while extra tubules (Z tubules) may traverse at the Z-line (but not in fast myofibres). Fast (phasic) myofibres, on the other hand, have a more compact internal structure with thick and thin myofilaments in a hexagonal array (six thin around one thick), short sarcomeres, and extensive sarcoplasmic reticulum. Differences between fast and slow myofibres are reflected in histochemical tests for ATPase and aerobic enzyme activity (strong ATPase and weak aerobic activity in fast myofibres, and vice versa). Whereas the histochemical properties of vertebrate myofibres are influenced by trophic effects from the motor innervation, this control is far less important in crustacean myofibres, where control of myofibre diameter appears to originate from the degree of passive tension on the myofibre. Thus, a tonic neuron may innervate both phasic and tonic myofibres.

Although toughening may be a problem with myosystems from larger crustacea, a loss of initial structural integrity seems to be the major problem with myosystems from smaller crustacea. Thus, frozen products readily become stringy or mushy. The basic cause is most likely biological, in that crustacean muscles are protected by a strong exoskeleton and operate in a weightless, aquatic environment. Thus, apart from → apodemes, crustacean myosystems may lack most of the connective tissues above the level of the endomysium that protect and bind together myofibres of animals exposed to gravity and bruising. To protect vertebrate muscles from overextension, there are strong internal connective tissues in parallel with the myofibres. Whereas, in crustacea, overextension is limited by the hinge-joint structure of the exoskeleton.

Although many crustacean muscles contain a high → glycogen concentration when alive, little or no glycogen survives post-mortem metabolism. Thus, for shrimp, a typical proximate analysis may be: 78% water, 0.5% ash, 20% crude protein, and 0% carbohydrate. Cholesterol is the dominant steroid, and originates from dietary sources. The characteristic aroma of various types of crustacea is determined by the relative amounts of 2,3-butanedione, 2-methyl-3-furanthiol, 2-acetyl-1-pyrroline, 3-(methylthio)propanol and 2-acetyl-2-thiazole.

(Sources: Ross, 1927; Getchell and Highlands, 1957; Peachey, 1967; Fahrenbach, 1967; Selverston, 1967; Gillies, 1971; Atwood, 1973; Bittner, 1973; Morin and McLaughlin, 1973; Boone and Bittner, 1974; Giddings and Hill, 1976; Govind *et al.*, 1978; Rossner and Sherman, 1978; Ogonowski and Lang, 1979; Moore and Eitenmiller, 1980; Chapple, 1982; Baek and Cadwallader, 1997a)

CSONTOS OLDALAS

In a → Hungarian beef carcass, the side bones or csontos oldalas is a long ladder of rib sections.

CUADRIL

A "quadrilateral" or box-like beef cut in → Argentina which includes the whole of the pelvis.

CUBA - BEEF CUTS

The pattern of beef cutting in Cuba is derived from that of Spain, as found in the Madrid region. Quarters of beef are completely boned, following the major muscles and their fasciae as detailed below.

Bola The *quadriceps femoris* group of muscles is called the bola or ball, rather like the boule d'aloyau (sirloin ball) in → France, or the kugel (ball) in → Germany.

Cañada The muscles medial to the femur: *adductor, pectineus* and *semimembranosus*.

Cogote A block of dorsal neck muscles.

Cohete The rocket-shaped (fusiform) *flexor digitorum superficialis* of the hindlimb.

Entraña Diaphragm muscle.

Falda de pecho The skirt muscle from the thorax is derived from prescapular muscles, including the *brachiocephalicus*.

Falda de morrillo o tapa de cogote A skirt of muscle containing *serratus ventralis* and adjacent muscles lateral to the ribs.

Falda de palomilla A skirt of muscle from the edge of the *gluteus* group.

Falda de vacío A skirt of *obliquus internus abdominis, obliquus externus abdominis* and *transversus abdominis*.

Falda real A skirt of muscle taken laterally to the ribs, including *latissimus dorsi*.

Filete Psoas major and *minor*, plus *iliopsoas*.

Filetillo A long filet of ventral neck muscles extending back to the sixth thoracic vertebra.

Jarretes Distal extensors and flexors of the hind and foreshank.

Lomo The *longissimus dorsi* and associated muscles of the loin.

Mariposa A butterfly-shaped cut of pelvic muscles including *obturatorius internus* and *externus*.

Paleta A mass of shoulder muscles including *infraspinatus* (punta de paleta), *supraspinatus* (boliche francés), plus *biceps* and *triceps brachii* (yema de paleta).

Palomilla The *gluteus* muscles, which when boned out may have a v-shape like a moth.

Pierna con boliche A mass of leg muscles including *biceps femoris* (pierna), *semitendinosus* (boliche blanco) and *gastrocnemius* (sapo).

Riñonada A rack of rib steaks extending back to the kidney (riñonada).

Tapa de paleta The "lid" of the "shovel" is the *subscapularis*.

(Sources: Norma Cubana, 1985; Fornias, 1998)

CUBE ROLL

In → Australian and → Irish beef cutting, a cube roll extends from ribs 4 to 13 and contains *longissimus dorsi, spinalis, semispinalis* and *multifidus dorsi*.

CUISSE

Hip of beef in Quebec.

CUISSE PONDEROSA

Quebec, a Ponderosa hip of beef.

CULOTTE

In → French beef cutting, the culotte is the posterior portion of the romsteck in the gluteal region. It contains sacral vertebrae three to five, plus the first caudal vertebra.

CUPIM

The *trapezius* and *rhomboideus* hump of a → Brazilian beef carcass, cut from the paleta.

CURING

Treatment of meat with sodium chloride and nitrite, plus other ingredients such as sugar (to modify taste and reduce the availability of water for bacterial growth), ascorbic acid (to stabilize product colour), and alkaline phosphates (to increase the → water binding capacity). Curing ingredients may be injected with a battery of hollow needles or through major arteries. Some meat products are smoked with hardwood sawdust (typically hickory or oak) on a heating element or treated with a liquid concentrate of smoke ingredients. This modifies the taste of the product and retards oxidative rancidity of fat and bacterial spoilage.

CUSHION

In → North America, a shoulder of lamb may be deboned and sewn tightly around the edges with string to produce a cushion-shaped cut of meat. In → England, a cushion is more likely to be of veal, composed of the topside muscles medial to the femur, neatly trimmed and with all the fat removed.

CUSK

A cod-like bony fish of northern seas, *Brosme*, → Gadiformes.

CUTANEOUS MUSCLE

This very thin but extensive muscle is located under the hide in beef, pork and lamb, functioning to twitch the skin. It forms the conspicuous area of red muscle seen on a side of beef (*cutaneous trunci*), with a smaller anterior patch on the shoulder (*cutaneous omobrachialis*). Myofibre hyperplasia in double-muscled cattle causes finger-like thickenings which often can be seen under the hide in live animals. In pork, the *cutaneous* muscle has two layers crossing obliquely.

CUTLET

From its origin from costelette or côtelette, one might expect a cutlet to be a slice of meat from the rib region. But this is seldom the case, as in the common North American usage for a pork cutlet, which is from the sirloin to include ilium rather than rib. Cutlet could be the most vague term in the whole of this book. Almost anything can be a cutlet.

CUY

Andean cavies and guinea pigs (Family Caviidae) are used as a meat source - primarily *Cavia aperea*. Imported to Europe in the sixteenth century they gave rise to *C. porcellus*, now a pet and a laboratory animal. Cuy is typically marinaded in wine or beer overnight.

CYPRINIFORMES

This order of bony fish contains the carp family (Cyprinidae) and the catfishes (Siluroidei). The family Cyprinidae includes large numbers of different species of carp, distinguished by carrying their teeth on pharyngeal bones instead of on the jaws. Adult carp are often bottom feeders able to tolerate warm water (15 to 30°C) and very low levels of oxygen, features that are advantageous in some types of aquaculture. The fry feed on zooplankton but later become able to digest vegetable matter.

For centuries, European and Asian carps (such as *Cyprinus* in Figure 47) have been bred as a food fish for aquaculture. Currently, the major producers are Russia, China, Japan and other Asian countries. The Crucian carp, *Carassius*, includes

important food fishes, plus a well known domestic pet, the goldfish. Catfishes are distributed through quite a few zoological families. The European catfish is *Silurus*, while the North American catfish is *Ictalurus*. Most catfish are naked and lack any well developed scales. Many catfish have several pairs of barbels around the mouth. Some have a spine on the leading edge of their fins, and a few catfish have poisonous spines.

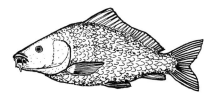

Figure 47. *The carp,* Cyprinus.

Catfish are relatively easy to rear in captivity because they tolerate warm water with a low oxygen concentration. Their skeletal musculature contains few bones. In a typical catfish processing operation the fishes are brought live to the plant. After being drained in a steel mesh basket, they are electrically stunned and skinned. The fish are then decapitated, eviscerated, washed, wrapped, and frozen. Mechanical skinning is possible after fish have been scalded. The carcass yield is about 50% for hand-dressed fishes, with final meat yields of 43% by mechanical deboning and 40% by hand. The average composition of catfish muscle is 75% water, 4% fat, 1% ash, and 20% protein. The muscle protein has an emulsifying capacity similar to that of pork and beef lean. The myosystem can be isolated mechanically from the skin and bones by shearing action between a conveyer belt and a drum rotating at a different speed.

(Sources: Tamura, 1961; Thomas, 1971)

CYTOSKELETON

Many different types of animal cells maintain their shape and internal structure by means of protein filaments with a diameter of 10 nm. → Myofilaments and all the filamentous components involved in muscle → contraction are viewed as a highly evolved and elaborate development of the ordinary cytoskeleton found in all cells.

The cytoskeleton includes microfilaments (composed of actin), as well as larger microtubules (composed of tubulin). Microtubules have a diameter of about 20 to 27 nm with a wall 5 to 7 nm thick. They are common in cells, particularly near the centrosome, and they form the spindle that separates the chromosomes during mitosis. In → myofibres, microtubules are well developed in the sarcoplasm around the nuclei. Both microfilaments and microtubules may be polymerized rapidly and, in some situations, may also be rapidly depolymerized when no longer needed. Mechanoreceptors such as integrin at the cell surface are connected with the cytoskeleton, thus explaining how cells react with their mechanical environment to maintain or alter their shape. A subunit of integrin occurs at the → myotendon junction and may be involved in the anchoring of → myofibrils.

AI/AU Tampa Library
4401 N. Himes Ave.
Tampa, FL 33614

DAB

A medium-sized flatfish, → Pleuronectiformes.

DAGADÓ

The belly of a → Hungarian pork carcass.

DAISY HAM

A fanciful name, sometimes used in North America for a smoked roll of pork shoulder.

DALAA

Ribs of a beef carcass in → Morocoo.

DARK, FIRM, DRY (DFD)

This widely used term, synonymous with dark-cutting, is used to indicate meat with a high ultimate pH (typically > 5.9) usually caused by low glycogen levels at the time of slaughter. The meat is dark because it has reduced light scattering, so incident light is transmitted deep into the sample and becomes trapped. The meat is firm and dry because myofilaments maintain their separation, thus retaining water within the → myofilament lattice.

DFD meat is typically caused by preslaughter stress or muscle activation. Transport exhaustion, hunger, fear, climatic stress or aggressive behaviour cause depletion of muscle glycogen and this limits the amount of lactate that can form postmortem. In beef, climatic stress may create a seasonal rhythm in the incidence of DFD meat, and aggressive behaviour is more common between males than females. Thus, bulls have the worst reputation for producing DFD beef.

(Source: Bendall and Swatland, 1988)

DASTI

Forelimb of a carcass in → Pakistan.

DECKLE

In → North America, the deckle muscles are located laterally in a primal rib cut, essentially covering it, as indicated by the English or German origins of the word (e.g., the deck covers a boat). The large deckle muscle is the *latissimus dorsi*, while the small deckle muscle is the *trapezius*. However, sometimes the tissues around the ribs, sternum and *pectoralis* muscles of a beef brisket are called the deckle. In → Australia, for example, the deckle is fat and connective tissue which is removed from the brisket. In → Swiss beef cutting, the Schulterdeckel is in the dorsal part of the shoulder.

DEER

→ Venison.

DELICATED

This term is used when several meat fragments have been folded and knitted together to make a cutlet (as in a delicated veal cutlet).

DELMONICO STEAK

In → North American beef cutting, the primal rib cut may be separated into rib-eye or delmonico steaks dominated by *longissimus dorsi* but also containing sections of *spinalis dorsi*.

DELTOIDEUS

A small, two-headed muscle on the lateral surface of the mammalian shoulder (Figure 48), although undivided in pork. The acromial head originates from the acromion process, while the scapular head originates from the posterior border of the scapula. Both are attached to the deltoid tuberosity of the humerus. In pork, the muscle originates from the aponeurosis over the *infraspinatus* and is inserted onto the deltoid ridge of the humerus.

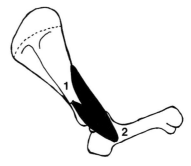

Figure 48. Lateral view of beef deltoideus *(solid black) showing its origin from the scapula (1) and insertion on the humerus (2).*

DEMI-MEMBRANEAUX

French for *semimembranosus*.

DEMI-TENDINEUX

French for *semitendinosus*.

DENMARK - BEEF CUTS

Denmark has two quite different patterns for beef cutting (Figures 49 and 50). The first is used internationally for exports and is based on division of the carcass into a pistola (axially from the thoracic vertebra 7 and including all the bones and muscles of the hindlimb) to leave a forequarter with a matching shape, as shown in Figure 49. The export cuts may be known by their English or Spanish names. The topside with fat (tapa con grasa) includes the *pectineus, adductor, semimembranosus, gracilis* and *sartorius* and is located medially. The silverside (contra con redondo) is dominated by *biceps femoris* and *semitendinosus*. From the round, this leaves the knuckle (babilla sin grasa) formed by the *quadriceps femoris*

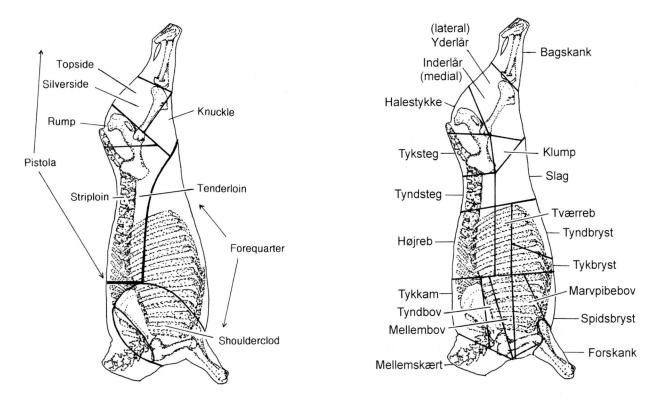

Figure 49. Denmark - pistola beef cuts.   *Figure 50. Denmark - domestic beef cuts.*

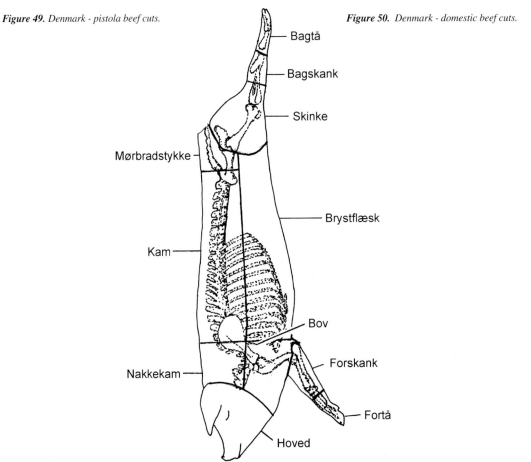

Figure 51. Denmark - pork cuts.

group of muscles. *Longissimus dorsi* may be isolated as a striploin (lomo con grasa), and the *psoas* muscles as a tenderloin (solomillo sin cordon). The shoulderclod (paleta) contains the main muscles of the upper forelimb from the *supraspinatus* down to the *triceps brachii*.

Domestic beef cutting in Denmark separates the round into a large inside round or inderlår, and a large outside round, the yderlår, with the back-shank or bagskank remaining distally. The ischium is named a tail-cut or halestykke. Thick (tyk) and thin (tynd) roasts are cut through the sacral and lumbar vertebrae to produce, respectively, the tyksteg and tyndsteg. More anteriorly, the prime rib roast is known as a high rib, or højreb, ventral to which is a cross-rib roast or tværreb. The *tensor fascia lata* is known as the klump, literally the lump. The hindflank is the slag, something to do with slaughtering. The shoulder region (bov) is cut from anterior to posterior to produce a dorsal thick crest or tykkam, a thin-shoulder cut or tyndbov, an intermediate shoulder cut or mellembov through the distal humerus, and a marrow-bone shoulder or marvpibebov exposing the marrow of the distal humerus. The brisket or breast is divided, from anterior to posterior, into the point of the brisket (spidsbryst), thick brisket (tykbryst), and thin brisket (tyndbryst). The neck is called the intermediate cut or mellemskært, and the front-shank is the forskank.

(Sources: ESS-Food; OEEC, 1961)

DENMARK - PORK CUTS

The ham or skinke may be cut with or without the sacral region, to separate a back-shank or bagskank, and back-toes or bagtå (Figure 51). The sacral region (rather than the more usual *psoas* muscles) receives the high appellation of tenderpiece or mørbradstykke, while the loin is called the kam or crest. The shoulder or bov is ventral to the nape of the neck or nakkekam, leaving a foreshank or forskank, and foretoes or fortå.

(Sources: ESS-Food; OEEC, 1961)

DENVER ROAST

In the USA, a Denver roast is a boneless oven roast of beef cut from the bottom round, and dominated by *semitendinosus* and *biceps femoris*.

DERRIÈRE DE PALERON

A → French cut of beef through the proximal part of the scapula.

DESMIN

A cytoskeletal protein linking the → Z-lines of adjacent → myofibrils.

(Source: Cullen *et al.*, 1992)

DEUKARI

The ham of a → Korean pork carcass.

DEVON ROAST

Flank and brisket in → Proten beef in England.

DEWLAP

Folds of skin and soft tissue over the sternum, between an animal's forelimbs.

DHUB

Lizard meat in Saudi Arabia. The essential amino acid profile surpasses dietary requirements for children and adults.

(Source: Abu-Tarboush *et al.*, 1996)

DIALA

The tail region of a lamb or beef carcass in → Morocco.

DIAMOND ROUND

A primal cut of beef in → North America, essentially the hip plus the sirloin tip from the sirloin (Figure 52).

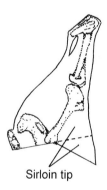

Sirloin tip

Figure 52. North American diamond round of beef.

DICED

Cut into bite-sized chunks.

DICKER SPITZ

A ladder of rib sections from an → Austrian beef carcass.

DICKES BUGSTÜCK

In → German beef cutting, the Dickes Bugstück or thick shoulder-piece is from the posterior part of the Schulter and contains much of *triceps brachii*.

DICKE SCHULTER

In → Austrian and → Swiss beef cutting, the thick shoulder is taken through the humerus, or posterior to it in an → Austrian pork carcass.

DIKKE LENDE

The thick loin or gluteal region of a beef carcass in the → Netherlands.

DIKKE RIB

The thick rib from the mid-thoracic region of a → Netherlands beef carcass.

DIKLIES

The diklies or thick flank of a → South African beef carcass contains the *quadriceps femoris* group of muscles.

DIKRIB

The thick-rib or dikrib of a →South African beef carcass contains the anterior thoracic vertebrae, ribs, and associated muscles which remain after removal of the blad

DILKAKJÖT

Lamb in → Iceland.

DIXIE STEAKS AND ROASTS

In North America, retail cuts of beef taken from the sirloin tip of a → diamond round.

DOGFISH

A small shark-like cartilaginous fish, → Chondrichthyes.

DOLLY VARDEN

A literary salmonid of the North Pacific, named after a colourful coquette in the novel "Barnaby Rudge" by Charles Dickens, *Salvelinus malma* → Salmoniformes.

DOUBLE CHOP OF LAMB

In North America, lamb carcasses may be cut with transverse cuts across the vertebral column so as to leave right and left sides of the carcass together. Thus, right and left loins of lamb together form a saddle that can be cut into double chops, each chop having left and right sides and two eyes of *longissimus dorsi*.

DOUBLE MUSCLING

A genetic condition in cattle (rarely in sheep) whereby the muscle mass is greatly increased by the prenatal formation of extra myofibres (myofibre hyperplasia). The cause is probably a gene that switches precursor cells away from becoming fibroblasts and towards becoming myoblasts. Fibroblasts form connective and adipose tissues. Myoblasts form muscle. Thus, double muscled beef develops extra muscle, but has less connective and adipose tissue than normal.

DROIT ANTÉRIEUR

French for *rectus femoris*.

DROIT INTERNE

French for *gracilis*.

DROUA

Camel hump in → Morocco.

DUBBELBIF

A lumbar steak from a → Swedish beef carcass showing two eyes of meat (double beef: i.e., *longissimus dorsi* and *psoas major*).

DUBLIN BAY PRAWN

A clawed → lobster, *Nephrops norvegicus*.

DUCK

Ducks, geese and swans are grouped together in the Order Anseriformes. Ducks comprise the Family Anatidae. Ducks are poor walkers but good swimmers, which means their legs are set far back in the body and are well muscled. Ducks are hunted extensively for their meat. The main types are the surface-feeding ducks such as the wild mallard (*Anas platyrhynchos*), → teal and widgeon (*Mareca americana*); the diving ducks such as the redhead (*Aythya americana*), canvasback (*Aythya valisineria*) and ring-necked duck (*Aythya collaris*); the sea ducks, which seldom provide meat with a pleasant taste; stiff-tailed ducks such as the ruddy duck (*Oxyura jamaicensis*); and the mergansers, whose meat is seldom palatable.

Thus, wild ducks produce meat with a wide range from delectable to unpalatable. How much of the range is hereditary and how much is nutritional? Breed differences in the taste of poultry meat are slight to undetectable, so I would join most others in supposing that the range in palatability of duck meat is largely nutritional. Thus, the unpopular taste of meat from sea ducks and mergansers reflects what the ducks have been feeding on. At the other end of the range, the canvasback is rated very highly for the taste of its meat, but is instantly disqualified if it has been feeding on rotting salmon. Probably one of the key features of this dietary effect relates to the digestion of fats and oils in the bird's diet. Fats and oils are formed from triglyceride - three fatty acids bonded to a glycerol backbone like the three arms of a capital letter E. As the triglyceride is digested and moved around the body to be deposited in the duck's own fat, nothing happens to the structure of the fatty acids, they just get uncoupled and recoupled to a glycerol backbone. Thus, a fatty acid with an unpleasant taste from rotting salmon can move, unchanged, from the edge of the sea to the edge of your plate.

Two types of ducks have been domesticated and are extensively farmed, the mallard (*Anas platyrhynchos*) and the Muscovy duck (*Cairina moschata*). The Muscovy may be identified by its claws (it is able to perch) and a red caruncle or knob between the beak and the eyes. Breast meat from female Muscovies may be tougher, drier and stronger in taste than that from males. The main changes produced by domestication have been to increase growth rates and reduce the colouration of feathers. Carcass conformation is often not much different from a muscular wild duck (at least, nothing like the contrast between a Pietrain pig and a wild boar). Apart from numerous duck breeds of layers and ornamentals, there are many meat breeds around the world. The top ratings might be the Aylesbury in England; the Rouen, Nantes and Barbary in France; and the Long Island in the USA. But this would probably be argued by fanciers of other breeds such as the Blue Swedish, Crested, Pekin, and Black Cayuga. The muscular Muscovy is the winner if meat yield and leanness are the main criteria. In North America, ducks are separated into a broiler-fryer category, less than 8 weeks of age, weighing 1.8 to 2.8 kg, either males or females; as distinct from the larger roaster, up to 16 weeks of age, either male or female. In Muscovies, the male is larger than the female. Youthfulness in ducks is detected by softness of the bill, sternal cartilage and trachea.

Feather removal is still a problem with ducks, and the final stages require dipping in hot wax. The breast meat is dark and tasty, like the leg meat. The skin on a duck is quite substantial and makes a good contribution to the overall meal. There are two basic approaches to cooking ducks. When they are small and not yet too fat, they may be roasted to leave subcutaneous fat in situ. Alternatively, for large or fat birds, the skin is slashed or lifted by inflation to facilitate the loss of fat by liquification during roasting. Traditionally, the large volume of easily liquified lard so produced is used for frying other components of the meal. Like most game, the taste of duck meat is greatly enhanced by aging, and modern consumers who take a duck straight from the freezer to the oven are missing the rewards of patience and culinary wisdom.

(Sources: Bertin, 1967; Rue, 1973; Land and Hobson-Frohock, 1977; Batty, 1979; Baeza *et al.*, 1998; McAndrew, 1990)

DUGONG

→ Manatee

DUNGENESS CRAB

A shore crab of the North Pacific, *Cancer magister*, → Crabs.

DUNGSIM

In → Korea, a long cut of beef including all the thoracic vertebrae. In pork, it is shorter and excludes the anterior thoracic vertebrae which are in the moksim.

DUNLIES

The dunlies is the abdominal muscle flank of a → South African beef carcass.

DUNNE BORST

In a beef carcass in the → Netherlands, the thin part of the breast, namely the postero-ventral rib region.

DUNNE LENDE MET HAAS

The thin loin in the lumbar region of a → Netherlands beef carcass.

DÜNNE SCHULTER

Thin shoulder of an → Austrian pork carcass.

DY

A lateral cut from the round of a → South African beef carcass containing the *biceps femoris* and *semitendinosus*.

DYSTROGLYCAN

Located outside the → myofibre, dystroglycan is part of the dystrophin-glycoprotein linkage between the → cytoskeleton and the → basement membrane.

(Source: Henry and Campbell, 1996)

DYSTROPHIN

Dystrophin is one of a family of proteins that helps to anchor the → cytoskeleton to the inside of the plasma membrane of a → myofibre. The dystrophin molecule has a rod-like shape, 110 nm long x 2 nm wide, with a sphere at each end. Dystrophin is named from the fact that it is defective in children with muscular dystrophy, where it was first discovered as a DNA mutation, then tracked back by finding what protein the DNA normally produced. In skeletal myofibres, most of the dystrophin is located at the → costameres although, in cardiac muscle cells dystrophin is more widely distributed along the plasma membrane. Other proteins in the dystrophin family include DRP-2 (dystrophin-related protein-2) and α- and ß-dystrobrevin. Dystrophin, spectrin and α-actinin may all have a common evolutionary origin.

(Source: Tidball and Law, 1991; Wakayama and Shibuya, 1991)

ÉCHINE

In → French pork cutting, the échine is the proximal part of the shoulder.

ECKSTÜCK

In → Swiss beef cutting, this is a postero-medial cut from the hindlimb.

ECUADOR - BEEF CUTS

Pulpa, in Latin-American Spanish, refers to a mass of boneless meat. In the hindquarter, the medial muscle mass is the pulpa negra, while the lateral mass is pulpa blanca (Figure 53). The pulpa negra contains the blackened exposure of the adductor muscle created when the carcass is split into sides. The rounded muscle mass, the pulpa redonda, is the → *quadriceps femoris* group of muscles. Relative to some other Latin-American countries the hip or cadera is fairly small, only the ilium and associated *gluteus* muscles. The posterior part may be called a Swiss steak or steak suizo. *Tensor fascia lata* may be separated as colita de cadera and *semitendinosus* as the salon (saloon steak). How the steak al minuto composed of relatively tough *gastrocnemius* gets to be called a minute steak is a mystery. Similarly, the hindlimb extensors and flexors are called rodaja, but whether this is from their lateral shape like sliced lemon (radaja) or their cross-sectional appearance like wheels (radaje) is debatable.

The prime roasts and steaks from anterior to posterior along the vertebral axis are lomo de asado (roast), club, and T-bone. The *psoas* is lomo fino. The rib or costilla is split from the flank or falda following the line of the last rib. *Trapezius* and *rhomboideus* contribute to the lomo de aguya Milanesa (schnitzel), *supraspinatus* is removed as steak especial, the remaining scapular muscles contribute to carne de paleta (blade meat), and the *triceps brachii* to pulpa de brazo (arm meat). The neck is cogote-nuca and the lizard-like extensors and flexors of the foreshank form the lagartillo. The brisket is the pecho.

(Source: Salazar, 2001)

EDGE-BONE

An ancient beef cut from → England.

EEL

An elongated teleost fish, → Anguilliformes.

ELADLAA

Posterior rib cuts of a beef carcass in → Morocco.

ELAND

The eland, *Taurotragus oryx*, is the largest of the African

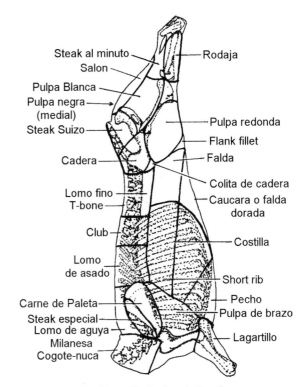

Figure 53. Beef cuts in Ecuador

antelopes. Being a muscular animal, it has been over-hunted for its meat, but has great potential for sustained meat production because it can be kept as a domestic animal, although growth rates may be low relative to cattle. Eland are ranched both within their natural territory, as in Zimbabwe, as well as in faraway places like the Ukraine. The eland has long, strong, forwardly directed horns with a spiral twist which pose a formidable problem in animal transport and abattoir handling. The carcass weight is about 300 kg with a 59% dressing percentage. The cutting pattern is shown for → South Africa.

(Source: Lightfoot, 1977; Eltringham, 1984)

ELASTICITY

Muscles and meat contain elastic structures both in series and in parallel (in other words, like links in a chain, as well as like strings on a guitar). The largest structures are the connective tissues visible by eye alone, while the smallest structures are at the molecular level. They are all very important in determining the texture of meat and muscle foods.

At the molecular level, the series elastic component is spread through the system of linkages whereby myosin molecule heads exert tension on a distant muscle attachment. Thus, as a muscle starts a normal → contraction, it must first take up the internal slack in its own linkage system. Even myofilaments themselves are extensible and contribute to series elasticity. For meat scientists, however, parallel elastic components such as titin are the most interesting because these may form part of the system that survives cooking to hold the sarcomere together, thus contributing to meat toughness.

(Source: Wakabayashi *et al.*, 1994)

ELASTIN FIBRES

Elastin fibres are microscopic, elastic fibres composed of the protein elastin. They are pale yellow in colour and birefringent when viewed with polarised light. Whereas → collagen fibres only lengthen by about 5% when stretched, elastin fibres may be stretched to several times their unloaded length, then they rapidly resume their starting length when released. Elastin resists severe chemical conditions, such as the extremes of alkalinity, acidity, and heat that destroy collagen fibres. Fortunately, there are relatively few elastin fibres in meat, otherwise cooking would do little to reduce meat toughness. The elastin fibres in muscles used frequently for locomotion are larger and more numerous than those of less frequently used muscles. Elastin fibres in the → epimysium and → perimysium of beef muscles range from 1 to 10 μm in diameter, but they are much smaller if they occur within a muscle (typically 0.2 to 5 μm).

Elastin fibres are composed of bundles of small fibrils (diameter 11 nm) embedded in an amorphous material. In the bovine → ligamentum nuchae, elastin fibrils may be constructed from smaller units or filaments approximately 2.5 nm in diameter. Elastin filaments are bound by non-covalent interactions to form a three-dimensional network and elastic fibres are assembled in grooves on the fibroblast surface where initially rope-like aggregations of fibrils become infiltrated with amorphous elastin. The elastin of the arterial system occurs in sheets that condense extracellularly without forming fibrils.

Elastin resembles → tropocollagen in having a large amount of glycine, but is distinguished by the presence of two relatively rare amino acids, desmosine, and isodesmosine. Elastin contains hydroxyproline, although it may not function to stabilize the molecule as it does in collagen. Tropoelastin, the soluble precursor molecule of elastin (70 to 75 kD), is secreted by fibroblasts after it has been synthesized by ribosomes of the rough endoplasmic reticulum and processed by the Golgi apparatus. In the presence of copper, lysyl oxidase links together four lysine molecules to form a desmosine molecule. Isodesmosine is the isomer of desmosine.

(Source: Bendall, 1967; Debelle and Tamburro, 1999)

EL DLOUE

Rib region of a lamb or mutton carcass in → Morocco.

ELECTRICAL STIMULATION

This is a procedure applied to the carcass after an animal has been stunned and exsanguinated. An alternating current is applied via electrodes which may penetrate or touch the carcass. The purpose is to cause muscle → contraction, which then accelerates → glycogen depletion and the onset of → rigor mortis. This reduces the risk of → cold shortening and allows the carcass to be rapidly refrigerated without risk of causing meat toughness. The increased rate of glycolysis caused by electrical stimulation increases the rate of → pH decline, so that the meat may appear slightly brighter or have slightly higher fluid losses (for example, during cooking). If muscle contractions are strong, mechanical disruption of the meat microstructure may enhance tenderness. However, strong contractions may also cause damage to the skeleton which can create problems during meat cutting.

The voltage required to achieve a suitable balance between the good and bad effects of electrical stimulation depends on a number of factors. The electrical continuity between electrodes and the carcass is important: large electrodes embedded in muscle or wet tissues are far more effective than electrodes with a small contact area or with a contact area insulated by fat. The electrical current is more effective when its frequency is matched to the relatively slow response time of muscle contraction. The typical 50 and 60 Hz frequencies used by the electrical supply grid are generally four or five times faster than the optimum for electrical stimulation of muscle contraction. The longer the delay between exsanguination and electrical stimulation, the higher the voltage required to obtain a suitable response. Early stimulation is particularly effective because it is mediated via the nervous system.

ELEPHANT

The international ban on the ivory trade has helped preserve the African elephant (*Loxodonta africana*, order Proboscidea) to some extent, although numbers are still decreasing. However, in Kruger Park, South Africa, numbers sometimes reach a point where culling is performed and the meat is canned. Birth control and translocation might be preferable, but have their own risks and complications.

(Source: Chadwick, 1996; De Boer *et al.*, 1998)

ELK

This depends on where you are. In Europe, an elk is *Alces alces*, which to the average butcher is indistinguishable from the beast Americans and Canadians would call a → moose. Some sources separate the two animals, *A. alces* for the European elk and *A. americana* for the North American moose. In North America, on the other hand, an elk is *Cervus elaphus*, which Europeans would call a red deer - a classic source of meat called → venison.

EL KAEBA

Hindshank of lamb or mutton in → Morocco.

EL KAIMA ELAMAMIA

Foreshank of lamb or mutton in → Morocco.

EL KATNA

Lumbar region of a beef or camel carcass in → Morocco.

EL SADR

Breast of lamb or mutton in → Morocco.

EL SALSOUL

Loin of lamb or mutton in → Morocco.

ELÜLSŐ CSÜLÖK

The foreshank and foot of a → Hungarian pork carcass.

ELÜLSŐ LÁBSZÁRHÚS

The foreshank of a → Hungarian beef carcass.

ELVER

An immature eel, → Anguilliformes.

EMU

The emu, *Dromaius novaeholandiae*, originating in the Australian desert, is now ranched in several countries. It produces dark meat with a low fat content. The distribution of muscles and primal cuts is similar to that of the → ostrich, except that *obturatorius medialis* is not large enough to rate being a separate tenderloin. Lateral limb muscles and their North American names are shown in Figure 54. Not shown are the medial cuts: the inside strip (*iliofemoralis*) and inside drum (medial part of *gastrocnemius*). The tenderness and juiciness of emu meat varies between muscles and types of birds.

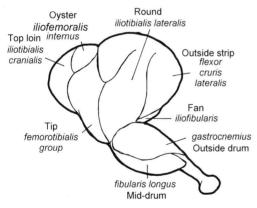

Figure 54. *Emu leg.*

(Source: Minnaar, 1998; Fitzgerald *et al.*, 1999).

EMULSION

The microstructure of some meat products (frankfurters, bologna, etc.) resembles an emulsion with an interfacial film and physical entrapment of fat in a matrix (Figure 55). Typically, fat droplets bounded by a protein membrane are trapped in a meshwork of myofibrillar proteins in water. The fat forms the discontinuous phase of the emulsion, and the aqueous proteins form the continuous phase. Actin and myosin become soluble in dilute salt solutions, so that the salt in a formulation strengthens the continuous phase of the emulsion. The continuous phase forms a gel-type matrix which, unlike oil-water emulsions, traps fat droplets and allows them to coalesce.

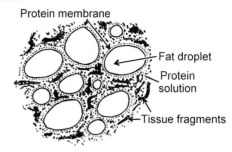

Figure 55. *Section through a meat emulsion.*

The emulsifying capacity of meat is the ability of dissolved meat proteins to bind fat droplets in an emulsion. When the emulsifying capacity is exceeded, the fat separates into large masses in the product. The emulsifying properties of myosin are superior to those of actin or the sarcoplasmic proteins. Connective tissue adds nothing to emulsifying capacity and, in excess, may lead to pockets of gelatin in the final cooked product. Extenders are proteins added to increase emulsifying capacity. Fillers such as starch may be added to improve cooking yields and to strengthen the product for slicing.

(Source: Lee, 1985)

ENDOMYSIUM

Fibrous connective tissue around individual skeletal → myofibres. Thus, the endomysium is located between the myofibres when the myofibres are in a → fasciculus. The endomysium is dominated by → reticular fibres formed from Type III collagen. The endomysium is embedded in the → basement membrane of the myofibre and helps to transmit forces from myofibrillar contraction to the connective tissue framework of the muscle. The orientation of reticular fibres changes when a muscle contracts or is stretched (Figure 56). At a stretched length, the protection provided by the endomysium prevents further elongation and is important in the prevention of myofibrillar damage.

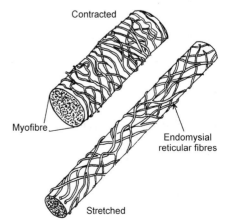

Figure 56. *Changes in the configuration of endomysial reticular fibres from the contracted to the stretched length of a myofibre.*

(Source: Orcutt *et al.*, 1986)

ENGLAND - BEEF CUTS

Many different styles of beef cutting are used in England. The dominant pattern adopted for uniform purchasing standards seems to be fairly close to that used in London and surrounding counties (Figure 57). Here, the forequarter may be separated from the hindquarter by a fairly straight cut between ribs 10 and 11, thus leaving more ribs in the hindquarter than would occur in North America. The forerib is the most posterior rib cut and may be cut in either of two ways, either as a forerib with four ribs (ribs 7 to 10), or as a Scotch-cut forerib with five ribs (ribs 6 to 10) and part of the blade of the scapula. Anterior to a 4-rib forerib is located a 4-rib middle ribs cut, and a 2-rib steakmeat cut. The dorsal part of the middle ribs containing the scapula is separated as a back ribs cut, while the more ventral part containing sections of ribs becomes the top ribs cut. Thus, the designation top is misleading because the top ribs are ventral to the back ribs. The dorsal part of the steakmeat is separated as the chuck and blade (scapula and associated muscles), while the remaining ventral part is called the leg of mutton cut, possibly because its globular triangle of *triceps brachii* is shaped like a leg of mutton. Middleribs and steakmeat together before separation are called a pony. The brisket contains the sternum and the ventral parts of the first six ribs. The shank is removed through the joint between the humerus and the ulna and radius. The clod contains most of the humerus, while the sticking contains most of the cervical vertebrae.

In breaking the hindquarter (still retaining the last three ribs), the top piece (hip) is detached from the rump, loin and hindquarter flank by a line through the fifth sacral vertebra. A comparison of English with → North American cutting reveals a major difference in terminology relating to the rump and sirloin. Before boning, a → rump in Great Britain initially contains the sacrum and ilium and overlaps a region which would be called the sirloin in North America. The English sirloin contains lumbar and posterior thoracic vertebrae and straddles the loin and rib cuts of North America. There is considerable variation within England, and what follows is a generalisation rather than a linguistic law, but the sirloin does appear to have changed ends of the loin in crossing the Atlantic. In England, the sirloin is usually an anterior cut of prime beef, while in North America it is a posterior cut. Another complication is that many English butchers will identify the primal cut containing the ilium along to the posterior ribs as a rump *and* loin, but then will proceed to break it into a rump (posterior) *and* sirloin (anterior). This may be explained as an ellipsis, in that rump *and* loin really mean rump *and* sirloin. Thus, the important distinction between loin and sirloin in North America may be blurred in England where, usually but not always, loin may be a verbal contraction of sirloin. Thus, from → Scotland down to the West of England, the lumbar and posterior thoracic cut known as sirloin, used to be called a loin in London. Possibly, at a time when many butchers believed that the loin was knighted to become the sirloin by that appreciative meat-eater King Henry VIII, the honour was not taken as seriously in London.

In England, the separation between rump (posterior) and sirloin (anterior) is made between the last lumbar and the first sacral vertebra, taking the anterior face of the ilium at a tangent, not far from where a North American butcher would separate sirloin (posterior) from loin (anterior), although the angles may differ a little.

In breaking the English top piece, another major point in terminology relates to what is meant by the top of the hip. If a whole English top piece or hip of beef is dropped on a cutting block, it is convenient to drop it with the lateral surface downwards onto the block, thus leaving the aitch bone (pubis) and chine bone (vertebrae) exposed as landmarks for cutting. Hence, the medial surface of the hip defines the English → topside. Quite literally, it is on top. The *semimembranosus* is located medially in the hip and is, therefore part of the topside. Conversely, the *semitendinosus* (eye of the round in North America) is located laterally in the hip and has a natural, silvery seam of epimysium along which to define the start of the English → silverside, which is lateral in position. Thus, the topside is medial to the silverside in England, just as the inside round is medial to the outside round in North America. The top piece with the leg (tibia) removed through the stifle (femoro-tibial) joint may be called a round or buttock of beef. The *quadriceps femoris* group of muscles inserting onto the patella are removed as toprump (often written top rump) or thick flank with a cut parallel to the femur.

In attempting to explore how methods of beef cutting evolved and travelled geographically, the resources are fairly restrictive. Smith (1876), the pioneer nutritional biochemist, lived in London and published an early beef cutting chart (Figure 58). The plan was based on a carcass that clearly had not been suspended from its tarsal region, and was very lightly muscled in the forequarter. Thus, there is considerable ambiguity in transferring Smith's (1876) plan to a contemporary skeletal outline, but the names of the meat cuts are extremely interesting.

The most striking name is the spaud, used for the *triceps brachii* in the shoulder region, which unites us through Middle English to old French, through spauld, espalde, and espaule, to both modern French, épaule, and to the Latin, spatulae, for the shoulder blades. Hence, in modern Italy, there are cuts through the shoulder of the beef carcass bearing names such as taglio di sottospalla (cut from the under-shoulder) and sezione e muscolo di spalla (section of shoulder muscle). While in Spain, the shoulder of a beef carcass is called the espalda. The linguistic leap from scapula (shoulder blade) to spatula, via spatha for double-edged broad sword likely would not worry ancient centurions or modern butchers, although it might trouble a professorial linguist. I imagine that the continuity of the pectoral muscle was continued from the fairly well defined long brisket, medial to the spaud, through to the thick brisket at the base of the neck. Naming the neck as the vein might relate to slaughtering (near where neck veins were cut) and, or have some connection to French beef cutting, where the collier is divided into a dorsal veine maigre and a ventral veine grasse.

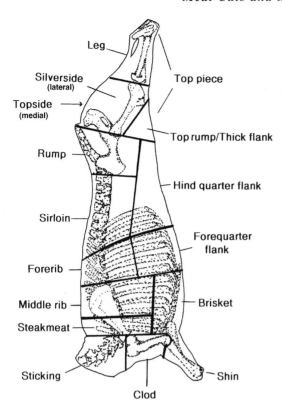

Figure 57. *England - beef cuts.*

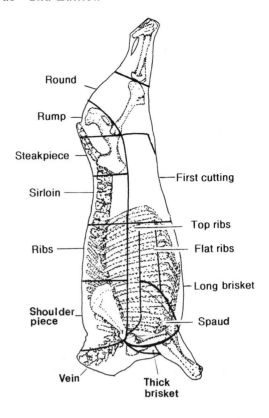

Figure 58. *England - beef cuts in 1876.*

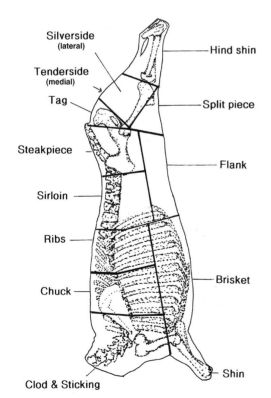

Figure 59. *West of England - beef cuts.*

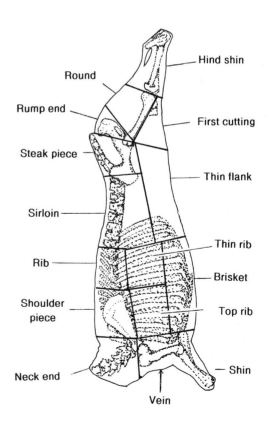

Figure 60. *Liverpool - beef cuts.*

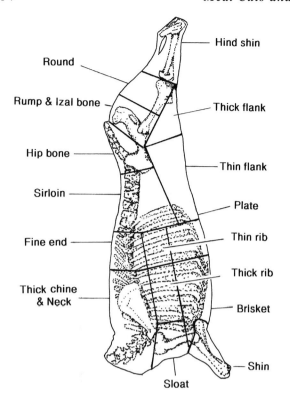

Figure 61. Northeast England - beef cuts.

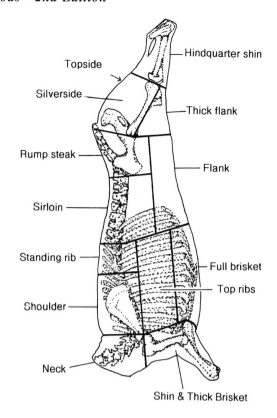

Figure 62. Manchester - beef cuts.

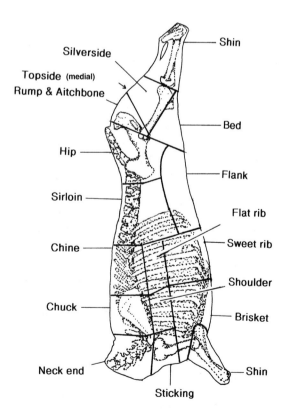

Figure 63. English Midlands - beef cuts.

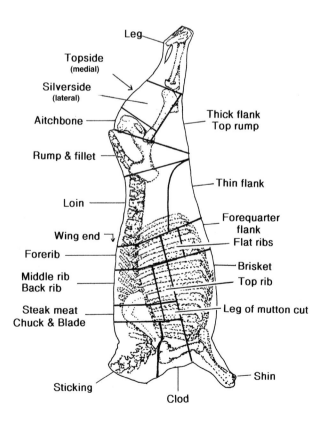

Figure 64. London and Home Counties - beef cuts.

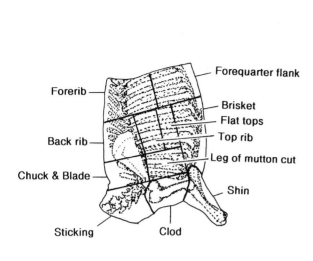

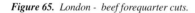

Figure 65. London - beef forequarter cuts.

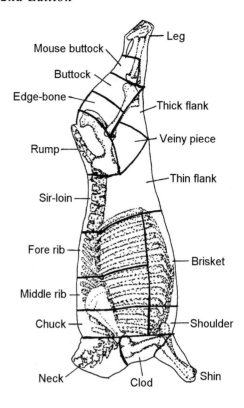

Figure 66. English beef cuts in 1816.

There are other salient points in Smith's (1876) beef cutting chart which help us bridge the Atlantic, from English to American beef cutting. Thus, the London rump of 1876 matches the standing rump of North America rather than the current English rump (Figure 57), and the beef carcass of Victorian London had a round with a deflected antero-ventral tip through the *tensor fascia lata*, just like the American → Diamond round. Thus, as has "gotten" to be the case with so many other words, modern American usage is closer to old English than is modern English. The London sirloin of 1876 may have been longer than that shown in Figure 58, but it is clear that sirloin and loin meant the same thing, because there is no loin *per se* and the sirloin is centrally located, whence it may have moved posteriorly (as in current North American usage) or extended anteriorly (as in current English usage). The steakpiece as a term for the gluteal muscles over the ilium is still retained in the West of England.

The West of England beef cutting pattern (Figure 59) reported by Gerrard (1947) is transitional between that of Victorian London and the currently dominant English cutting pattern (Figure 57). Thus, the Victorian London steakpiece was maintained as the West of England steakpiece, while the round has been replaced by silverside (lateral) and tenderside, for medial top side. Although the limits of the chuck, clod and sticking in Figure 59 are different from those of Figure 57, the naming pattern is the same, because clod and sticking have always gone together.

The survival in the West of England of the name tag for a cut in the ischial region is worthy of note. Tags are familiar to us now, as small, pendant tails to a garment or fishing fly. To the shepherd, however, they were more likely to be tags of dirty wool below the sheep's tail - hence the name tag near to where the tail is located on a beef carcass (a young sheep also was called a teg or tag).

The "first cutting" known by Smith (1876), survived in the Liverpool area and was reported by Gerrard (1949), but in a more posterior location (Figure 60). This is not such a major change for English cutting, where thin flank (abdominal muscles) and thick flank (*tensor fascia lata* and *quadriceps femoris* group) often go together. Thus, first cutting moved from thin flank to thick flank.

Other interesting continuities from Smith (1876) to Gerrard's (1949) Liverpool cutting pattern are the vein (moved ventrally down the neck), the shoulder piece in the scapular region, the posteriorly located steak piece (ilium), rump (ischium) and round (femur). Bearing in mind the great importance of Liverpool as a point of emigration from the Old World to the New World, perhaps this continuity of rump and round from Victorian London (Figure 58), to Liverpool (Figure 60), and then to New York is no mere coincidence. If this speculation is correct, then Liverpool to New York represents a major route for the migration of meat cutting methods.

Keeping a round of beef in the American style, rather than splitting off a medial top side and a lateral silverside, seems to have been a northern cutting pattern in England. Thus, the round seen in Liverpool cutting (Figure 60) also occurs on the other side of England in the Northeast (Figure 61), as well as even farther north in Scotland.

The izal bone associated with the rump is rather a mystery, but the eye of the pizzle is not far away, and eye of the pizzle might have been contracted down to izal (the eye of the pizzle indicates where the root of the penis was detached in a male carcass). Explaining the sloat is little easier, but still not certain. One possibility is the word slot (from the Old French, esclot), which refers to the slight depression above the sternum in the human chest. Another more likely possibility is the Middle English use of slot, meaning to pierce through. Used in this way, the sloat might be regarded in the same way as the vein or sticking - near where cattle are exsanguinated in the abattoir. Also of great interest in Northeast English cutting is the location of the plate as a postero-ventral rib cut, just as in America. Being just east of Liverpool, the cutting pattern in Manchester (Figure 62) closely resembles that of Liverpool.

North of the border in → Scotland, there are some unique names used in cutting the beef hindquarter, with the hindlimb having something of a North American style round. But we find topside and silverside again as we work our way southwards back into the Midlands of England (Figure 63). Why the postero-ventral rib cut is regarded as sweet in the Midlands is uncertain. Possibly, it relates not to sweetness in taste, but to the distinction between fresh or sweet water versus brine or salt water. Brine curing is the fate of most beef briskets, whereas the more posterior cut makes a better fresh roast and is far less often salted. It is more difficult to explain the bed cut anterior to the femur. In cross section, the seams of epimysium separating the *rectus femoris* from the adjacent *vastus* muscles are often quite rectilinear and can produce a shape like a side-view of a bed. Who knows?

Having worked our way back to London (where we started English beef cuts with Figure 58), we find that both names and cutting pattern had all undergone major changes by the time they were recorded by Gerrard in 1949 (Figure 64).

Thus, just like historical artefacts such as pottery and furniture, both chronology and geography are needed to define a particular style. During the 1960s, forequarter cutting continued to change, with Figure 65 showing the pattern taught at the Smithfield College, as reported by Rixson (1970). And it is appropriate to leave the last word on this subject to the late Derrick Rixson who, as well as being a great expert on meat cutting, proved himself a superb social historian and archaeologist. Rixson (2000) discovered an English meat cutting pattern dating back to 1816 (Figure 66). It confirms the ancient English sirloin as a lumbar cut of meat anterior to the rump, and sirloin is hyphenated as sir-loin, reminiscent of the old butcher's joke about King Henry conferring a knighthood on his favourite cut of beef. A baron of beef was sir-loin + rump of both left and right sides unseparated. Distal to the ischium the cut had a variety of names: edge-bone,

ridge-bone, → aitch-bone or round. Rather surprisingly the buttock contained the distal part of the femur rather than the proximal part. The uniquely named mouse buttock presumably denoted the smaller, distal end of the buttock containing the *gastrocnemius*. Its alternate name was the bed. The veiny piece was more likely named from the *fascia lata* forming a layer like a vein of quartz in a rock, rather than from any veins of the vascular system in this region. The thick and thin flanks and rib cuts were very close to some of our contemporary cutting patterns, including the shoulder whose alternate name was the leg-of-mutton piece. Given the alternate name of the neck was the sticking, then the clod and sticking was essentially the same as cut now.

(Sources: Smith, 1887; Gerrard, 1947; Moore *et al.*, 1983; Unger and Wilson; Rixson, 1970, 2000)

ENGLAND - LAMB CUTS

English lamb cutting is quite variable, but the basic pattern is fairly simple, as shown in Figure 67, which is similar to the London and Home Counties pattern of Gerrard (1949) and Gerrard and Mallion (1977), and not much different from the standardised cutting for caterers (Moore *etal.*, 1983). The most important feature is that the shoulder (the scapula and associated limb muscles), is released from the underlying vertebral column and ribs, leaving the cervical vertebrae as scrag, and four thoracic vertebrae and dorsal ribs as middle neck. The scrag and middle neck shown in Figure 67 may be called a neck and middle. The best end neck contains no neck at all, but is all thoracic. Left and right loins from an unsplit carcass form a saddle of lamb. Also on an unsplit carcass, a chine and end is loin plus best end neck plus middle neck.

ENGLAND - PORK CUTS

A general pattern for pork cutting in England is shown in Figure 68. There is considerable variation in the separation of the leg, depending on whether it follows a curved line cut by knife and hand-saw, or a straighter line cut with a band-saw. The posterior end of the loin may be separated as a chump cut, while the posterior end of the belly may be separated as a flank cut. A major point to note is the location of the English sparerib of pork, which corresponds to something like a North American blade or Boston shoulder. In England, ribs and intercostals also may be cut American-style from the belly, but then they are designated as barbecue spareribs. It is difficult to find the source of the spring in the hand and spring. It is a very old name, and may have some connection to the shape of the ribs as they approach the sternum. A similar shape presented by the ribs of a wooden ship approaching the keel was called the spring.

(Sources: Gerrard, 1949; Gerrard and Mallion, 1977; Moore *et al.*, 1983)

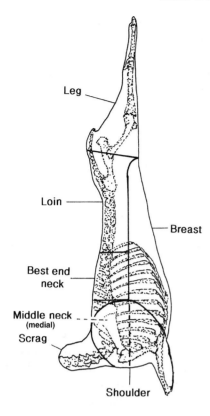

Figure 67. England – lamb cuts.

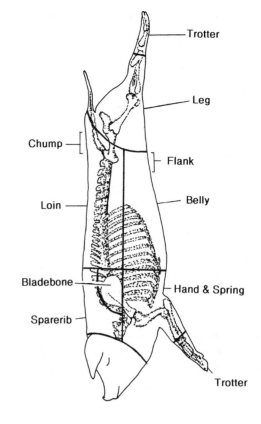

Figure 68. England – pork cuts.

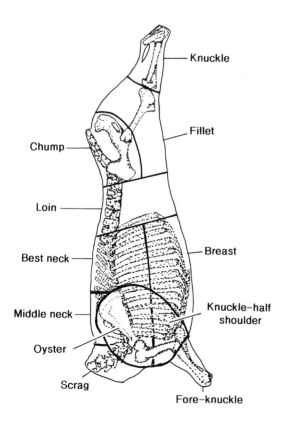

Figure 69. England – veal cuts.

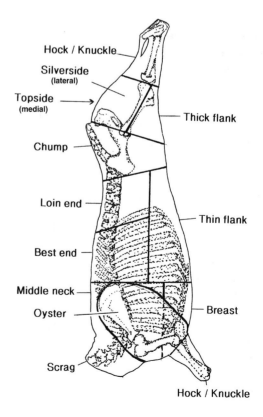

Figure 70. England - white veal cuts for caterers.

ENGLAND - VEAL CUTS

Veal carcasses differ considerably in size from small carcasses retailed locally as a byproduct of dairy farming, through to larger carcasses raised specially on a low iron ration for white veal. Figure 69 (from Gerrard, 1949, and Gerrard and Mallion, 1977) follows a pattern for a small carcass that has many similarities to lamb cutting. Thus the oyster (scapula, humerus and associated muscles) is separated from the underlying medial middle neck, and the distal part of the oyster may be separated as a knuckle-half shoulder.

Large, white veal carcasses for caterers and the hotel trade are broken to allow a greater subdivision of cuts (Figure 70), which resemble an amalgamation of beef and small veal carcass cuts (Moore *et al.*, 1983). The loin end and best end may be cut together as a loin and end. A primal cut including the chump and posterior cuts may be called a haunch. A cushion of veal is composed of the topside muscles medial to the femur, neatly trimmed and with all the fat removed. Many of the names used for unsplit carcasses (such as chine and end, and saddle) also may be applied to comparable veal cuts. Thin escalopes of veal are cut from either topside or silverside, while transverse sections across the tibia, fibula and tarsal bones become osso bucco.

ENGLISCHER

Just as many continental European names for the finer steaks of beef are widely used in English-speaking countries, the compliment is returned by the finer roasts being described as English on the continent of Europe. Hence, the Englischer is a primal cut containing the best roasts on an → Austrian beef carcass.

ENGLISH CHOP OF LAMB

In North America, lamb carcasses may be cut with transverse cuts across the vertebral column so as to leave right and left sides of the carcass together. Thus, right and left loins of lamb together form a saddle that can be cut into double or English chops, each chop having left and right sides, and two *longissimus dorsi* and two *psoas major* muscles. The lumbar vertebrae may be removed, leaving the subcutaneous fat intact to hold the two sides together, with the help of a skewer.

ENGLISH CUT ROAST OR ROLL

On the American east coast, an English cut roast is taken at the edge of the *triceps brachii*, ventral to the scapula in the cross rib region of the chuck, while an English cut roast (also known as a Boston cut roast) would be located in the chuck over the ventral part of rib 5.

(Source: Bull, 1951)

ENKELBIFF

In → Sweden, a simple (enkel) beefsteak has one eye of meat (*longissimus dorsi*) and originates from the posterior rib region.

ENROLLADA

Spanish for a cut of meat that has been boned and rolled as, for example, pecho enrollada from a Mexican lamb carcass.

ENTE

German for → duck. Entchen is a young duck.

ENTRAÑA

Beef diaphragm muscle in → Cuba or → Chile.

ENTRECOSTO

The ventral rib region of a → Portuguese pork carcass.

ENTRECOTE KAM

A midthoracic roast from a → Norwegian beef carcass.

ENTRECÔTES

Entrecôtes steaks are taken through the vertebral muscles in the thoracic region, not literally between the ribs as the name might suggest (Figure 71). In → France, entrecôtes premières are restricted to thoracic vertebrae 5 to 11, while entrecôtes secondes are more anterior and originate from thoracic vertebrae 1 to 5. The name entrecôtes is also used in → Sweden and → Mexico.

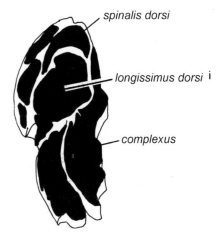

Figure 71. An entrecôte steak.

ENTRECÔTES AVANT

Rib of beef in → Belgium.

ENTREDEUX

Literally between the two - the cut between the filet and queue of a pork carcass in → Luxembourg.

ENTRELARDÉ

Lard maigre or belly of pork in → Belgium.

ENTREMEADA

The belly of a → Portuguese pork carcass.

ÉPAULE

A → French shoulder of veal or lamb, or shoulder of beef or pork in → Luxembourg.

ÉPAULE PICNIC

A picnic shoulder of pork in Quebec.

EPAXIAL MUSCLE

The muscle mass above the horizontal connective tissue septum in the musculature of a fish (→ myomere).

EPIMYSIUM

Connective tissue on the surface of a whole muscle. The epimysium is dominated by collagen fibres formed from Type I → collagen.

ESCALOPE

French equivalent of a cutlet. In England, thin escalopes of veal are cut from either → topside or → silverside.

ESPALDA

A scapular cut from a → Spanish beef carcass.

ESPALDILLA

While espaldilla is Spanish for the scapula, the cut of this name on a → Mexican pork carcass involves the more distal part of the forelimb. The pork espaldilla y codillo includes the ventral neck and jowl. In a Mexican lamb carcass, the espaldilla may be a large primal, or a smaller shoulder roast.

ESSEX ROAST

Top rump in British → Proten beef.

ETUPOTKA

In → Finland, the foreshank of a beef or pork carcass.

ETUSELKÄ

In → Finnish beef cutting, the etuselkä remains in the anterior thoracic region after the shoulder meat is removed. In Finnish pork cutting, it is square-cut and more substantial, like a Boston butt in → North America.

ETUSORKKA

The forefoot of a pork carcass in → Finland.

EULACHON

An oily Pacific salmonid, *Thaleichthyes*, → Salmoniformes.

EWE

A female sheep that has produced a lamb.

EXTENDER

Proteins added to a processed meat product to increase emulsifying capacity.

EXTENSOR CARPI OBLIQUUS

A small forelimb muscle, → *abductor pollicis longus*.

EXTENSOR CARPI RADIALIS

A medium-sized muscle located anteriorly and medially on the mammalian foreshank. It originates from the lateral condyloid crest and coronoid fossa of the humerus and its metacarpal insertion is lost in a dressed beef carcass.

EXTENSOR CARPI ULNARIS

This is somewhat of a trick muscle in the mammalian foreshank. Morphologically it is an extensor, but it may function as a flexor. In a dressed beef carcass, only the origin on the lateral epicondyle of the humerus survives, and the insertion to the accessory carpal bone is severed.

EXTENSOR DIGITORUM BREVIS

A small muscle of the hindshank in beef and lamb, but larger in pork. It originates on the tibial tarsal bone.

EXTENSOR DIGITORUM COMMUNIS

One of the major forelimb extensors, located laterally. Two fleshy parts usually may be seen laterally in a cross section of a beef shank (*extensor digitorum communis = extensor digitii tertii proprius* and *extensor digitorum medialis*) while, in pork, five parts may appear (to digits II, II + III, V + VI, IV, and V). There is considerable confusion between authors about this muscle, which is not worth any effort on our part because, in beef, the muscle ends up in hamburger or beef shank slices where its identity is thoroughly lost.

EXTENSOR DIGITORUM LATERALIS

In the mammalian forelimb, a muscle originating with separate heads on the ulna and radius and inserting onto the third and fourth digits. The muscle belly is central and lateral in the foreshank. In the hindlimb, it is located laterally from the stifle (knee) to the hock.

EXTENSOR DIGITORUM LONGUS

One of the mammalian hindshank muscles, originating from the extensor fossa of the femur and subdivided into several parts to the third and fourth digits.

EXTÉRIEUR DE RONDE

In Quebec, outside or bottom round of beef.

EYE OF THE ROUND

In → North America, → Australia and → Ireland, the eye of the round is a beef *semitendinosus* muscle from the outside or bottom round.

EZELTJE

The *tensor fascia lata* of a beef carcass in the →Netherlands, alternatively called the liesstuk.

FABRICAGEHAM

The whole or manufacturing ham in → Netherlands pork cutting.

FACEIRA

The jowl of a → Portuguese pork carcass.

FAGIANO

Italian for → pheasant.

FAISAN

French for → pheasant.

FAKHDA

Posterior hindlimb muscles of a beef carcass in → Morocoo.

FALDA

Skirt meat in some Spanish-speaking countries. In Costa Rica, falda is beef *transversus abdominis* and *obliquus abdominis*. In Cuba, there are several types of beef skirt: falda de pecho (prescapular muscles including *brachiocephalicus*), falda de morrillo o tapa de cogote (*serratus ventralis* and adjacent muscles lateral to the ribs), falda de palomilla (from the edge of the *gluteus* muscles), falda de vacío (*obliquus internus abdominis, obliquus externus abdominis* and *transversus abdominis*) and falda real (*latissimus dorsi*). In → Ecuador, the falda is composed of abdominal muscles but, in → Argentina, the falda is more in the sternal region.

FALKA

Lumbar region of a camel carcass in → Morocco.

FALSCHES FILET

In → German beef cutting, the false filet or Falsches Filet (sometimes Falschesfilet) is the *supraspinatus*.

FAN

The *iliofibularis* of → ostrich and → emu meat is often called the fan.

FANCIFUL NAME

The conflict between the need for standardized names and innovative names for local marketing was solved by ICMIST (1973) in the USA by designating the former names as recommended names and the latter as fanciful names. Thus, a fanciful name may follow a recommended name on a label. Fanciful names listed by ICMIST (1973) for the USA include California roast, patio chops, pop-up steak, paradise roast, chuck-wagon cut, smoked callie and London broil.

FANTAIL SHRIMP

A shrimp tail with the shell removed from all the abdominal segments except the last, thus leaving the fan or telson as a finger hold.

FARAONA

Italian for → guinea fowl.

FÄRS

Ground or minced meat in Sweden.

FARTŐ

A lateral cut from the beef round in → Hungary, sometimes translated as the rump.

FARTŐHEGY

A cut containing the ilium in → Hungarian beef cutting.

FASAN

German for → pheasant.

FASCICULUS

Fasciculus is the Latin name for a bundle of myofibres, as in a bundle of skeletal → myofibres forming meat or a food myosystem. The plural is fasciculi.

FAUX-FILET

Rib eye (*longissimus dorsi*) in Quebec.

FEATHER BLADE

For beef cutting in → Northern Ireland, the inside feather blade is *subscapularis* and the outside feather blade is *infraspinatus*.

FEDERSTÜCK

In → Switzerland, the feather piece is a ladder of ribs from a beef carcass.

FEHÉRPECSENYE

Eye of the round or *semitendinosus* in → Hungarian beef cutting.

FELDERSTRUKTUR

In a cross section of a → myofibre under the microscope, if the → myofibrils are large and are grouped together in clumps, the arrangement is described as felderstruktur.

FESA

A medial cut from the hindlimb of an → Italian beef carcass, including *semimembranosus, adductor, pectineus, sartorius* and *gracilis*.

FESONE DI SPALLA

The *triceps brachii* of an → Italian beef carcass.

FESSIER

French for the gluteal muscles (*gluteus medius, gluteus accessorius* and *gluteus profundus*).

FETTES MEISEL

An → Austrian beef cut at the base of the neck.

FIANCHETTO

An → Italian cut of beef based on the *tensor fascia lata* at the junction between the flank and the hindlimb.

FIANCO

Posterior abdominal muscles on an → Italian lamb carcass. Fianco is Italian for the flank.

FIBRILLENSTRUKTUR

In a cross section of a → myofibre under the microscope, if the → myofibrils are small and separate the arrangement is described as being fibrillenstruktur.

FIBRONECTIN

A high molecular weight glycoprotein found in the → basement membrane of → myofibres.

(Source: Kannus *et al*., 1998)

FIJNE RIB

The prime rib region of a → Netherlands beef carcass.

FILEESELKÄ

In → Finnish beef cutting the fileeselkä is a back filet in the lumbar region.

FILEH

Psoas muscles of a beef carcass in → Jordan.

FILET

In most countries the filet is the *psoas major* (plus the much smaller *psoas minor*), but occasionally the filet is the *longissimus dorsi* or another muscle. There is little or no standardisation of the names for parts of the filet along its length. Rombauer and Becker (1975) suggest the following

sequence of names from posterior to anterior: Chateaubriand, filet steak, tournedo, filet mignon, and finally the tip sections cubed for Stroganoff, brochettes or kabobs. The traditional English spelling is fillet.

FILET D'ANVERS

Located in the posterior region of the beef hindlimb in → Luxembourg or → Belgium.

FILETE

Psoas major and *minor*, plus *iliopsoas* in beef cutting in → Cuba , → Chile and → Mexico. In Mexico, slices are filete mignon.

FILETILLO

In → Cuban beef cutting, a long filet of ventral neck muscles extending back to the sixth thoracic vertebra.

FILET-KAM

A lumbar cut of a → Norwegian beef carcass.

FILET MIGNON

In North America, filet mignon may be steaks of beef *psoas major* and *minor* from the short loin, or even the sirloin, as indicated on older Live Stock and Meat Board charts. In Europe and elsewhere, filet mignon is far more likely to be small steaks taken across the anterior thin end of *longissimus dorsi*, medially to the scapula. In → France, for example, filet mignon may originate in the surlonge, medial to the scapula, with larger tournedos originating more posteriorly in the train-de-côtes.

FILETO

The *psoas* muscles in → Greek meat cutting.

FILETTO

Psoas muscles on an → Italian beef carcass.

FILI

The *psoas* muscles of a beef carcass in → Morocco.

FILLER

An ingredient such as starch added to a processed meat product to improve cooking yields and strengthen the product for slicing.

FILLET

English spelling of filet - usually denoting *psoas major* for beef. However, a rump and fillet of veal in → England may be a large cut extending right across the femur.

FINE END

A cut of beef in the posterior thoracic region in Northeast → England and → Ireland.

FINGERBONE RIB

An entrecote beef steak in → Northern Ireland.

FINLAND - BEEF CUTS

The pattern of beef cutting in Finland (Figure 72) is now merging towards an international European style, but the unique language preserves many traditional names. The shanks are removed from fore and hindlimbs as etupotka and takapotka, respectively. As in other cuts of meat (paisti), etu is front or anterior, while taka is rear or posterior. The whole round, called a long roast or pitkä paisti, is divided into a medial part or sisäpaisti (mostly *semimembranosus, adductor* and *pectineus*), and a lateral part or ulkopaisti (mostly *semitendinosus* and *biceps femoris*). As used for naming other cuts as well, sisä means inside while ulko means outside. The *quadriceps femoris* group forms a corner or angle cut called the kulmapaisti. For Finland, the classical "roast beef cut", paahtopaisti, is taken through the gluteal muscles overlying the ilium.

The filet or filee gives its name to the fileeselkä, or back filet, with the grilling filet (parifileeselkä) being located posteriorly and the outside filet (ulkofileeselkä) anteriorly. A primal shoulder cut (lapa) containing the scapula, humerus and associated muscles is lifted from the anterior ribcage to leave the anterior back or etuselkä. The neck is called the niska, and the brisket is called the rinta. The kylki is the plate or flank, but also gives its name to the general name for a chop or cutlet, a kyljys. The gap-flank, välikylki contains the middle part of a six-rib forerib of beef.

(Sources: Lihateollisuuden Tutkimuskeskus, 1989)

FINLAND - PORK CUTS

As shown in Figure 73, the straight ham or suora kinkku is a primal cut that may be subdivided into a roasting joint (paahtopaisti) similar to that cut from a beef carcass in Finland. Removal of the paahtopaisti from the suora kinkku then leaves a round ham or pyöristetty kinkku, a hindshank or takapotka, and a hind hoof or takasorkka.

The primal cut along the posterior thoracic and lumbar vertebrae is the back, or Selkä, from which the *longissimus dorsi* and *psoas* muscles may be removed as ulkofilee and sisäfilee, respectively. The postero-ventral rib region plus abdominal muscles is the side of pork (kylki), from which a muscle layer may be removed as the barbecue flank or grillikylki. The anterior thoracic region forms an anterior back cut or etuselkä, separated from the shoulder or lapa ventrally. The etuselkä may be subdivided into pork neck (sian niska) or economy cutlets (talouskyljys). More distally down the forelimb are the foreshank (etupotka) and fore-hoof (etusorkka). The head (pää) is removed with a straight cut, as shown in Figure 73 A number of generic items are named

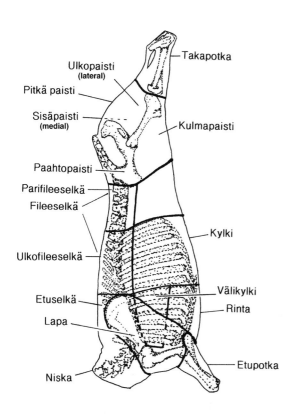

Figure 72. Beef cuts in Finland.

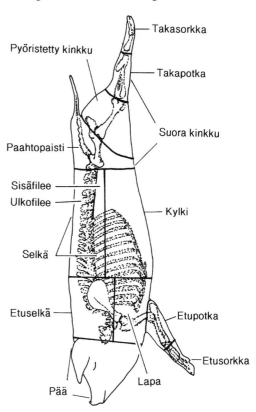

Figure 73. Pork cuts in Finland.

with the prefix sian (pig or pork) such as: siankyljys, a pork chop; sianliha, fresh pork; sianliikkiö, cured ham; and sianpaisti, roast pork.

(Sources: Lihateollisuuden Tutkimuskeskus, 1989)

FINNAN HADDIE

A lightly salted, split, and cold-soaked haddock. The name is derived from a Findon haddock, named from Findon, just south of Aberdeen on the east coast of Scotland.

FISH

Divided into cartilagenous fish such as sharks, skates and rays (→ Chondrichthyes) and bony fish (→ Osteichthyes).

FISK

Swedish for → fish. In Icelandic, fish is fiskur, but the prefix fisk- is used for a range of fish products such as fiskhakk (minced fish), fiskpylsur (fish sausage), etc.

FJÄRILSKOTLETT

Swedish for → butterfly cutlets.

FKHAD

Leg of lamb or proximal leg of poultry in → Morocoo.

FLÆSK

Danish for pork.

FLANC

In Quebec, flank of beef or lamb, or belly of pork. In → Belgium, the flank of a beef carcass.

FLANCHET

Flank of beef in → Luxembourg or → Belgium.

FLANCHIS

Flank of a beef carcass in → Belgium.

FLANCO

Abdominal muscles of a → Brazilian beef carcass. An alternative name for the vazio.

FLANKEN RIBS

In → North American beef cutting, flanken ribs or brust flanken, are cut across the ribs, posterior to the humerus and dorsal to the scapula. The dominant lateral muscle is *pectoralis profundus*, while medially may be found *rectus thoracis*, *serratus ventralis cervicis* and intercostal muscles.

FLANK OF BEEF

A primal cut of beef in → North America, dominated by abdominal muscle (*obliquus abdominis internus, obliquus abdominis externus* and *transversus abdominis*) which usually are either removed as thin steaks, cubed for stewing beef or ground to make minced beef. In a similar general position, a flank or thin flank is found in most → English and → Irish beef cutting methods.

FLANK PLATE

In → Australian beef cutting, the flank plate contains *obliquus internus abdominis* and *obliquus externus abdominis*.

FLAP

In → New Zealand lamb cutting, the flap contains the postero-ventral rib region plus abdominal muscles.

FLATBIFF

The medial part of the hindlimb in a → Norwegian beef carcass.

FLATFISH

→ Pleuronectiformes.

FLAT RIB

A cut of beef in the postero-dorsal region of the ribcage in the Midlands of → England.

FLATRIBBE

The postero-ventral rib region of a → Norwegian pork carcass.

FLATSTEIK

In → Iceland, an alternative to → ytralæri, indicating a cut taken from the lateral part of the hindlimb.

FLAT TOPS

A London cut of beef in the postero-dorsal region of the ribcage (→ England).

FLAVOUR

Water soluble components are responsible for the general meaty taste of a muscle food and are concentrated in the → sarcoplasm of → myofibres, although extra components may be added from the → myofibrils as the meat is aged. Water soluble components purified by dialysis may be used experimentally to recreate the general taste of meat when they are heated. Some of the most active compounds when heated with glucose and phosphate are inosinic acid, glycoprotein and amino acids. There is a progressive degradation of nucleotides post-mortem, from ADP to AMP to inosinic acid to ribose + hypoxanthine which contributes to the increase in flavour that occurs as beef is aged or → conditioned. Pentose sugars such as ribose may be more important in the generation of a meaty flavour than hexose sugars such as glucose.

The species-specific taste of a muscle food (whether it is beef or lamb, for example) originates from volatile components present in lipid components. Differences in fatty acid composition are involved, as in the distinction between lamb and mutton. The pattern of lipid autoxidation also is important, particularly lactones, ketones, alcohols and lower fatty acids. In the oxidation of unsaturated fatty acids, the 18-carbon group is most important (alk-2-ones, alkanols, alk-2-dienals, etc). Major aroma sources are 3-methyl butanol for beef and acetyl thiazole for pork. The flavour of grass-fed beef originates from diterpenoids created by the breakdown of chlorophyll by ruminal microorganisms.

(Source: Melton, 1990; Griebenow *et al.*, 1997)

FLEISCHDÜNNUNG

In → German beef cutting, the flesh-end thin flank or Fleischdünnung contains the abdominal muscles posterior to the ribs.

FLEISCHWURST

A creamy German emulsion → sausage flavoured with pistachio.

FLESH END

A → Scottish cut of beef containing the *quadriceps femoris* group anterior to the femur.

FLETCH

→ Flitch.

FLEXOR CARPI RADIALIS

A medial muscle of the mammalian foreshank, originating on the medial condyle of the humerus and inserting onto the third metacarpal bone.

FLEXOR CARPI ULNARIS

A medial, posterior muscle of the mammalian foreshank. The ulnar head of the muscle originates from the olecranon, while the humeral head originates from the epicondyle of the humerus. The insertion is on the accessory carpal bone. The ulnar head may be absent in pork.

FLEXOR DIGITORUM LONGUS

Part of *flexor digitorum profundus* in the hindlimb.

FLEXOR DIGITORUM PROFUNDUS

In the mammalian foreshank, *flexor digitorum profundus* is a three-headed muscle with origins on the humerus, radius and ulna, and insertions on third and fourth digits. Together, the three heads (humeral, radial and ulnar) are quite substantial (> 0.6% total muscle weight). In the hindshank, the corresponding muscle is fusiform in shape and also has three heads with common insertions on third and fourth digits: *tibialis caudalis* from the lateral tibial condyle, *flexor hallucis longus* from the lateral part of the tibia, and *flexor digitorum longus* from a ridge on the posterior part of the tibia.

FLEXOR DIGITORUM SUPERFICIALIS

In the mammalian foreshank, this muscle has two parts (superficial and deep). It originates from the medial epicondyle of the humerus and inserts onto third and fourth digits. In the hindlimb is another muscle with the same name. It is quite a large muscle (0.4% total muscle weight) with an origin from the supracondyloid fossa of the femur and insertions to third and fourth digits.

FLEXOR HALLUCIS LONGUS

Part of *flexor digitorum profundus* in the hindlimb.

FLITCH

Something cut from the side of an animal, such as a flitch of side bacon or of halibut.

FLOMEN

In → German pork cutting, the Flomen is between the loin or Kotelett and the belly or Bauch.

FLOUNDER

A medium-sized flatfish, → Pleuronectiformes.

FOLALDA-

In → Iceland, a prefix indicating a cut of horse meat, such as folaldasnitzel (snitzel) , folaldagúllas (gulash), folaldastrimlar (horse meat strips), or folaldabógsneiðar (blade steaks).

FOOT OF PORK

In → North American pork cutting, the hindfoot is removed by cutting through the tuber calcis, and the front foot is removed just distally to the ulna and radius. The foot contains a lot of bone and very little muscle. The gelatinized collagen of skin and tendon is its main attraction when cooked.

FOREKNUCKLE

In → English veal cutting, the foreknuckle is a cut based on the radius and ulna.

FOREL

Danish for trout, with the plural being forelørred, →
Salmoniformes. The Swedish is forell.

FOREQUARTER FLANK

→ English cuts of beef involving the postero-ventral region
of the rib cage, originating from the London area.

FORERIB OF BEEF

In → England, the forerib is the most posterior rib cut of the
forequarter of beef and may be cut in either of two ways,
either as a forerib with four ribs (ribs 7 to 10), or as a Scotch-
cut forerib with five ribs (ribs 6 to 10) and part of the blade of
the scapula. A six rib forerib may be cut in → Northern
Ireland.

FORESADDLE OF LAMB

A North American primal cut of lamb that avoids splitting
the carcass, right from left sides. It includes the rib (rack),
shoulder, breast and foreshank of both sides, still joined in
the midline.

FORESHANK OF LAMB

In → North America, the foreshank may be removed with
the breast using a single cut from the anterior of the sternum
to the ventral part of rib 12.

FORESHIN

The ulna, radius and associated muscles in beef cutting in
→Northern Ireland.

FORESKANK

In → Danish meat cutting, a foreskank is a foreshank of beef
or pork.

FOR-KNOKE

The foreshank (radius and ulna) of a → Norwegian beef,
veal or pork carcass.

FORMALDEHYDE

Apart from being a common laboratory reagent used to
preserve meat for microscopic examination (10% formalin
≈ 4% formaldehyde), formaldehyde may be produced as a
post-mortem metabolite (→ trimethylamine oxide) that affects
fish muscle texture, primarily in the cod family (→
Gadiformes).

FORTÅ

Front trotter or toes of a → Danish pork carcass.

FOT

→ Swedish for the foot of a carcass, as in pork.

FRALDA

The flap of abdominal muscle in a → Portuguese lamb
carcass. In a → Brazilian beef carcass it is the *obliquus
abdominis internus*, part of the vazio.

FRAMDEL

A front or forequarter of lamb in → Sweden.

FRAMHLUTI

Neck and antero-dorsal thoracic region of an → Icelandic
beef carcass.

FRAMHRYGGUR

An → Icelandic primal cut from the thoracic region of beef
or horse.

FRAMLÄGG

The foreshank of a → Swedish beef carcass.

FRAMPARTUR

The forequarter of a carcass in → Iceland.

FRANCE - BEEF CUTS

The major primal cut of the hindlimb is the globe (Figure
74), which is split to give a medial tranche or tende de tranche
(*sartorius, gracilis, pectineus, adductor* and
semimembranosus), a lateral semelle or gîte (*biceps femoris*
and *semitendinosus*), and an anterior tranche grasse
(*quadriceps femoris* group of muscles).

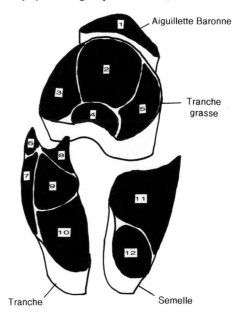

Figure 74. *French beef cuts from the globe (tranche, semelle and tranche
grasse) plus the Aiguillette baronne from aloyau. Muscles are tensor fascia lata
(1), rectus femoris (2), vastus medialis (3), vastus intermedius (4), vastus
lateralis (5), sartorius (6), gracilis (7), pectineus (8), adductor (9),
semimembranosus (10), biceps femoris (11) and semitendinosus (12).*

In some French cutting patterns, the jarret or shank is extended more proximally into the globe and split into distal (bout de gîte) and proximal (milieu de gîte) slices across the tibia and distal femur, respectively. Another variation is to split the muscles into an anterior (gîte-gîte) and posterior (gîte-noix nerveaux). The romsteck is dominated by the gluteal muscles and overlapping *biceps femoris*, and the posterior portion of the romsteck may be separated as a culotte (including sacral vertebrae three to five, plus the first caudal vertebra). Together, the romsteck (*gluteus* muscles) and aiguillette baronne (*tensor fascia lata*), plus the filet (*psoas major* and *minor*) and faux-filet (contre-filet or *longissimus dorsi*) before being boned out, form a primal cut called the aloyau. In Quebec, the aloyau is far more restricted and refers to the porterhouse or posterior lumbar region alone. The adjacent bavette d'aloyau (*obliquus internus abdominis* and *obliquus externus abdominis*), is part of a different primal, a panneau which extends anteriorly to include the plat-de-côtes couvert and découvert. These two terms, couvert (covered) and découvert (uncovered), relate to the removal of the shoulder. Thus, while the deep muscles posterior to the scapula remain covered, those beneath the scapula are uncovered by removal of the shoulder.The remaining cuts identified in Figure 75 (which only shows the deep cuts made medially to the scapula and humerus) include the collier (veine maigre and grasse) and the pis-de-boeuf (poitrine, tendron and flanchet). The train-de-côtes is a primal cut that includes the surlonge and both train-de-côtes couvert and découvert.

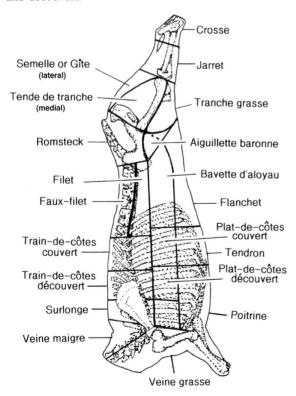

Figure 75. *France - beef cuts.*

Of the cuts obtained from the forelimb (Figure 76), the derrière de paleron may be the most proximal, but the dominant

separation is into the anterior jumeaux (*supraspinatus* and *biceps brachii*) and the posterior macreuses (*triceps, deltoids* and *subscapularis*). The insertions of all these muscles may be taken as a transverse cut, the charolaise, distal to which is the jarret and crosse. All these forelimb cuts are contained in a primal called the paleron.

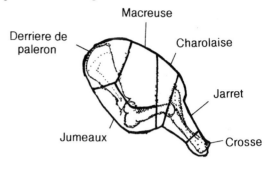

Figure 76. *France - forelimb beef cuts.*

Not much beef remains on the bone in France, and many of the larger muscles are individually layered with fat and made into long, tied rolls, from which sections of any length are cut for customers.

(Sources: OEEC, 1961; Gerrard and Mallion, 1977; Dumontiele *et al.*, 1981; Déterville, 1982; Monin, 1997)

FRANCE - LAMB CUTS

The shoulder is removed as the épaule, which may be subdivided into a proximal palette containing the scapula and a distal half, the coude (Figure 77). This uncovers the deeper vertebral muscles of the carré découvertes. The côtelettes are subdivided into posterior premières and anterior secondes. As in many countries, the lumbar, sacral and femoral regions are cut squarely to give the côtelletes filet, selle and gigot raccourci, respectively.

When left and right sides remain together on an unsplit carcass, the selle and côtelettes filet form the baron. A forequarter may be called a casque. And the papillon is the collet plus épaule.

(Sources: Dumontiele *et al.*, 1981; Déterville, 1982, Monin, 1997)

FRANCE - PORK CUTS

The bardière or back-fat is removed as a sheet which is used for wrapping other rolled joints of meat (Figure 78). A long length of the vertebral column from the posterior edge of the scapula back to the tail is cut as a loin (longe), from anterior to posterior, carré de filet, milieu de filet and pointe de filet (Figure 79). The pointe de filet resembles a sirloin steak of pork in North America, and is named from the *iliacus*, the end of the *psoas major* filet. The jambon or leg may be split at the femoro-tibial joint to separate the jambonneau de derrière. The travers is a long strip of flank and ribs, sometimes taken as far forward as the neck, but not if the distal shoulder region is removed as a square-cut hachage. Similarly, the

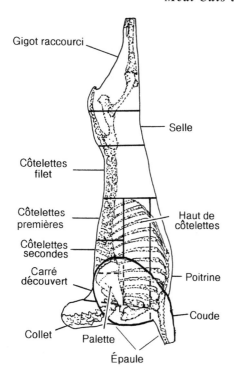

Figure 77. France – lamb cuts.

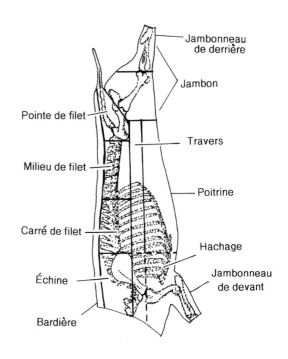

Figure 78. France – pork cuts.

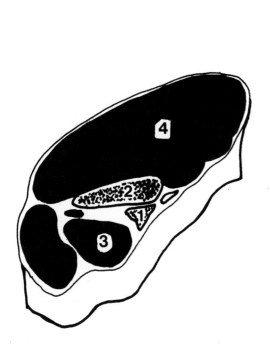

Figure 79. Pointe de filet showing sacrum (1), ilium (2), iliopsoas (3) and gluteus medius (4).

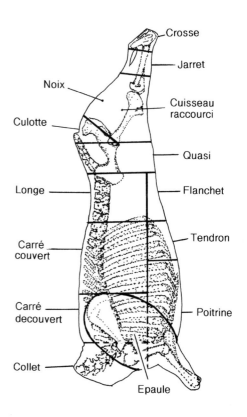

Figure 80. France – veal cuts.

poitrine may run the whole length of the ventral region of the carcass, or be cut short if a hachage is cut. Another variation instead of a square cut échine and hachage, is to remove the shoulder as a sequence, from proximal to distal, of échine (vertebrae and proximal scapula), palette (distal scapula) and épaule (*triceps* down to the ulna), in which case the sternal region remains as a cut called plat-de-côtes. Jambonneau de derrière may be called jarret arrière, and jambonneau de devant may be called jarret avant. Carré de filet may be called carreé de côtes, and its anterior end may be called the grillade. The *psoas* may be removed as a pork filet mignon.

(Sources: Gerrard and Mallion, 1977; Dumontiele *et al.*, 1981; Déterville, 1982; Monin, 1997)

FRANCE - VEAL CUTS

After removal of the shoulder (épaule), the remainder of the carcass is cut in a fairly rectilinear manner, as shown in Figure 80. From the épaule, the distal muscles around the ulna and radius may be removed as a jarret-avant, in which case the jarret of the hindlimb would be called jarret-arrière. Collier may be used as an alternative name for the collet in the neck region. The ventral part of the carré couvert may be removed as a series of flat ribs called haut-de-côtes.

There is considerable variation in hindquarter cutting, especially in the shape of the culotte which may be cut something like a North American sirloin of veal and called the quasi-culotte. Sometimes the *quadriceps femoris* group is removed as a noix patissière, in which case the noix shown in Figure 80 would become the sous-noix. Possibly the noix, or nut, relates to the patella, as is evident in nearby → Luxembourg. Bearing in mind the international variation in the location of the sirloin, it is of particular interest to find the loin being called the longe in veal, but not in beef or lamb.

(Sources: Dumontiele *et al.*, 1981; Déterville, 1982; Monin, 1997)

FRANKFURTER

An emulsion → sausage. The name originates from Frankfurt am Main in Germany. In North America, there is now little, if any, difference between frankfurters, wieners and hot dogs.

FRANSYSKAN

The *quadriceps femoris* muscle group of a → Swedish beef carcass. The literal translation is Frenchwoman. Guess why. The same name may be used for slices from the corresponding region of a leg of Swedish lamb.

FRENCHED

In North America, when a cut of meat such as a lamb rib chop is Frenched, the tissues are trimmed away from a protruding bone such as a rib so that it can be given a frill of paper by which to hold it at the dinner table.

FRENCH ROLLS

A North American cut of beef in which the tip from a → diamond round or hip is boned and rolled. It usually contains some portion of *rectus femoris* and one of the *vastus* muscles.

FRESH

For muscle foods, fresh implies simply that the product in question has not been frozen, cured, processed, or cooked. It has no connotations relating to the time lapse post-mortem. Thus, fresh beef is best after aging or conditioning for a few weeks in a meat cooler. Fresh frozen does, however, imply a short delay between slaughter or capture and freezing.

FRICANDEAU

A lateral cut from the hindlimb of a pork, lamb or veal carcass in → Austria. In → Belgium, the fricandeau may be the pointe au filet, a cut of pork containing sacral and posterior lumbar vertebrae. However, fricandeau (with the plural being fricandeaux) also may be used in a more general way to indicate almost any sliced or cubed cut of meat.

FRICASSEE

Following from the general use of → fricandeau, the same word root gives rise to a number of culinary terms to describe sliced, cubed, diced or comminuted meat dishes - fricassee, fricandelle, etc.

FRIZZES

A highly-seasoned dry → sausage containing pork in a natural → casing.

FROG LEGS

The large hindlimb muscles of frogs provide a myosystem with a taste and texture not unlike that of chicken, which is curious to the meat scientist because many of the myofibres have a very slow contraction speed relative to those of poultry. Thus, the slow myofibres of the frog have a → felderstruktur pattern of myofibrils, thick Z-lines, no M-line or membrane action potentials, and numerous mitochondria.

Frog legs are usually purchased with the skin off, with the muscles set in rigor mortis and ready to use. Fresh material may be more difficult to catch hold of - as Luigi Galvani discovered just before he discovered bioelectricity. The frog hindlimb is fairly flat and kicks laterally. The individual muscles have their own anatomical nomenclature. Using their mammalian equivalents for convenience, the most notable feature is that the equivalent of the *biceps femoris* is very thin and that the main muscles seen from above the frog are the *vastus lateralis* and *semimembranosus*. From a ventral view, the *adductor* is the largest muscle.

Wild frogs are getting scarce in many parts of the world as acid rain, draining of wetlands and the global spreading of amphibian diseases take their toll. There is much to commend farming rather than hunting, as in the Brazilian farming of the bullfrog, *Rana catesbeiana*,.

(Sources: Marshall, 1900; Rombauer and Becker, 1975; Franzini-Armstrong, 1973); Ramos et al., 2003).

FUGLAKJÖT

Icelandic for → chicken.

FYNE RIB

Prime rib roast of beef (posterior thoracic region) in the → Netherlands.

GADIFORMES

Cod-like bony fish of the families Gadidae and Merluciidae (Figure 81) are the second most important source of fish myosystems after the Clupeiformes. The cod fishery is an ancient but abused industry. Cod and their relatives are demersal (bottom-living) and have three dorsal fins and two anal fins. Many have a sensory barbel on the lower lip. Adults feed on other fish and their spawn.

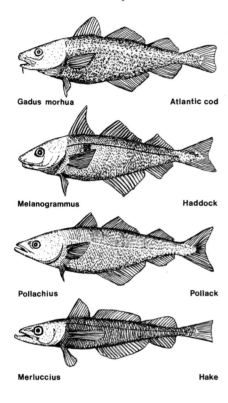

Gadus morhua   **Atlantic cod**

Melanogrammus   **Haddock**

Pollachius   **Pollack**

Merluccius   **Hake**

Figure 81. Fishes of the cod family.

Gadus morhua, the Atlantic cod (Figure 81), is or was of major economic importance on both sides of the Atlantic, usually weighing > 2 kg. When cooked, the myosystem separates easily into large, moist, white flakes. Pacific cod, *G. macrocephalus*, is extensively fished by Asian countries, being processed to fillets and fish fingers aboard factory ships. *Melanogrammus*, the haddock (Figure 81), distinguished by its black, thumb-shaped blotch on the side of the body behind the head, produces a fine white myosystem that may be consumed fresh, or lightly salted and smoked. The typical weight is from 0.5 to 1.8 kg. *Pollachius*, the pollock (Figure 81), usually weighs up to 6 kg and is sold fresh, frozen or dried. *P. virens* is the Boston bluefish, although it is greenish in colour. It has a deep body with a firm, white myosystem. Apart from its green colour, it may be distinguished from cod by its forked tail, pointed snout, projecting lower jaw, and reduced or absent chin barbel. *Brosme*, the cusk, has continuous dorsal, caudal and anal fins, and considerable variation in colour, from red to brown to yellow. Cooked cusk is similar to cod.

Merluccius, the hake (Figure 81), is in a separate family, the Merluciidae. The myosystem of the silver hake (*M. bilinearis*) rapidly softens post-mortem and the fillets are best cooked directly from the frozen state. For *M. merluccius*, the proximate analysis is water, 79.04%; total nitrogen, 2.9%; ash, 1.31%; lipid, 2.24%; and non-protein nitrogen, 0.304%. Seasonal variation occurs in composition. The white hake, *Urophycis tenuis*, and very similar red hake, *U. chuss*, both have a high, triangular first dorsal fin, followed by a continuous second dorsal fin extending to just before the tail fin.

Also classified within the Gadiformes are the grenadiers or rattails (family Melanoididae). They have a worldwide distribution as bottom-dwellers. The head is large but, after that, the body weirdly tapers down to nothing. Despite its odd shape, *Macrourus rupestris*, the roundnose grenadier, can be filleted to obtain a myosystem like that of cod, but sweeter. The livers are used as a source of vitamin A.

(Sources: Hardy, 1959; O'Boyle, 1985; Lamb and Edgell, 1986; Markle, 1989; Atkinson, 1993; Bishop, 1993; McGlade, 1993)

GAIMA

Hindlimb of camel in → Morocco.

GALLIFORMES

This order of birds includes the most important commercial species reared or harvested for their meat: → chickens, → turkeys, → guinea fowl, → pheasant, → grouse, → partridge, and → quail. They tend to be fairly large, heavy birds which walk a lot and have well muscled legs. They fly primarily as a means of escape and, being heavy, tend to have weight-lifter's flying muscles over the breast. Hence, they make the ideal meat birds. This natural propensity for muscle development, combined with selective breeding and superior nutrition, has resulted in the astounding growth rates of modern chickens and turkeys in the developed countries.

GALLINA DE GUINEA

Spanish for → guinea fowl.

GAMBA

Respectively, the gamba anteriore and posteriore are the fore- and hindshanks of an → Italian beef carcass.

GAM CORDS

Tendons of the hock used to hang a pork carcass from a gambrel.

GANASCIA

Italian for the throat of an animal.

GANDOUZ

Forelimb of a camel carcass in → Morocco.

GANS

German for → goose.

GANSO

Ganso is a goose in Spanish, but also the *biceps femoris* in → Chilean beef cutting.

GAP FILAMENTS

In electron micrographs of highly stretched muscle, it is sometimes possible to see very thin → myofilaments extending across the gap between the ends of thick and thin myofilaments. These are called gap filaments and most likely are → titin filaments.

(Source: Trombitás *et al.*, 1991)

GARDEN

The neck of a carcass in → Pakistan.

GARGRA

Ribs on a camel carcass in → Morocco.

GÅS

Swedish for → goose.

GASTROCNEMIUS

A large muscle of the mammalian hindlimb (> 2% total muscle weight) contributing to the Achilles tendon at the hock, from which beef carcasses traditionally are suspended (Figure 82). Unlike the human calf muscle, which protrudes from the back of the leg behind the shin, the *gastrocnemius* muscle in beef, pork and lamb is covered by other muscles from the upper part of the hindlimb. The *gastrocnemius* has two large heads, medial and lateral, both originating proximally on the posterior part of the femur.

GAUR

Bos gaurus, the largest living, wild bovine, from which was derived the semi-domesticated → mithan. Bulls may weigh 940 kg and may reach 1.9 m at the shoulder, although larger animals have been recorded. There is massive development around the *trapezius* muscle, leaving a marked drop toward the lumbar region. Gaurs need conservation to protect them from hunting and habitat erosion.

(Source: NRC, 1983)

Figure 82. *Lateral view of beef* gastrocnemius *(solid black) showing its origin from the femur (1) and insertion on the tuber calcis (2).*

GAYAL

→ Mitha.

GEFILTE FISH

A ball or cake of comminuted fish muscle with spices.

GELBWURST

Veal and pork → luncheon meat coloured pale yellow by milk and egg.

GELINOTTE

French for → grouse.

GELSOLIN

A protein involved in → actin polymerisation.

GEMELLUS

A small, thin, triangular muscle in the mammalian pelvic girdle, posterior to the distal part of the femur (Figure 83).

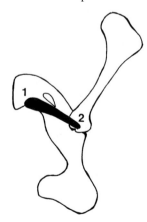

Figure 83. *Lateral view of beef* gemellus *(solid black) showing its origin from the* ischium *(1) and insertion on the* femur *(2).*

GEMSBOCK

→ Oryx

GENOA SALAMI

A dry → sausage.

GERMANY - BEEF CUTS

As in many European countries, beef cutting in Germany is extremely variable with countless regional names. Honikel (1996) suggested the pattern shown in Figure 84 for an overall view.

Separation of the round into a top dish or Oberschale and a bottom dish or Unterschale is a similar pattern to that used in England, with the hindquarter placed on the block ready for cutting with the lateral surface downwards so that the medial surface becomes the top. Thus, Oberschale compares with English → topside, and Unterschale with → silverside. This leaves most of the pelvis in the haunch or Hüfte. Reminiscent of the French ball of the loin, boule d'aloyau, the *quadriceps femoris* group of muscles plus *tensor fascia lata* also may be called a ball, the Kugel. The Blume refers to the body part near the tail (as shown in Figure 85), but also may be a cut straight across the carcass like a sirloin in North America. There is a long cut of loin and ribs called the Roastbeef, with the anterior end called the peak or tip of the ribs, Hohe Rippe, and being identified with the entrecôte. The lumbar region also may be called loin, in German, Lende. The shoulder

may be lifted from the forequarter leaving a tongue-piece or Zungenstück connected to the neck or Kamm. As in the English sticking cut, the neck also may be called a Stich cut. *Supraspinatus* is removed as a false filet or Falsches Filet, with all three heads of *triceps brachii* forming the shoulder cut or Schulter. The breast or Brust underlies the Schulter, and is one of the German names familiar in North America (→ brust flanken ribs). The foreshank is known by names such as Vorderhesse or legmeat, Beinfleisch.

Another cutting pattern outlined by OEEC (1961) and shown in Figure 85, removes a leg or keule which is split four ways into a (1) tail-piece or Schwanzstück containing *biceps femoris* and *semitendinosus*, (2) ball or Kugel containing the *quadriceps femoris* group, (3) medial Oberschale as in Figure 84, and (4) hindshank or Hinterhesse. The roasts in the prime rib region are called Hochrippe rather than Hohe Rippe, but with a similar meaning as the high-end of the rib cut. From the forelimb detached from the rib-cage, the shoulder-blade cartilage (Schulterblattknorpel) is removed proximally, to leave from anterior to posterior, the false filet or Falschesfilet, the shovel-piece or Schaufelstück, and the Dickes Bugstück (thick shoulder-piece) containing much of *triceps brachii*. And "spanning the ribs" is the Spannrippe, sometimes called the cross-rib cut or Querrippe. The vertebral region medial to the removed scapula is the Fehlrippe, or short-rib (where the implication of short is of a cut with something missing rather than a cut with minimal length). The brisket may be divided into the anterior breast-bone (Brustbein), middle breast (Mittelbrust), and after-breast (Nachbrust). The thin flank is

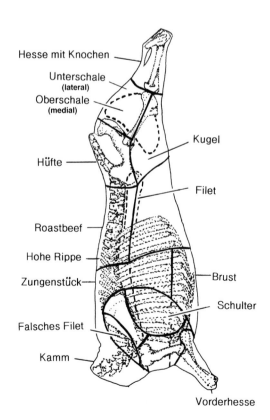

Figure 84. Germany – beef cuts.

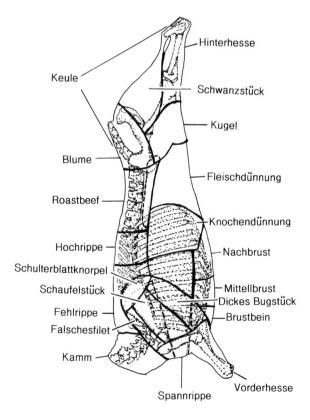

Figure 85. German standard method for beef cutting.

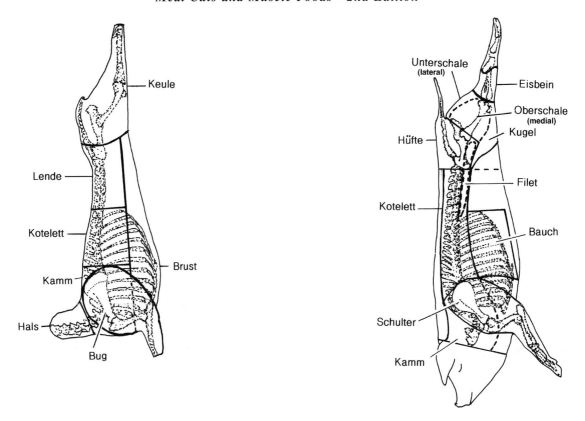

Figure 86. German lamb cuts.

Figure 87. Germany – pork cuts.

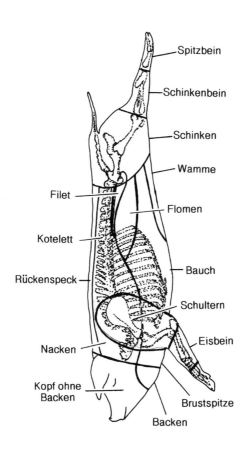

Figure 88. Germany – pork cuts.

divided into the bone-end thin (Knochendünnung) and flesh-end thin (Fleischdünnung).

(Sources: OEEC, 1961; Honikel, 1996)

GERMANY - LAMB CUTS

The Bug or shoulder contains the scapula and is lifted laterally off the anterior thoracic region, which is called the Kamm or neck (Figure 86). The cervical part of the neck is called the Hals or throat. The Brust or breast is cut the full length of the carcass. The lumbar and posterior thoracic regions are split into the Lende or loin, and Kotellet or cutlets, to leave a full leg or Keule.

(Source: Gerrard and Mallion, 1977).

GERMANY - PORK CUTS

Two of many different cutting patterns for pork carcasses in Germany are shown in Figures 87 and 88, one suggested by Honikel (1996) and one by OEEC (1961), respectively.

In the first pattern (Figure 87), the ham is split several ways like a miniature German beef carcass. The haunch or Hüft contains most of the pelvis, the Unterschale is lateral, the Oberschale is medial, and the anterior *quadriceps femoris* group forms a Kugel. Alternatively, the ham may be cut whole, as a Schinken (Figure 88). After removal of the back-fat or Rückenspeck, the whole loin is called a Kotelett (this term also may be used for a single chop or cutlet). It may be boned out as a Kotelett ohne Knochen (loin without bone). The Bauch is the belly, which may be separated from the paunch or Wamme. The breast point, or Brustspitze, is adjacent to the jowl, called the bake or Backen, which may be removed to leave a head without the bake cut (Kopf ohne Backen). Despite its literal translation as ice-foot, an Eisbein is a pig foot salted in jelly (Breul, 1952). The Schinkenbein is the foot of the ham. A cut of meat from the neck of a carcass in German meat cutting may be called the Kamm or Nacken. For the whole animal, both terms relate to the dorsal part of the neck.

(Sources: OEEC, 1961; Honikel, 1996)

GIGOT D'AGNEAU

Leg of lamb in → France and Quebec. Gigot d'agneau raccourci is a short cut leg of lamb, composed of parts 2 to 4 in Figure 128. Gigot also may be used for a leg of lamb in → Scotland.

GILT

A female pig that has not yet farrowed (produced a litter of piglets).

GIMMER

A female sheep between first and second shearing.

GIRELLO

An → Italian cut of beef based on the *semitendinosus*.

GIRELLO DI SPALLA

The *supraspinatus* of an → Italian beef carcass.

GÎTE NOIX

Biceps femoris in the → French beef cut, the tranche.

GLAZED

A frozen product, quickly dipped in water and re-frozen to create a surface layer of ice to prevent freezer burn.

GLOBE

In → French beef cutting, the globe is a hindlimb primal cut containing the femur. It is split to give a medial tranche or tende de tranche (*sartorius, gracilis, pectineus, adductor* and *semimembranosus*), a lateral semelle or gîte (*biceps femoris* and *semitendinosus*), and an anterior tranche grasse (*quadriceps femoris* group of muscles).

GŁOWA

The head on a → Polish meat carcass.

GLUTAEUS

Gluteus muscles.

GLUTEUS ACCESSORIUS

A deep muscle in the mammalian pelvic girdle, lateral to the ilium from which it originates. It inserts on the head of the femur.

GLUTEUS MEDIUS

A major muscle of the mammalian pelvic girdle (≈ 3.8% total muscle weight). It originates from much of the lateral surface of the ilium and inserts onto the greater trochanter of the femur (Figure 89).

Figure 89. *Lateral view of the main part of beef* gluteus medius *(solid black, excluding an anterior extension to the* longissimus dorsi *aponeurosis), showing its origin from the ilium (1) and insertion on the femur (2).*

GLUTEUS PROFUNDUS

A medium-sized muscle (> 1% total muscle weight) located over the mammalian hip joint and under the *gluteus medius*. It originates from the ilium and associated ligaments and inserts onto the femur (Figure 90).

Figure 90. *Lateral view of beef* gluteus profundus *(solid black) showing its origin from the ilium (1) and insertion on the femur (2).*

GLYCOGEN

Glycogen (Figure 91) is a storage carbohydrate in the → myofibre sarcoplasm where it forms granules visible by electron microscopy. Glycogen granules are ellipsoidal with a laminar structure which indicates they may grow from one edge rather than a central point. The starting point for a glycogen granule is a protein called glycogenin. In longitudinal sections of skeletal muscle, glycogen and its associated enzymes are concentrated at the → I-bands.

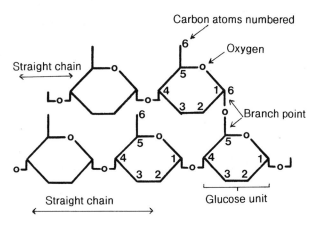

Figure 91. *Chemical structure of glycogen.*

Glycogen is a polysaccharide formed from large numbers of D-glucose units. However, the glycogen from some animal tissues is a proteoglucan (glycoprotein) that may contain other monosaccharides apart from D-glucose as well as phosphate ester groups. The "D-" prefix indicates that the glucose is a member of a family of carbohydrates related to one of the mirror-image forms of a parent molecule. Straight chains are formed by 1-4 linkages while branch points are formed by 1-6 linkages.

GLYCOLYSIS

Glycolysis is the enzymatic release of energy from the oxidation of glucose units. In meat, post-mortem glycolysis starts with the enzyme phosphorylase eroding straight chains of glucose units in → glycogen granules, starting at their non-reducing ends (at carbon 4) and attaching a phosphate group (to carbon 1) as it removes the glucose unit. Phosphorylase keeps eroding all the accessible straight chains on the outside of the glycogen granules until it comes to the fourth glucose unit preceding a branch point. The three glucose units before the fourth which forms the branch point are removed together, and are added to an adjacent free straight chain so that the 1-6 linkage thus exposed at the branch point becomes accessible (Figure 91). Another enzyme, debranching enzyme, removes the glucose unit forming the branch point. Instead of being released as glucose-1-phosphate, this glucose unit from a branch point remains as free glucose. Thus, total glycogenolysis liberates glucose-1-phosphate and glucose in a ratio related to the mean length of straight chains versus the number of branch points.

Farther along the glycolytic pathway, six-carbon molecules derived from the glucose units of glycogen are split to produce two molecules of pyruvate, each with three carbon atoms. Most of the glycolytic enzymes are concentrated in the I-band of the → sarcomere. In the living animal, under aerobic conditions, pyruvate formed in the cytosol of a myofibre normally enters a mitochondrion. Pyruvate then is converted to acetyl-CoA which becomes fused to oxaloacetate to form citrate. The citrate is oxidized in the Krebs' cycle, which is completed by the regeneration of oxaloacetate. Continuous activity of the Krebs' cycle is fuelled by a range of carbohydrates, fatty acids and amino acids, and is the primary system for the aerobic generation of energy. Large numbers of molecules of ATP are produced from ADP by oxidative phosphorylation in the mitochondrial membrane.

However, the bulk of the muscle in a recently slaughtered animal is anaerobic, and the production of two pyruvate molecules from a glucose-1-phosphate molecule results in the reduction of 2NAD$^+$. Thus, somewhere else in the myofibre, NADH must be re-oxidized for glycolysis to continue. In the live animal, this is done by the mitochondrial Krebs' cycle. After slaughter, however, the Krebs' cycle is halted, and NADH is re-oxidized in the → sarcoplasm by lactate dehydrogenase (LDH) during the conversion of pyruvate to lactate. Pyruvate is of no immediate use anaerobically since mitochondrial oxidation has ceased, but its conversion to lactate ensures a continued supply of NAD$^+$ for the continuation of glycogenolysis and anaerobic glycolysis. These events are only the initial stages of complete carbohydrate oxidation, and they do not regenerate much ATP. Post-mortem, the net gain of ATP is reduced to only two molecules of ATP per molecule of glucose-1-phosphate. Molecules of glucose released from glycogen branch points generate a total net gain of 3 ATP.

In fish, glycogenolysis usually is extremely rapid once the fish is out of water. Fish may develop rigor mortis in less than an hour. Thus, they may be frozen rapidly after capture.

GNAGI

Pork trotters in → Switzerland.

GOAT

Domestication of the goat, *Capra hircus,* started at least 10,000 years ago. In countries where retailing is dominated by supermarkets rather than by farmer's markets, most commercial goat meat is obtained from kids (young animals) produced as a by-product of milk production. Most of the consumers in these countries do not like the strong smell and taste of meat from older goats, and only the meat from kids fed milk or a replacer diet is available. This is generally pale and has a low fat content. For milk-fed animals, the target slaughter weight would be about 30 kg at 18 to 19 weeks. For animals fed concentrate, the target would be 30 kg at 23 to 24 weeks. Dressing percentages may be as high as 60% for young milk-fed kids, but they decrease towards 45% for progressively older animals, especially if forage fed. As goats become older, their meat gets darker and tougher, and their fat gets yellower. The lean meat yield from carcasses is relatively high (around 68%) because there is seldom any subcutaneous fat. Carcass cutting generally follows the local procedure for lamb. In countries where strong-tasting meat is acceptable or even preferred, meat from older slow-growing goats is the main product. The main meat-breeds of goats are found in South America, China, Africa, and India. Mature carcasses may produce about 64% lean muscle. Breeds such as Angora and Cashmere are kept primarily for textile fibre production. Cashmere kids have dark meat.

(Source: Wilkinson and Stark, 1987; Snell, 1996; Simela *et al.*, 1999; Dhanda *et al.*, 1999; Zeder and Hesse, 2000)

GOLEŃ PRZEDNIA

The foreshank of a → Polish veal or lamb carcass.

GOLEŃ TYLNA

The hindshank of a → Polish veal or lamb carcass.

GOLONKA PRZEDNIA

A foreshank of a → Polish pork carcass containing the distal ulna and radius.

GOLONKA TYLNA

A hindshank of a → Polish pork carcass containing the proximal tibia and fibula.

GÖMBÖLYŰ-FELSÁL

In a → Hungarian beef carcass, the *quadriceps femoris* group of muscles is called the ball-round or gömbölyű-felsál which

makes an interesting parallel to similar names for this muscle group in French (→ boule d'aloyau), Spanish (→ bola) and German (→ Kugel).

GOOSE

Ducks, geese and swans are grouped together in the Order Anseriformes. Geese comprise the Family Anserinae. Most geese are larger than most ducks and, thus, attract most of the attention from most of the wildfowl hunters. Some geese are so common as to be a problem, such as the common Canada goose (*Branta canadensis*), while others totter sadly on extinction. Canada geese produce good meat, which is probably why they congregate in cities where firearms are traditionally used for killing people not procuring dinner. Some geese produce highly rated meat, such as the American brant (*Branta bernicla*), the white-fronted goose (*Anser albifrons*), and the blue goose (*Chen caerulescens*); while the meat from others, such as the emperor goose (*Philacte canagica*) is reputed to have a vile taste. Meat from the snow goose (*Chen hyperborea*) is rated as intermediate. The cause is likely to be nutritional rather than hereditary (→ duck). In Europe, the grey-lag goose (*Anser anser*), the ancestor of domestic geese, is the classical quarry for the hunter, as well as several other species of *Anser* and *Branta*. The lag in grey-lag relates to this goose being somewhat of a migratory laggard.

The meat breeds of domesticated geese include the Embden (distinguished by its white feathers and blue eyes), Toulouse (grey feathers with a pronounced gullet pouch below its beak), Chinese (white, grey or brown feathers plus a pronounced knob or caruncle above the beak), Roman (like a small Embden), and African (like a large Chinese). The best meat is from goslings up to 6 months, so selection of good produce should be on signs of youthfulness. In North America, young Goose is from birds 15 to 20 weeks, with carcass weights of 2.5 to 6.5 kg, either males or females. The beak is still relatively flexible, as are the tracheal cartilages.

(Sources: Vesey-Fitzgerald, 1946; Rue, 1973; Batty, 1979)

GOOSE SKIRT

In English beef cutting, sheets of stewing muscle trimmed from the subpelvic region.

GORGE

French for the throat, as in a pork carcass in → Luxembourg.

GÓRKA

In → Poland, the posterior thoracic region of a beef or veal carcass or the anterior thoracic region of a lamb carcass.

GÖTEBERG

This Swedish city has given its name to a type of semidry → sausage which, if made traditionally, is coarsely chopped, soaked in brine, then smoked.

GOVEDINA

Beef in → Serbia & Montenegro.

GOLYASHKA

Fore- or hindshanks from a beef or pork carcass in → Russia.

GRACILIS

An expansive, thin muscle on the medial surface of the mammalian hindlimb (Figure 92). Its origins on the subpelvic ligament and prepubic tendons do not survive splitting of the carcass into right and left sides. Distally, the *gracilis* tapers to a fascia over the stifle joint.

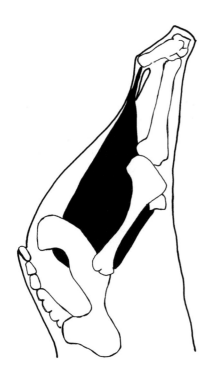

Figure 92. *Lateral view of the extent of beef* gracilis *(solid black, behind the skeleton).*

GRAISSE DE ROGNON

Panne or kidney fat from a → Belgian pork carcass.

GRASSCUTTER

→ Cane rat

GRASSELLA

Italian for the posterior flank of an animal.

GREECE - BEEF CUTS

After removing the tibial shank (opisthio kotsi) and the kiloto containing the gluteal muscles, the main cutting pattern of a beef hindquarter in Greece is into the trans, the noua kai oura, and the strogilo (Figure 93). The trans is located medially to the femur and contains *semimembranosus, adductor* and the other medial muscles. The noua kai oura is located laterally to the femur and contains the *biceps femoris* and *semitendinosus*. The strogilo is located anteriorly to the femur and contains the *quadriceps femoris* group of muscles. In the lumbar region, the *psoas* muscles are removed as a fileto, leaving the *longissimus dorsi* as contra brizoles in the lumbar region and brizoles in the posterior thoracic region. The stithoplevres contains a long ladder of ribs (used for rib rolls), leaving the abdominal muscles in the lapa and the pectoral muscles in the stithos. The spala contains the scapula, humerus and associated muscles which are lifted off from the underlying rib cut, the kapaki. The *biceps brachii* is removed from the spala as the pontiki (mouse). The cervical region is called the lemos.

(Sources: Vareltzis, 1998)

GREECE - LAMB CUTS

The bouti is the full hindlimb posterior to the ilium, from which may be detached the shank or kotsi (Figure 94). After removal of the shoulder or spala, the vertebral column is separated into the neframia in the lumbar region and the paidakia in the thoracic region. The breast and flank, or stithos and lapa, are removed together with a straight cut, leaving only the neck, or lapa.

(Sources: Vareltzis, 1998)

GREECE - PORK CUTS

After trimming off any appropriate back and belly fat, the hindlimb is cut whole as a bouti, and the hindshank or kotsi is removed (Figure 95). Shoulder meat is removed in the spala, from which the foreshank or omprosthio kotsi is detached. The major vertebral cut is the brizoles which follows the length of the *longissimus dorsi* from its origin medial to the scapula to its insertion on the anterior face of the ilium. The belly is the pansteta.

(Sources: Vareltzis, 1998)

GREEN COD

Pollachius virens, more a pollock than a cod, although with three dorsal fins and two anal fins just like the cod, → Gadiformes.

GRENADIER

A tapered fish of the cod family, → Gadiformes.

GRÍSA-

In → Iceland, a prefix denoting a range of pork cuts such as grísarif (pork sparerib), grísakótilettur (pork cutlets), grísalæri med beini (bone-in leg of pork), etc.

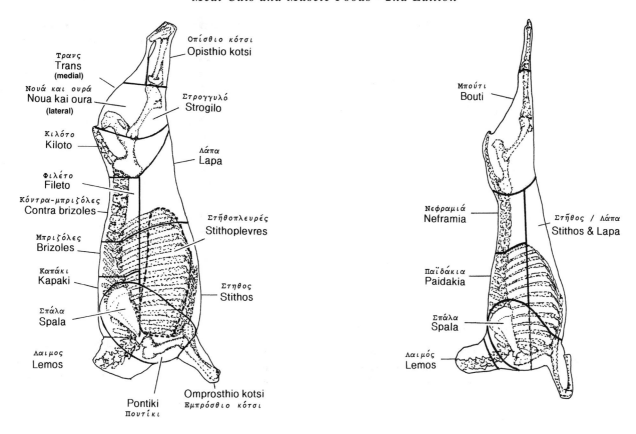

Τρανς
Trans
(medial)

Νουά και ουρά
Noua kai oura
(lateral)

Κιλότο
Kiloto

Φιλέτο
Fileto

Κόντρα-μπριζόλες
Contra brizoles

Μπριζόλες
Brizoles

Καπάκι
Kapaki

Σπάλα
Spala

Λαιμος
Lemos

Οπίσθιο κότσι
Opisthio kotsi

Στρογγυλό
Strogilo

Λάπα
Lapa

Στηθοπλευρές
Stithoplevres

Στηθος
Stithos

Omprosthio kotsi
Εμπρόσθιο κότσι

Pontiki
Πουτίκι

Figure 93. *Beef cuts in Greece.*

Μπούτι
Bouti

Νεφραμιά
Neframia

Στῆθος / Λάπα
Stithos & Lapa

Παϊδάκια
Paidakia

Σπάλα
Spala

Λαιμός
Lemos

Figure 94. *Lamb cuts in Greece.*

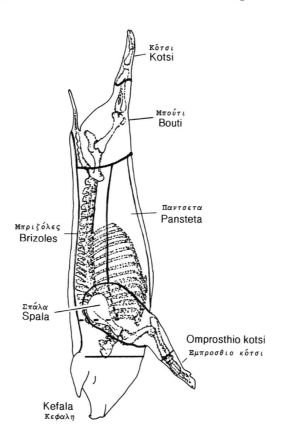

Κότσι
Kotsi

Μπούτι
Bouti

Παντσετα
Pansteta

Μπριζόλες
Brizoles

Σπάλα
Spala

Omprosthio kotsi
Εμπροσθιο κότσι

Kefala
Κεφάλη

Figure 95. *Pork cuts in Greece.*

GROS CROISÉ

Anterior part of the brisket on a → Belgian beef carcass.

GROSSE CROISURE

The grosse croisé, or anterior part of the brisket on a → Belgian beef carcass.

GROSSE CUISSE

The pièce levée of a → Belgian beef carcass.

GROSSE TÊTE

Tranche grasse of a → Belgian beef carcass.

GROUDINKA

The sternal region of a → Russian carcass.

GROUPER

A serranoid bony fish, → Perciformes.

GROUSE

Grouse are birds in the order → Galliformes. The red grouse, *Lagopus scoticus*, is one of the foremost game birds of the British Isles, living in hills and bogs where there is an abundance of its primary food, heather and various upland berries. The grouse produces a meat with a stronger taste than that of pheasant and does not require extended aging for full flavour development (as does pheasant). Three days of aging would not be unreasonable for a grouse (but would not do much for a pheasant). Grouse can be distinguished from other game birds by their feathered legs and toes. Young birds may be identified by a fragile beak and skull, the third primary flight feather being shorter than the others, and the first two primary flight feathers being pointed rather than rounded.

The largest grouse is the capercaillie, *Tetrao urogallus*. In Canada and the USA, the ruffed grouse, *Bonasa umbellus*, does a marvellous imitation an old tractor starting up as part of its mating ritual. It must be confusing for the courting couple if a real tractor starts up. The confusion is worsened by hunters who call this bird a partridge. Other North American grouse are the dusky blue grouse (*Dendragapus obscurus*) of the Pacific Northwest and Rockies, which produces highly prized meat except when on its winter diet of resinous berries; the spruce grouse (*Canachites canadensis*) of Canada, which is best on its late summer diet of berries, but to be avoided when on its winter diet of conifer needles;

the sharp-tailed grouse (*Pediocetes phasianellus*) of the central and northern brushlands of North America; and the sage grouse (*Centrocercus urophasianus*), of the sagebrush areas of the Rockies, which comes ready spiced from its diet. Other North American birds hunted as grouse include the ptarmigans (willow, rock and white tailed; *Lagopus lagopus*, *L. mutus* and *L. leucurus*) and greater and lesser prairie chickens (*Tympanuchus cupido* and *T. pallidicinctus*).

(Sources: Vesey-Fitzgerald, 1949; Bertin, 1967; Rue, 1973; McAndrew, 1990)

GRUDI

Sternal region in → Serbia & Montenegro.

GUILLEMOT

→ Svartfugl

GUINEA FOWL

There are many types of guinea fowl around the world. They lack feathers on their head and neck. The domestic guinea fowl, *Numida meleagris*, has red wattles like a turkey. Guinea fowl meat is dark, approaching that of the pheasant, something like a chicken-pheasant mixture. Guinea fowl are produced commercially for both their meat and eggs. A typical carcass might be slightly over 1 kg from a bird of 8 to 9 weeks. Above this the meat tends to become too dry.

GULASCHFLEISCH

Neck and distal limb cuts of beef in → Austria.

GURNARD

A bony fish with spines and a mailed head, → Scorpaeniformes.

GUSTOSTÜCKE VOM KNÖPFEL

The proximal hindlimb of a beef carcass in → Austria.

GUSTOSTÜCKE VOM VORDERVIERTEL

The proximal forelimb of a beef carcass in → Austria.

GUTTOR

Thigh muscles around the femur in beef, veal or camel in → Pakistan.

HAAS

Dutch for a → hare, but it may also be a cut of beef or pork from the lumbar region.

HABRAT AL-DAHER

A lumbar cut of veal in → Jordan.

HACHAGE

In → French pork cutting, the hachage is a square cut of pork including the distal part of the scapula, the humerus and the ventral parts of the first four ribs.

HACHÉ

French for ground, minced or comminuted meat.

HADDOCK

A small cod-like fish of the North Atlantic, *Melanogrammus*, → Gadiformes.

HAGGIS

This Scottish delicacy is composed of coarse fragments of skeletal and/or cardiac muscle which are flavoured with fragments of liver and/or spleen. Other key ingredients are fat, oatmeal, pepper and salt. Traditional haggis is cooked in a sheep's rumen.

HAIE

Another name for the pièce levée of a → Belgian beef carcass.

HAKE

A cod-like fish, *Merluccius*, → Gadiformes. It produces a high quality myosystem but is now overfished.

HALESTYKKE

In → Danish beef cutting, the halestykke or tail cut is a roast from the proximal part of the hindlimb.

HALIBUT

A large flatfish, → Pleuronectiformes.

HÄLLEFLUNDRA

Swedish for halibut, → Pleuronectiformes.

HALS

The hals is the neck of the carcass in many different countries. For example; in → Sweden, where it may also be called the halskött in beef; in the → Netherlands; in → Switzerland and → Austria.

HALSBIT

A neck cut from a → Swedish pork carcass.

HALSKARBONADE

A neck cut from a → Netherlands pork carcass.

HAM

Unchanged much from its Old English source word, hamm, this cut of meat generally relates to the muscle mass around the femur and tibia, and is most frequently used in relation to a cut of pork including most of the hindlimb. Because of the chubby shape of the pork hindlimb, the *quadriceps femoris* muscle group is reduced in exposed area relative to the beef round shown in Figure 213, but otherwise the muscle groups are very similar, as seen in Figure 96.

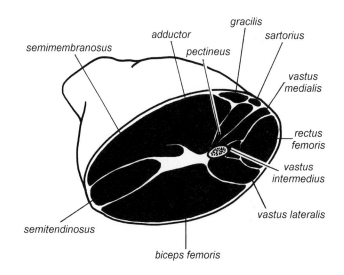

***Figure 96.** A shank end of ham.*

Smoking a ham may not necessarily involve heating much above an internal temperature of 60°C, which may leave the consumer with the responsibility to cook the ham to 70°C.

HAMBURGER

Hamburger is comminuted (ground or minced) beef, named from the German city of Hamburg. The name was introduced into the USA by German immigrants around 1850. Hamburg steak became hamburger steak by 1889, which was shortened to hamburger by 1908 and to hamburg by the 1950s.

(Source: Merriam-Webster, 1991)

HAM HOCKS

In North America, the tibial end of a smoked ham of pork.

HAMPE

Hampe is a → French term for the muscle of the diaphragm (skirt). It is also used for the flanc of a beef carcass in → Belgium.

HANCHE

A → French cut of beef including both the romsteck and aiguillette baronne.

HAND

The term hand is commonly used to indicate the shoulder region of a pork carcass in → England, but in → Sri Lanka the term is also used in beef and lamb, as well as for pork.

HAP

Another name for the flanc of a → Belgian beef carcass.

HARD SALAMI

A dry → sausage.

HARE

Being more solitary animals than rabbits, and living on the surface instead of in crowded underground warrens, hares did not offer the same possibilities for intensive husbandry that made rabbits so attractive as meat animals in days gone by. When available, hares are the product of hunting, which is generally welcome because the hare may be a fast-breeding nuisance like the rabbit in agricultural areas. Hares are generally larger than rabbits. A young hare is called a leveret. In Europe, the brown hare (*Lepus europaeus*) is the most common species. The jack-hare (male) is slightly smaller than the doe (female), and reaches about 3.5 kg live weight. Irish and Scottish hares are subspecies of *Lepus timidus*, known as the blue or varying hare from its change to a white coat in the winter. Across Canada and the mountains of the USA, the common hare is another varying hare, *Lepus americanus*, also called the snowshoe rabbit. A typical live weight is 1.8 kg. Its meat is lean and dry. In northern Canada, the arctic hare, *Lepus arcticus*, is relatively large (reaching 5.5 kg live weight) and forms an important meat source for Inuit hunters. Various species of *Lepus* are important small game animals of the plains of North America, but they are mostly called jackrabbits, even though they are not burrowing animals.

Jugged hare is a traditional method of casseroling hare. It requires the collection of as much blood as possible when the animal is eviscerated. The blood is pureed with the liver as the basis for a rich sauce (which is spoiled if allowed to boil).

(Sources: Burton, 1962; Burt and Grossenheider, 1976; McAndrew, 1990; Rue, 1981)

HASE

German for → hare.

HÁTSZALONNA

Pork backfat in → Hungary.

HÁTSZÍN

Steaks through the loin of a →Hungarian beef carcass.

HÁTULSÓ CSÜLÖK

The hindfoot of a → Hungarian pork carcass.

HÁTULSÓ LÁBSZÁRHUS

The hindshank of a beef carcass in → Hungary.

HAUNCH

In → England, a haunch may be a pair of legs of lamb from an unsplit carcass, or a unilateral primal cut of veal including the chump and posterior cuts. The term is widely applied to hindlimb of → venison in English-speaking countries.

HAUT DE CÔTELETTES

In → French lamb cutting, the haut de côtelettes is a long series of midlength rib cuts.

HAWAII - PORK CUTS

Pork cutting in Hawaii is similar to the typical American pattern, but the separations are made at different positions to produce a very large ham, wide belly and simple shoulder (Figure 97).

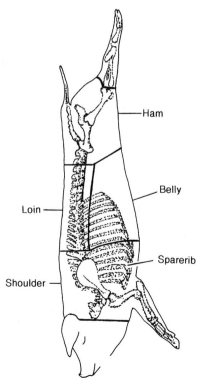

Figure 97. *Hawaii – pork cuts.*

(Source: Gómez *et al.*, 1992).

HAXL

Haxl vorne and Haxl hinten are the fore and hind-trotters of a pork carcass in → Austria.

HEAD CHEESE

Meat fragments set in a thick gelatin that can be thinly sliced.

HEEL OF ROUND

In → North America beef cutting, the heel of the round is the posterior part of the shank cut from a hip of beef (Figure 98). The most characteristic muscle is a horseshoe of *gastrocnemius*, surrounded by the distal extremities of *biceps femoris* (lateral), *semitendinosus* (posterior), *semimembranosus* and *gracilis* (medial). In heel slices taken level with the femur-tibia joint other muscles appear. Between the *gastrocnemius* and the femur may appear *flexor digitorum superficialis* and *popliteus*. The heel muscle is a cut of beef in → Northern Ireland composed of *gastrocnemius* and *flexor digitorum superficialis*.

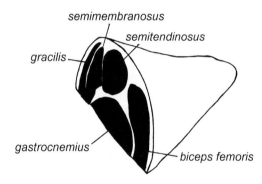

Figure 98. *North American heel of round of beef.*

HEIFER

A female beef animal that has not yet had a calf.

HERRING

A pelagic bony fish, such as *Clupea harengus*, → Clupeiformes.

HESSE MIT KNOCHEN

In → German beef cutting, the hindshank may be called the Hesse mit Knochen.

HILT

A female pig that has not yet farrowed (produced a litter of piglets).

HINDSADDLE OF LAMB

A North American primal cut of lamb that avoids splitting the carcass, right from left sides. It includes the loin, flank and legs of both sides, with the loin still joined in the midline.

HINDSHIN

A primal cut of beef in → Northern Ireland including the tibia, tarsals and associated muscles.

HINTERER WADSCHINKEN UND WADELSTUTZEN

Hindshank of an → Austrian beef carcass.

HINTERES

Rib and sternal cuts on a beef carcass in → Austria.

HINTERES AUSGELÖSTES

Trapezius and *rhomboideus* on an → Austrian beef carcass.

HINTERHESSE

In → German beef cutting, the hindshank may be called the Hinterhesse.

HIP

A primal cut of beef in North America (Figure 99), essentially the same as a Chicago round. It may be broken into shank, round, rump and sirloin tip.

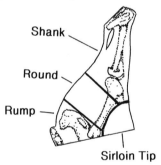

Figure 99. *North American hip of beef.*

HIP BONE

Apart from being a general name for the pelvic bone (ilium, ischium and pubis), the hip bone also is an English cut of beef from Northeast → England.

HISKARA

A retail cut through the neck of a pork carcass in → Sri Lanka.

HISTOCHEMICAL TYPES OF MYOFIBRES

Individual → myofibres of a variety of animals show a physiological specialization for either fast or slow contraction. In many of the invertebrates and lower vertebrates, slow myofibres are neurally activated via a relatively slow, progressive change in membrane resting potential along the myofibre, whereas fast myofibres may be activated by a rapid, self-propagating action potential. In the higher vertebrates (birds and mammals), nearly all myofibres are activated by action potentials, but still with a range from fast to slow. Slow myofibres generally have features appropriate to the sustainable utilisation of blood-borne energy sources, such as: (1), a large contact area with numerous capillaries; (2), a high concentration of myoglobin to facilitate the inward transport of oxygen; (3), numerous large mitochondria, especially under the plasmamembrane; (4), evenly distributed myofibrils with a relatively small cross-sectional area separated by abundant sarcoplasm (→ Fibrillenstruktur pattern); and (5), a short-term reserve supply of stored energy in the form of lipid droplets. Fast myofibres, on the other hand, generally have myofibrils with a relatively large cross-sectional area clumped together (→ Felderstruktur pattern) and exhibit features more suited to the anaerobic utilization of stored glycogen granules, such as numerous glycogen granules and high levels of phosphorylase, and they lack the aerobic features of red myofibres. Slow (red) myofibres have a slow isoenzyme of myosin, while fast (white) myofibres have a fast form. However, appreciable numbers of myofibres may show intermediate conditions where fast contraction speeds are combined with both the aerobic features of slow (red) myofibres and the anaerobic features of fast (white) myofibres. Thus, these intermediate myofibres contract rapidly, and may use either aerobic or anaerobic energy sources.

Myofibre histochemistry is important because of its commercial impact on meat. Briefly, slow (red) myofibres tend to have slow growth but confer taste and succulence to the meat, while fast (white) myofibres grow rapidly in response to genetic selection or superior nutrition, but they may produce meat that is rather dry and bland.

HLAVA

The head of pork carcass in → Slovakia.

HOCHRIPPE

In → German beef cutting, roasts from the prime rib region may be called either Hohe Rippe or Hochrippe.

HOCK

Hock has two meanings. For a live animal, the hock is a backwardly projecting joint of the hindlimb such as the tibio-tarsal joint in beef, from which the carcass is suspended. For a pork carcass, however, the hock (Figure 100) may be the distal part of the forelimb (distal to the picnic).

HODE

The head of a → Norwegian pork carcass.

HOGGET

A yearling sheep, typically a female.

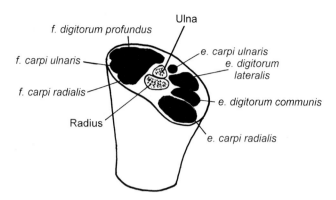

Figure 100. *North American pork hock, showing some of the flexor (f) and extensor (e) muscles.*

HÖGREV MED BEN

In → Swedish beef cutting, this cut contains the anterior thoracic vertebrae after removal of the bog or shoulder. In a Swedish veal carcass, after the shoulder or bog is lifted from the ribcage, the corresponding cut in the antero-dorsal thoracic region is the high-rib back or högrev/rygg. Similar terminology is used for → reindeer in Sweden.

HOHE RIPPE

In → German beef cutting, the Hohe Rippe may be part of the Roastbeef in the posterior thoracic region.

HØJREB

In → Danish beef cutting, the højreb or high rib is the prime rib roast from the posterior thoracic region.

HOLSTEINER

A semidry → sausage.

HONEY CUT

A North American beef chuck roast taken ventrally to the posterior part of the scapula.

HÖRPUSKEL

Icelandic for → scallop.

HORSE MACKEREL

A tuna-like bony fish of the family Carangidae, → Perciformes.

HORSE MEAT

As well as meat from horses, horse meat also includes meat from various species of donkey or ass around the world, plus meat from hybrids like mules (offspring of a male donkey and a female horse) and hinnies (offspring of a female donkey and a male horse). The total world consumption of horse meat is low relative to beef, pork, lamb and poultry, but a high proportion of what is consumed is traded internationally. In the USA and Canada, for example, there is a surplus of old horses and a minuscule domestic demand, which matches nicely to countries such as France, Japan, and Italy where demand exceeds supply. The horse would not fare well as a primary meat producer because, relative to the competition, it has low fertility, a long gestation period and a low feed conversion efficiency.

Horse meat has a taste intermediate between that of beef and venison, and is somewhat sweet. It has very low levels of fat and a high concentration of myoglobin making it dark in colour. The ultimate → pH may be relatively high (pH ≈ 5.7), especially in meat from older horses. The pH may be especially low in meat from Thoroughbred horses and high in meat from colts (poulain). Horse muscles may contain residual glycogen, especially at a high ultimate pH. In some countries, horse meat is sold at specialised retail outlets, such as the boucherie chevaline of France. In other countries it may appear as another vacuum-packed specialty item in a supermarket. A few countries have bone-in primal cuts for horse meat (→ Iceland) but in most other countries there are no bone-in cuts of horse meat. Muscles are removed intact from the whole carcass to be stripped individually of their → epimysial connective tissues. Once this is done, horse meat is relatively tender (bearing in mind the age and life history of much of the source material). The price differentials for horse meat muscles are similar to those found in a beef carcass, with top prices going for muscles suitable for rapid cooking with dry heat. Horse meat may be used to add depth of colour and taste to a variety of processed meat products such as mortadella and salami. Artificial marbling may be injected into horse meat for the Japanese market.

(Sources: Gade 1976; ITC, 1983; Lawrie, 1985; Weyermann and Dzapo, 1997)

HORSESHOE

Part of the *gastocnemius* muscle in the heel of a beef round.

HOSSZÚ-FELSÁL

A medial cut of the beef round in → Hungary.

HOSSZÚKARAJ

The lumbar region of a pork carcass in → Hungary.

HOT BONING

Removal of meat from the carcass immediately after slaughter, before the development of rigor mortis or the start of refrigeration. Hot boning reduces refrigeration and meat handling requirements, decreases shrinkage losses, improves shelf life and sometimes improves colour and tenderness.

However, hot-boning is a skilled operation and the shapes of many traditional meat cuts are lost.

(Source: Hamm, 1982)

HOT DOG

A hot dog is an emulsion → sausage, named from the frankfurters roasted by street vendors in the city of New York. In North America, there is now little, if any, difference between frankfurters, wieners and hot dogs.

HOTEL RACK OF LAMB

A North American cut of lamb that avoids splitting the carcass, right from left sides. It includes the rib (rack) cuts of both sides, still joined in the midline. It contains ribs 6 to 12 of both sides.

HOUGH

Alternative spelling of hock, used for → Scottish beef cuts level with the tibia, and with the radius and ulna cuts.

HOVED

→ Danish for head, as in the head of a pork carcass.

HØYRYGG

A shoulder cut from a → Norwegian beef carcass.

HREINDÝR

Icelandic for → reindeer.

HROSSAKJÖT

Horse meat in → Iceland.

HRUD'

The brisket of a beef carcass in → Slovakia.

HRYGGSNEIÐAR

Steaks through the *longissimus dorsi* in → Iceland.

HRYGGUR

The back or spine of a carcass in → Iceland, as in primal cuts of lamb and pork and retail cuts of other species.

HRYGGVÖÐVI

Longissimus dorsi, isolated and trimmed from a carcass in → Iceland.

HUACHALOMO

A rhomboidal cut of beef in → Chile containing dorsal neck

muscles extending over the thoracic vertebrae. Perhaps purchased by those who cannot quite afford real lomo, but wish to pretend otherwise.

HUDSON BROIL

In → North America, a retail cut of beef from the flank.

HÜFERSCHERZEL

An → Austrian beef cut containing the ilium.

HÜFERSCHWANZEL

Tensor fascia lata of an → Austrian beef carcass.

HÜFTE

In → German and → Swiss beef cutting, the Hüfte or haunch includes most of the pelvis.

HUHN

German for → chicken.

HUMAR

Icelandic for → lobster.

HUMMER

Danish or Swedish for → lobster.

HUMP

In → South Africa, a cut of beef dorsal to the scapula.

HUNGARY - BEEF CUTS

A major division of the hindquarter (hátulsónegyed) of beef (marha) is into the medial long round (hosszú-felsál) and lateral rump (fartő), separating the *semitendinosus* as the white-roast (fehérpecsenye) (Figure 101). The gluteal muscles are contained with the top of the rump (fartőhegy) while the *quadriceps femoris* group of muscles are known as the ball round (gömbölyű-felsál), as they are in French (→ boule d'aloyau), Spanish (→ bola) and German (→ Kugel) as well.

Along the vertebral column, in the lumbar region, the major division is into the dorsal hátszín (translated by most sources as sirloin, following the typical English designation of the location of the sirloin) and into the ventral vesepecsenye or *psoas* muscles. The more anterior bélszín may also be translated as sirloin or, more specifically, as bélszín angolos or English sirloin. In the midthoracic region, is located the regular roll of beef, rostélyos. And anterior to this is the tarja (chuck) and nyak (neck). The szegy or brisket is divided into

three parts, from anterior to posterior, these are the szegyfő, szegy and puha-szegy (soft brisket). Between the dorsal and ventral cuts is a long ladder of rib sections called the side bones or csontos oldalas, and posterior to this is the soft sirloin or puha hátszin. The shoulder is the lapocka, which is divided into thick shoulder (vastag lapocka), blade filet (oldal lapocka) and shoulder top (lapocka fedő). The fore and hindshanks, respectively, are called elülső and hátulsó lábszáhús.

(Source: Barnabas and Süth, 1998)

HUNGARY - LAMB CUTS

Bárány is lamb, and is used as a prefix for the lamb cuts. As in many other countries, the lamb carcass may be split either as a pistol cut of leg plus loin (pisztolyvágásu bárány) or with left and right sides remaining together to create a saddle in the lumbar region (báránygerinc) (Figure 102). The báránycomb is the leg. The shoulder (báránylapocka) is lifted from the ribcage to leave a short forequarter with breast, bárány eleje.

(Source: Barnabas and Süth, 1998)

HUNGARY - PORK CUTS

Sertés is pork, so that félsertés is a side of pork (Figure 103). After removal of the leg (sertéscomb), backfat (hátszalonna), shoulder (sertéslapocka), and head (sertésfej), the whole vertebral axis is divided into an anterior collar (sertéstarja), a rib chop region (rövidkaraj), and a lumbar chop region (hosszúkaraj or sertéskaraj). The sertés-szűzpecsenye is the filet of *psoas* muscles. Ribs and spareribs are called sertésoldalas, and the belly is the dagadő. The fore and hindshanks are, respectively, the elülső csülök and hátulsó csülök. The sertésfarok is the tail.

(Source: Barnabas and Süth, 1998)

HURE

An alternate name for the tête of a pork carcass in → Belgium.

HUVUD

→ Swedish for head, as on a pork carcass.

HYPAXIAL MUSCLE

The muscle mass below the horizontal connective tissue septum in the musculature of a fish (→ myomere).

H-ZONE

If a muscle is at its resting length, the gap between opposing thin → myofilaments at the midlength of the → sarcomere causes a pale H-zone in the middle of the A-band.

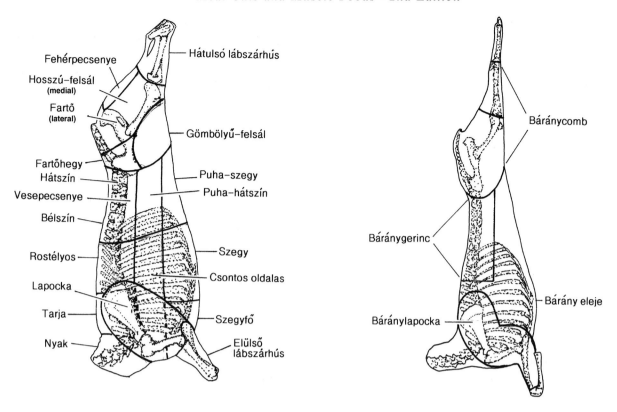

Figure 101. *Hungary – beef cuts.*

Figure 102. *Hungary – lamb cuts.*

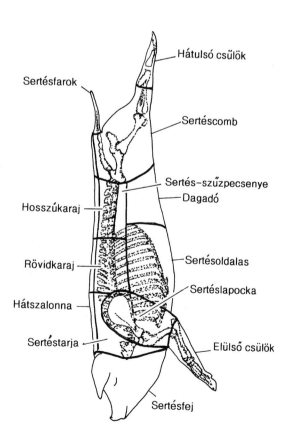

Figure 103. *Hungary – pork cuts.*

I-BAND

The I-band is one of the → transverse striations along the → myofibre. It gets its name from being optically isotropic - appearing very dim under a light microscope fitted with crossed polarisers.

ICELAND - BEEF CUTS

The primal cuts of beef (nautakjöt) in Iceland are shown in Figure 104. The lærleggur giving rise to the hindshank (lærskanki) includes the tibia, tarsal bones and associated extensor and flexor muscles. Only the most distal part of the *gastrocnemius* remains on the lærleggur. The round or læri is divided into the miðlæri (medial muscles) and the afturstykki (lateral muscles). Thus, the latter cut, dominated by the gluteals and *biceps femoris*, gives rise to steaks through the gluteal region, which are called mjaðmasneiðar or sirloinsteikur, following the North American rather than the English designation of the sirloin. Most of the other adopted names follow a similar North American pattern, such as, porterhouse steikur and T-bein steikur and klúbbsteikur. The *quadriceps femoris* group of muscles is removed as the grófsnyrtur klumpur. The trimmed medial muscle mass of *semimembranosus, adductor* and *pectineus* becomes innralæri or inner thigh. Grófsnyrt læristunga is the *semitendinosus*.

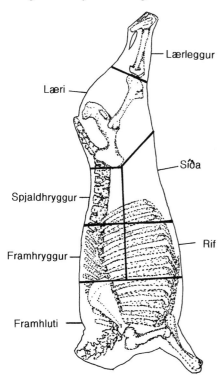

Figure 104. Iceland - beef cuts.

The lumbar (spjaldhryggur) and posterior thoracic (framhryggur) cuts are used as steaks and roasts, and both primals together before separation are called the whole loin or langhryggur. Hryggur is Icelandic for the back or spine. *Longissimus dorsi* and the *psoas* muscles may be separated and trimmed as, respectively, hryggvöðvi and lundir. Beinlauser framhryggsneiðar are rib steaks, called beinlauser

framhryggur when kept as a roast. *Longissimus dorsi* from the rib region is called framhryggsvöðvi if separated and trimmed. The flank or síða muscles which continue anteriorly over the ribs of the rif cut may be separated as the síða úr stuttum framparti, in which case, the posterior flank muscles are identified as síða úr afturpart. Like the nautical terms fore and aft on a ship, fram and aft give useful clues to meat cut locations in the Norse languages. To make a point, let us now move on from the afturpartur (hindquarter) to the frampartur (forequarter).

Herðar is Icelandic for the shoulders, so the scapula or shoulder blade is the herðablad, most of which is located in the framhluti or front part of the carcass. The most lateral part of the shoulder is the bógur, hence the humerus is the bógleggur. Like the → bog in Norway and Sweden, or the → bov in Denmark, this word root is used for many of the meat cuts around the shoulder region in Iceland. Bógur is the whole shoulder (scapula, humerus, ulna and radius, plus associated muscles), frampartur án bógs is the remainder of the forequarter after removal of the shoulder, bógvöðvi is *triceps brachii* and bógsneiðar are cross-cut arm steaks. From the ribcage underlying the bógur, the typical ladder of mid-rib sections, as found in many other countries, is simply called the beef-rib or nautarif. The posterior thoracic rib roasts are called framhryggur, or framhryggsneiðar með beini when cut as bone-in steaks.

(Sources: Þorkelsson and Hilmarsson, 1994; Hilmarsson, 1999, IAIS)

ICELAND - HORSE MEAT

Iceland has a cutting pattern for horse meat or hrossakjöt, sometimes called folald to indicate its origin from relatively tender foals (Figure 105). The nomenclature is similar to that for beef in Iceland. As in most other countries, however, many muscles are subsequently separated individually to facilitate stripping of surface connective tissues (→ horse meat). Some major muscle groups are innralæri (*semimembranosus, adductor* and *pectineus*), mjaðmasteik (gluteal muscles), ytralæri (*biceps femoris*), klumpur (*quadriceps femoris* group), hryggvöðvi (*longissimus dorsi*), lundir (*iliopsoas*) and bógvöðvi (*triceps brachii*). However, quite a few prime cuts are sold as bone-in steaks, such as T-beinsteikur (T-bone steaks) and even folaldabógsneiðar (blade steaks).

(Source: Þorkelsson and Hilmarsson, 1994; IAIS)

ICELAND - LAMB CUTS

The main primal cuts of lamb (dilkakjöt) are shown in Figure 106. The lambapistóla is composed of the læri plus hryggur of both sides. Lamba- is used as a prefix in combination with many of the names used in Icelandic beef cutting, such as lambalæri (leg of lamb) and lambahryggur (rib of lamb). The hindlimb may be cut into a series of transverse steaks called lærissneiðar, while the forequarter may be cubed to make súpakjöt or sectioned to make hálssneiðar. Shoulder sections

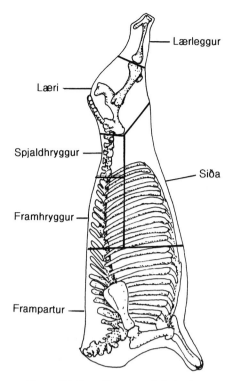

Figure 105. Cuts of horsemeat in Iceland.

are bógsneiðar. Bringur is breast of lamb. As well as typical unilateral and bilateral cutlets (kótilettur and tvöfaldar kótilettur, respectively), the rib region also may be used for something comparable to a → crown roast. But instead of being curved so that the ribs radiate outwards like the points of a crown, the democratic Icelanders go for a straight rack or lambakóróna (Figure 107). The Icelandic word kór denotes the chancel or choir of a church, traditionally with a series of arched ribs supporting the roof.

Figure 107. Icelandic lambakóróna.

(Sources: Þorkelsson and Hilmarsson, 1994; Hilmarsson, 1999, IAIS)

ICELAND - PORK CUTS

The primal cuts of pork (svínakjöt) are shown in Figure 108. The hindlimb may be used as a roast on the bone (grísalæri með beini) or sliced transversely to make steaks (grísalærissneiðar). The hryggur may be cut to give either 8 (ósnyrtur grísahryggur með 8 rifjum) or 11 ribs (ósnyrtur grísahryggur með 11 rifjum), with a corresponding change to the length of the siða (ósnyrt grísasiða). The prefix grísa-indicates pork. Thus, pork chops are grísakótilettur. The framhryggur is used in a variety of ways, either as a pork-back with shoulder-blade roast (grísaframhryggur án herðablaðs) or to make shoulder chops (grísaframhryggsneiðar). The siða is split into rib (grísarif) and belly meat (úrbeinuð grísasiða).

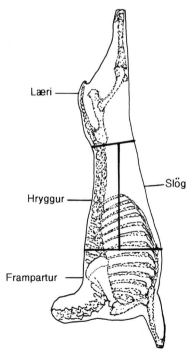

Figure 106. Iceland – lamb cuts.

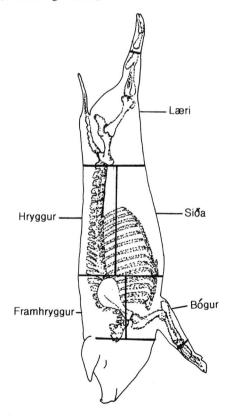

Figure 108. Pork cuts in Iceland.

(Sources: Þorkelsson and Hilmarsson, 1994; Hilmarsson, 1999, IAIS)

ICELAND - VEAL CUTS

The main nomenclature for veal (kálfakjöt) in Iceland follows that for beef except, of course, that there is less subdivision of the primal cuts A veal pistóla is composed of the lærleggur, læri, spjaldhryggur and framhryggur.

(Source: Þorkelsson and Hilmarsson, 1994)

IGE

Yoruba for the brisket in → Nigeria.

ILIACUS

A medium-sized, rounded muscle ventral to the ilium in mammals, where it contributes tender meat to an expensive part of the carcass. It originates mainly from the ilium, and inserts onto the trochanter minor and neck of the femur. It joins with the posterior part of the *psoas major* to form a compound muscle, the *iliopsoas*.

ILIOCOSTALIS

A long, thin, segmented muscle located dorsally over the mammalian ribcage (Figure 240). It is often called the *longissimus costarum*.

IMPALA

The impala, *Aepyceros melampus*, is a graceful African antelope reaching 45 to 82 kg live weight. The dressing percentage is relatively high (58 to 68%). The → South African cutting pattern is shown.

(Sources: Dorst and Dandelot, 1970; Eltringham, 1984)

IMPERIAL ROAST

A cut from the anterior scapular region in British → Proten beef.

INDERLÅR

Inside round in → Danish beef cutting.

INDREFILET

The *psoas* and *iliacus* muscles removed from a → Norwegian beef carcass.

INFRASPINATUS

A major muscle (≈ 2% total muscle weight) filling the infraspinous fossa of the scapula in mammals (Figure 109). It has a strong round tendon inserting on the lateral tuberosity of the humerus.

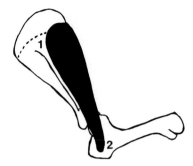

Figure 109. *Lateral view of beef* infraspinatus *(solid black), originating on the scapula (1) and inserting on the humerus (2).*

INNANLÅR

A medial cut from the round of a → Swedish beef carcass.

INSIDE ROUND

In →North American beef cutting, the inside or top round is taken from the medial face of the hindlimb and is dominated by *adductor* and *semimembranosus*. Also, in → Australian, → New Zealand and → Irish beef cutting, the inside round is a medial cut containing *semimembranosus*, *adductor* and *pectineus*.

INTEGRIN

Integrins occur on the surface of the cell and link the extracellular matrix to the → cytoskeleton within the cell. They are well developed at → myotendon junctions.

(Source: Bao *et al.*, 1993)

INTERCOSTALES

In the mammalian ribcage, the spaces between adjacent ribs are filled with two layers of muscle (*intercostales externi* and *interni*) which both act to move the ribs forwards and outwards during respiratory inspiration. Together they contribute a substantial amount (> 2.5% total muscle weight) of meat to the carcass, much of which is in premium cuts, such as ribs of beef and North American spareribs of pork.

INTÉRIEUR DE RONDE

Quebec, inside or top round of beef.

INTERTRANSVERSARII

These are a series of small muscles acting between the transverse processes of vertebrae when the mammalian vertebral column is fixed in shape or flexed laterally. The *intertransversales caudae* are in the tail, posterior to the last sacral vertebra, the *intertransversales cervicis* and *intertransversarius longus* are in the neck, and the *intertransversales lumborum* are in the lumbar region.

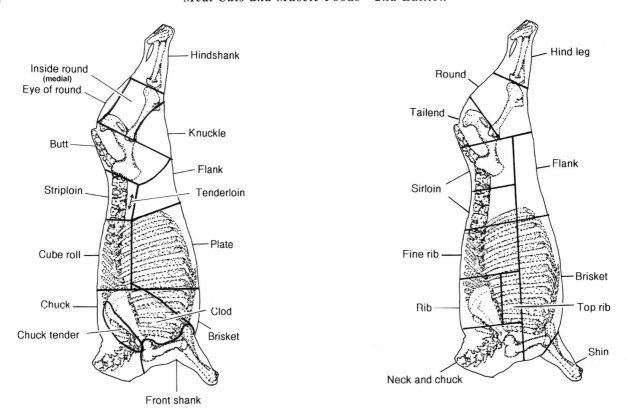

Figure 110. Ireland – beef cuts.

Figure 111. Ireland – beef cuts.

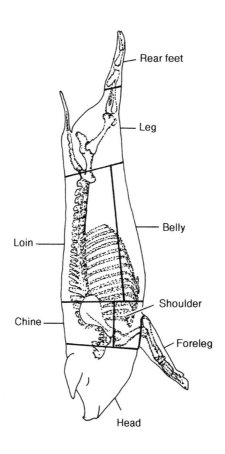

Figure 112. Ireland – pork cuts.

INTRAFASCICULAR

An adjective describing position, within a → fasciculus or bundle of myofibres. Intrafascicularly terminating myofibres are those which end along the length of a fasciculus, rather than where the end of the fasciculus is anchored in connective tissue.

IRELAND - BEEF CUTS

A cutting pattern for Irish beef (Figure 110) was collected in a survey by Pomeroy and Williams (1967). The round is separated into an eye of the round (*semitendinosus*), an inside round containing the medial muscles, and a knuckle containing the *quadriceps femoris* group. The fate of lateral muscles such as *biceps femoris* is uncertain. The gluteal muscles and ilium are contained in a large primal called the butt, and the lumbar region is divided into striploin (*longissimus dorsi*) and tenderloin (*psoas major*). The prime roasts of the mid and posterior thoracic region are contained within the cube roll, with the distal portions of the ribs contained in the plate. Cutting of the shoulder includes a chuck tender composed of the *supraspinatus*, a clod dominated by *triceps brachii*, and a chuck composed of the anterior thoracic vertebrae and ribs. The front shank includes the humerus as well as the ulna and radius.

Another quite different pattern was collected by OEEC (1961), as shown in Figure 111. It looks like a band-saw cutting pattern for bulk beef and has an unusual name, the tailend, for a cut containing sacrum and ischium. The name for the fine rib matches that used in Northeast England, the fine end, although the Irish cut is much larger.

(Source: OEEC, 1961; Pomeroy and Williams, 1976)

IRELAND - PORK CUTS

The age of the source material is a concern, and most likely reflects a traditional cutting pattern, because the rib sections on the loin are very long by contemporary standards (Figure 112). The shoulder is split in the same way as in North America but the Boston butt is called the chine, the picnic is called the shoulder, and the hock is called the foreleg.

(Source: OEEC, 1961)

ISCHIOCAVERNOSUS

In bulls, rams and boars, this muscle is inserted in the penis and pulls it anteriorly. The origin of the muscle is on the sciatic tuber of the ischium, and a severed cross section of the *ischiocavernosus* may be seen on sides of a carcass where its presence gives a useful clue to the sex of the carcass.

ISI

Ibo for the head in → Nigeria.

ISOMETRIC CONTRACTION

Isometric contraction is when a muscle contracts against a large load, which is unmoved so that tension rises. Thus, the shape of the muscle remains about the same because it does not shorten.

ISOTONIC CONTRACTION

Isotonic contraction is when a muscle shortens and moves a constant load. Thus, the tension on the load does not change much.

ISRAEL

Information on Israeli beef cutting presented here is only an approximation. The Israeli beef cutting charts reviewed were all based on the animal in a standing posture, and transliteration has been difficult. Figure 113 shows some of the English equivalent terms suggested by a commercial company, Marbek. The main feature of cutting the round is that it is split into medial and lateral primal cuts which are difficult to show on a lateral skeletal profile. Pelvic, lumbar and posterior thoracic regions are cut into steaks and roasts as in many other countries, with removal of the *psoas major* as a filet. Anteriorly, however, the shoulder is removed whole then subdivided along the length of the bones. *Supraspinatus* is removed as a shoulder filet, while *triceps brachii* is separated posteriorly.

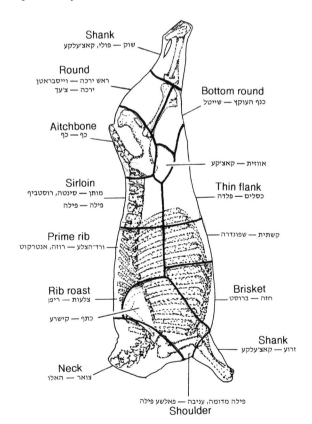

Figure 113. Israel – beef cuts.

(Sources: Marbek; Devear; Sirkis, 1975).

ISTER

Another name for the belly or buklist of a → Norwegian pork carcass.

ITALY - BEEF CUTS

There are many regional variations in Italian beef cuts and names. Figure 114 shows the pattern used by both OEEC (1961) and Favati (1971).

The flesh of the haunch or polpa di coscia is a large primal cut, including the gluteal muscles which may be called a thigh section or sezione di coscia. The steaks of the lumbar region are in a cut called either the bistecche di lombo or the braciole di lombo, with a similar alternate naming of the prime rib region, either as bistecche di costa or braciole di costa. The flank (pancia) and brisket (petto) are removed to leave a long cut level with the midlength of the ribs which may be known as the lista (stripe) or spuntatura di costa (tips of the ribs), with the most anterior part having a special name, the taglio di sottospalla (cut of the under-shoulder). The shoulder muscle (muscolo di spalla) and clod muscles (sezione e muscolo di spalla) are removed together with the foreshank (gamba anteriore) to leave the underlying ribs (braciole di sottospalla) and neck (collo).

Another Italian pattern of beef cutting with excellent documentation is from Chizzolini (1997) as shown in Figure 115. The main separation of the hindlimb is into the posterior girello (*semitendinosus*), medial fesa (*semimembranosus, adductor, pectineus, sartorius* and *gracilis*), lateral sottofesa (*biceps femoris*) and the anterior noce (*quadriceps femoris* group) and fianchetto (*tensor fascia lata*). The distal muscles of the hindlimb are divided into the posterior campanello (*gastrocnemius*) and anterior muscolo posteriore (distal extensors and flexors). The scamone is composed of the gluteal muscles and the filetto by the *psoas* muscles, while the major primal cut along the thoraco-lumbar region is the lombata, located dorsally to the flank or pancia. The shoulder muscles are divided into the medial copertina di sotto (*subscapularis* and *teres major*), the dorsal girello di spalla (*supraspinatus*),

the more ventral copertina (*infraspinatus*), the ventral fesone di spalla (*triceps brachii*) and the anterior polpa di spalla (*biceps brachii* and *pectoralis profundus*). The underlying cuts of the ribcage are the costate (anterior thoracic vertebrae and dorsal parts of ribs), sottospalla or reale (rib midlength), and petto (sternum and ventral parts of ribs).

(Sources: OEEC, 1961; Favati, 1971; Chizzolini, 1997).

ITALY - LAMB CUTS

The general regions for lamb cuts in Italy (Figure 116) are the cosciotto o quarto (haunch or quarter), lombata (loin), costolette (hotel rack), spalla (shoulder), collo (neck), fianco (flank), petto o pancia (breast or belly) and zampa (shank).

(Source: Favati, 1971).

ITALY - PORK CUTS

In Italy, the carcasses of heavy pigs (live weight 160 kg) have traditionally been cut while hot to ensure optimal quality for further processing (Chizzolini et al., 1992). The skull is separated from the cervical vertebrae but the *masseter* and cheek (testa senza guanciale) remain on the side of pork where they will later appear in the bacon (Figure 117). The hindlimb (prosciutto posteriore) is removed including the pubis and ischium, and the trotter or zampi is detached.

(Sources: OEEC, 1961; Favati, 1971; Chizzolini et al., 1992; Chizzolini, 1997).

ITAN

Yoruba for the hindquarter in → Nigeria.

IZAL BONE

A cut of beef in the ischial region in Northeast → England.

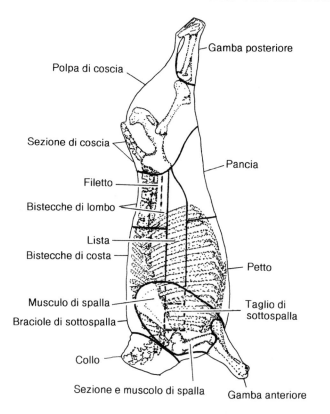

Figure 114. *Italy – beef cuts.*

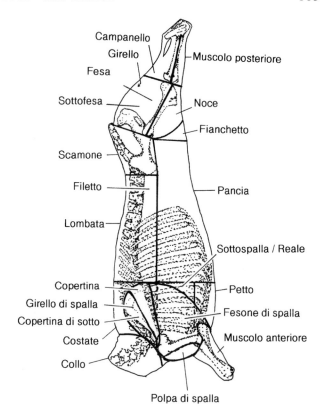

Figure 115. *Italy – beef cuts.*

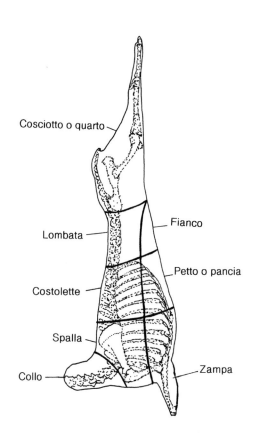

Figure 116. *Italy – lamb cuts.*

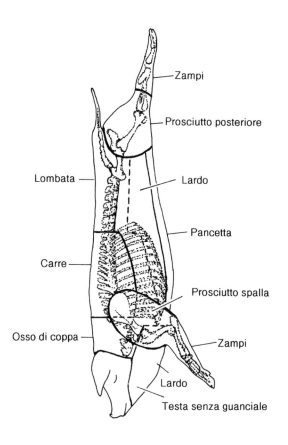

Figure 117. *Italy – pork cuts.*

JACK

A tuna-like fish of the family Carangidae, → Perciformes.

JACKSMELT

Atherinopsis, the California jacksmelt, → Mugiliformes.

JAGDWURST

Hunter's sausage contains ham, bacon and pork, and is only lightly smoked.

JAMBE

Primal cuts of → French beef in the distal regions of the limbs. Both the jambe de derrière (hindlimb) and jambe de devant (forelimb) are subdivided into a crosse and jarret.

JAMBETTE

The fore- or hindshank of a pork carcass in → Belgium otherwise known as the jarret.

JAMBON

Leg of pork in → France.

JAMBONNEAU

The shank (distal) end of a pork ham in Quebec, or the distal part of the pork hindlimb or forelimb in → France or → Belgium. An alternative name for pork jambonneau is jarret. In → Luxembourg, the radio-ulnar region of a pork carcass is the jambonneau, while the tibio-fibular region is the jambonneau de derrière.

JAMÓN REDONDO

The "round ham" - a beef cut in → Argentina including the *semitendinosus* and medial hindlimb muscles.

JANB

Camel flank in → Morocco.

JANJE

Lamb in → Serbia & Montenegro.

JAPAN - BEEF CUTS

Japanese meat cutting involves almost the complete removal of bones and anything else that is not fat or muscle - including all lymph nodes, periostium, sinews, skin, and ligamentum nuchae. Some types of Japanese beef are very fat with major seams of intermuscular fat, most of which is removed to leave highly marbled meat that is thinly sliced and rapidly cooked at the dinner table, holding it with chopsticks and dipping it into lightly spiced boiling water. Traditional Japanese breeds such as Wagyu produce very palatable meat for this *shabu-shabu* cooking method.

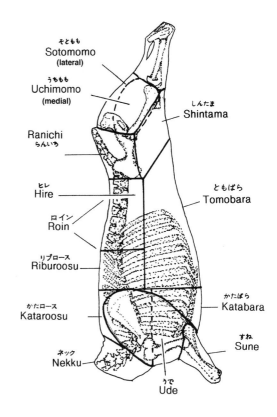

Figure 118. *Japanese beef cuts.*

The forequarter is removed from the hindquarter by a straight cut passing between ribs 5 and 6, although a more posterior separation may be made between ribs 6 and 7, or between 7 and 8 (Figure 118). The ude or shoulder containing the scapula, humerus, ulna, radius and associated muscles is separated from the ribcage to leave *serratus ventralis* on the ribcage as well as spread across the shoulder roast or kataroosu and the dorsal part of the brisket of katabara. The foreshank or sune is removed and contains the ulna, radius, extensors and flexors. The ude and sune are boned and trimmed of all fat and gristle. The brisket or katabara is removed from the ribcage with a cut parallel to the vertebral column, starting about one third the way along rib 5. The neck or nekku is separated from the shoulder roast with a cut between cervical vertebrae 6 and 7.

The flank or tomobara is separated from the hindlimb by a cut along the anterior edge of the *tensor fascia lata*, then by a second cut parallel to the vertebral column. After removal of the kidney fat, the *psoas major* and *minor* are removed from the loin as the hire, then the loin or roin is separated with a cut between the last lumbar and first sacral vertebra. The uchimomo containing *pectineus, adductor* and *semimembranosus* is removed medially, and the remainder is subdivided into the shintama (*quadriceps femoris* group), ranichi (gluteal muscles), sotomomo (*semitendinosus* and *biceps femoris*) and hindshank (*gastrocnemius* and distal extensors and flexors).

(Source: Japan Meat Grading Association, 1979; Irie, 1999)

JAPAN - PORK CUTS

The whole shoulder (limb bones and first four ribs, as in Figure 119) is removed perpendicularly to the vertebral column between thoracic vertebrae 4 and 5, although a separation between 5 and 6 is allowable. Similarly, the ham or momo may be removed leaving either the last lumbar vertebra on the loin or on the ham. At the junction between the belly and the ham, instead of a straight line (solid line in Figure 119), the separation may curve posteriorly (broken line in Figure 119). Anterior to the pubis, the *psoas* muscles are removed as the filet or hire. A long loin called the roast or roosu is removed from the bacon belly (bara) and ventral ribs by a line parallel to the vertebral column at about one third rib length. As all the cuts are trimmed, the shoulder is separated into arm (ude) and shoulder roasts (kata) near the top of the scapula.

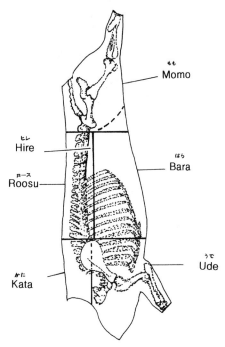

Figure 119. Japanese pork cuts.

(Source: Japan Meat Grading Association, 1979; Irie, 1999)

JARRET

The jarret is a distal cut of beef through the hindlimb in → France. It may be split into distal (bout de gîte) and proximal (milieu de gîte) slices across the tibia and distal femur, respectively, or split into an anterior gîte-gîte and a posterior gîte-noix nerveaux. Jarret also may refer to the forelimb shank, as in → Luxembourg. In → Belgium, jarret also is used for both fore- and hindshanks of beef and pork.

JARRET AVANT

Quebec, foreshank of beef or lamb, or hock of pork.

JARRETE

Jarrete is Spanish for knuckle or shank, as in the hock of a →

Spanish beef carcass. Also used for the distal extensors and flexors of the hind and foreshank in → Cuba.

JNIAH

Ulna and radius region of the poultry wing in → Morocco.

JONAH CRAB

A Western Atlantic shore crab, *Cancer borealis*, → crabs.

JORDAN – BEEF CUTS

The leg or Mozeh Khalfieh is removed by a separation between the tibia and femur, leaving a large round or Wazeh from which the posterior muscles of the Al-Shaikeh are removed (Figure 120). The Al-Khad is derived from the ischial region and the Al-Ekweh from the remaining sacral and ilial region. The prime steaks of the loin translate as "rose beef" while the psoas has almost a universal name – the fileh. Lacking a typical separation into forequarter and hindquarter through the posterior rib region, the flank or Al-Khaserah is very large. The thoracic vertebral region is Metleh. The brisket is divided into an anterior Al-Asse and a posterior Al-Sader. The Rakabeh is the neck, and the Mozeh Amamieh is the foreshank.

(Source: Abdallah, 2003)

JORDAN – LAMB CUTS

Excluding the cannon bone (Al-Zend), the leg of lamb (Al-Fakhed) is large and terminates just short of the ischium (Figure 121). The loin (Al-Thaher) is tapered anteriorly to leave meat on the rib (Al-Sader). The Al-Katef includes the anterior scapula plus distal limb bones and muscles. The neck is the Al-Rakabeh and the flank is the Al-Khaserah.

(Source: Abdallah, 2003).

JORDAN – VEAL CUTS

The shanks of both fore- and hindlimbs are removed as Al-Mozeh (Figure 122). The remainder of the hindquarter is divided into the posterior Al-Habra, the anterior Al-Fakheth, and the intermediate Al-Habra. The steakmeat along the vertebral column is Habrat Al-Daher in the lumbar region and Al-Daher in the thoracic region. The abdominal muscles are Al-Khaserah and the ribs are Al-Delea. The Al-Katef contains the scapula and humerus with associated muscles. The Rakabeh is the neck.

(Source: Abdallah, 2003).

JOUE

The jowl of a pork carcass in → Belgium.

JOWL OF PORK

In → North American pork cutting, the jowl is removed from the picnic with a cut that follows the crease lines in the skin.

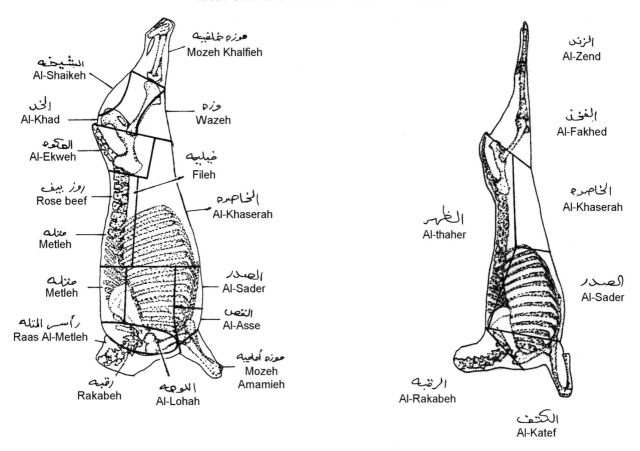

موزه خلفيه
Mozeh Khalfieh

الشيخه
Al-Shaikeh

الخد
Al-Khad

وزه
Wazeh

العكوه
Al-Ekweh

فيليه
Fileh

روز بيف
Rose beef

الخاصره
Al-Khaserah

متله
Metleh

منزله
Metleh

الصدر
Al-Sader

رأس المتله
Raas Al-Metleh

النقص
Al-Asse

موزه أماميه
Mozeh
Amamieh

رقبه
Rakabeh

اللوحه
Al-Lohah

Figure 120. *Beef cuts in Jordan.*

الزند
Al-Zend

الفخذ
Al-Fakhed

الخاصره
Al-Khaserah

الظهر
Al-thaher

الصدر
Al-Sader

الرقبه
Al-Rakabeh

الكتف
Al-Katef

Figure 121. *Lamb cuts in Jordan.*

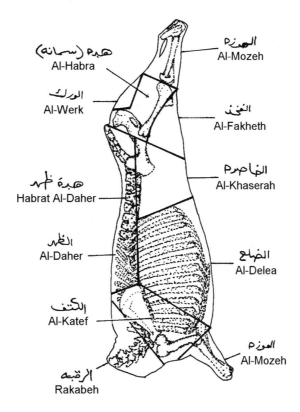

هبره (سمانه)
Al-Habra

الموزه
Al-Mozeh

الورك
Al-Werk

الفخذ
Al-Fakheth

هبرة ظهر
Habrat Al-Daher

الخاصره
Al-Khaserah

الظهر
Al-Daher

الضلع
Al-Delea

الكتف
Al-Katef

الموزه
Al-Mozeh

الرقبه
Rakabeh

Figure 122. *Veal cuts in Jordan.*

JUMEAU

→ *Gastrocnemius.*

JUMEAUX

Despite → jumeau indicating the *gastrocnemius* muscle of the hindlimb, the jumeaux is a → French cut of beef from the anterior part of the forelimb containing *supraspinatus* and *biceps brachii.*

KABANOSY

A dry pork → sausage which has been cured, smoked and stuffed into narrow casings.

KABOB

From an Arabic word, indicating a cube or chunk of meat suitable for cooking on a long skewer. Kebab is an alternative spelling.

KAFARGABA

Hausa for the forequarter in → Nigeria.

KAI

Hausa for the head in → Nigeria.

KAISERTEIL

A medial cut from the hindlimb of an → Austrian pork carcass.

KALBFLEISCH

German for veal.

KALBI

A long antero-posterior cut from a → Korean beef carcass including most of the ribs and some of the dorsal abdominal muscles. In a Korean pork carcass, the kalbi is shorter than in beef and restricted to the anterior rib region.

KÁLFA-

An → Icelandic prefix indicating a cut of veal, as in kálfasnitzel (veal snitzel), kálfagúllas (veal gulash), kálfahakk (minced veal), kálfabógsneiðar (veal shoulder slices), etc.

KALKOENSTUCK

Part of the leg of veal in the → Netherlands.

KALKÚNI

Icelandic for a → turkey.

KÅLLAP

The flank of a → Swedish beef carcass.

KALVKÖTT

Swedish for veal.

KAM

The crest of the neck and back in a → Danish pork carcass, giving its name to the major thoraco-lumbar cut. A similar nomenclature is used for the loin of a → Norwegian veal carcass.

KAMABOKO

A cooked → surimi gel with a globular or fibrous microstructure.

(Source: Alvarez et al., 1999)

KAMM

A cut of meat from the neck of a carcass in → German meat cutting. For the whole animal, the Kamm is the dorsal part of the neck, as distinct from the cervical part of the neck which may be called the Hals or throat, as in lamb cutting.

KAMMUSSLA

Swedish for → scallop.

KANGAROO

Development of sheep ranching in Australia increased the habitat available for kangaroos. Kangaroos compete for the same feed and water as sheep, but opinions vary between farmers and conservationists as to the extent and economic impact of the competition. Australians joke that they might have done better developing the kangaroo rather than introducing sheep. When one tastes fresh (rather than canned) kangaroo meat at a good restaurant in Australia it confirms that many a true word has been said in jest. But it has been difficult to transform the product image from "petfood obtained from vermin" to "gourmet food obtained from free-range wildlife". Kangaroo meat provided an important source of food for European pioneers in Australia, but became socially unacceptable as people grew wealthier and eating meat from wild animals became an indication of poverty. Kangaroo meat subject to proper evisceration, veterinary inspection and controlled refrigeration is, of course, as safe as any other source of meat. It is unfortunate that most scientific efforts have been directed at detecting kangaroo meat adulterating other products, rather than in characterising and developing kangaroo meat for its own sake.

Male and female red kangaroos (*Macropus rufus*) can reach 90 kg and 40 kg live weight, respectively, but most culled animals are about half this weight. Male eastern grey kangaroos (*Macropus giganteus*) may also reach 90 kg, but are more likely to be around 70 kg, with females reaching half that weight. To the untrained eye, the western grey kangaroo (*Macropus fuliginosus*) looks the same as the eastern grey. Animals are obtained by night hunting with a searchlight and are generally shot in the head. The gut is removed and carcasses are chilled and transported to an abattoir where dressing is completed. Dressing percentages are about 65% for males and females, and for red and grey kangaroos, within an empty body weight range from 15 to 40 kg. Carcasses produce about 80% muscle, 14% bone and 6% waste. When hopping, the body posture of the red kangaroo tends to be horizontal while that of the grey kangaroo tends to be more vertical. Since the tail counterbalances the body weight, red kangaroos have a higher proportion of tail meat than grey kangaroos. Males have a higher proportion of tail than females in the grey but not in the red kangaroo. The tail may be removed before abattoir dressing commences to make the hanging carcass easier to manipulate. There is essentially no subcutaneous fat. Carcasses from wild kangaroos have less then 1% fat, and a high proportion of this is polyunsaturated.

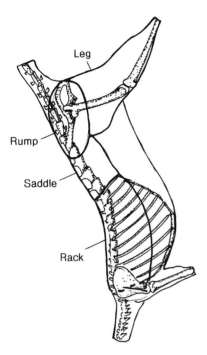

Figure 123. Australia - kangaroo cuts.

The major cuts for gourmet food are shown in Figure 123. Left and ride sides may remain together in the lumbar region to make a saddle. The leg is broken into medial, lateral and antero-distal parts known, respectively, by names such as topside, silverside and knuckle. A variety of fillets are isolated from the more tender muscles.

(Source: Kirkpatrick and Amos, 1985; Dawson, 1995; Sales and Dingle, 1998; SGM, 1999)

KANINCHEN

German for → rabbit.

KAPAKI

In → Greece, a cut of beef from the antero-dorsal thoracic region, located medially to the spala.

KARBONADE

A chop in the → Netherlands.

KARBONADESTRENG

The loin of a pork carcass in the → Netherlands.

KARÉ

The loin of a → Slovakian pork carcass.

KARFI

Icelandic for rosefish (→ Scorpaeniformes)

KARKÓWKA

The posterior cervical region of a →Polish beef carcass or the anterior thoracic region of a Polish pork or veal carcass. In a Polish lamb carcass, the karkówka is the cervical region.

KARP

Swedish for carp → Cypriniformes.

KARREE

In → Austria and → Switzerland, the Karree is a primal cut of pork including most of the vertebral column.

KARRÉ MED BEN

The anterior thoracic region of a → Swedish pork carcass after removal of the scapula and bog.

KATNA

Lumbar region of a camel carcass in → Morocco.

KAVALIERSPITZ

Subscapularis of an → Austrian beef carcass.

KEULE

→ German for leg, typically used for a leg of lamb.

KEY STEAKS

A poorly defined North American beef steak, typically *longissimus dorsi* with or without rib.

KFASS

A triangular cut of poultry in → Morocco.

KHAANG

The jowl of a pork carcass in → Thailand.

KHO KHAA LAANG

The hindfoot or trotter of a pork carcass in → Thailand.

KHO KHAA NAA

The forefoot or trotter of a pork carcass in → Thailand.

KIELBASA

Polish for → sausage, the most familiar type being pork-based, smoked, coarse-ground and heavy on the garlic. Kielbasa szynkowa is made with lean ham. Kielbasa serdelki is made with veal. Kielbasa cytrynowa is made with pork flavoured with lemon.

KILOTO

In → Greece, a cut of beef containing the sacrum, ilium and gluteal muscles.

KING CRAB

A long-legged deep-sea crab, *Paralithodes camtschatica,* → crabs.

KIPPER

A salted, smoked herring, *Clupea harengus,* → Clupeiformes.

KIRJI

Hausa for the brisket in → Nigeria.

KISZKA

Skeletal muscle may make only a minor contribution (from facial muscles of the head) to this cooked → sausage which is dominated by tripe and liver.

KJÚKLINGER

Icelandic for → chicken.

KKCF

An English abbreviation for kidney knob and channel fat on a beef carcass.

KLAPRIB

A ladder of beef ribs in a → Netherlands beef carcass.

KLAPSTUK

The abdominal muscles of a →Netherlands lamb carcass.

KLOBASA

Slovene for sausage.

KLUMP

A → Danish cut of beef around the *tensor fascia lata*.

KLUMPUR

Quadriceps femoris in → Iceland.

KNACKWURST

A smoked → sausage with a dark colour.

KNAP

→ Scottish cuts of beef in the most distal parts of the limbs remaining on the carcass.

KNOBLAUCH

Knoblauch is German for garlic, but is also used to indicate a garlic sausage.

KNOCHENDÜNNUNG

In → German beef cutting, the anterior part of the flank containing ribs is the bone-end thin flank or Knochendünnung.

KNOCKWURST

→ Knackwurst

KNUCKLE

For beef in North America, this may be the tip of a → diamond round composed of the three *vastus* muscles, plus *rectus femoris* and some of *tensor fascia lata*. A similar designation is used for → Australian, → New Zealand, export → Danish and → Irish beef. In → England, the knuckle may be the tibial cut of a veal carcass, with the ulna and radius cut as the fore-knuckle.

KOBASICA

Sausage in → Serbia & Montenegro.

KOBE BEEF

Japanese beef with a massive amount of → marbling, reputed to originate from cattle that drink beer and get a regular massage.

KOLENICA

Prednja kolenica and zadnja kolenica are fore and hindshanks, respectively, of beef, pork and lamb carcasses in → Serbia & Montenegro.

KOLJA

Swedish for haddock, → Gadiformes.

KOP

Dutch for the head, as in a → Netherlands pork carcass.

KOREA - BEEF CUTS

As shown in Figure 124, the major division of the hindlimb is into the smaller medial udun and the larger lateral suldo. The *psoas* filet or ansim is separated from the remainder of the short loin or chaekeut. After lifting the shoulder or abdari, the dungsim is a long cut including all the thoracic vertebrae. The moksim includes the neck and anterior of the chuck. The yangjee includes the entire ventral region from the sternum to the abdominal muscles, while the kalbi is an equally long cut including a ladder of rib sections plus dorsal abdominal muscles.

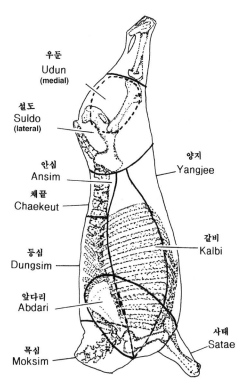

Figure 124. Korea – beef cuts.

(Source: Joo, 1999).

KOREA - PORK CUTS

Lamb and veal are not used in Korea, so the remaining meat cuts are for pork (Figure 125), and many of the same names as beef are in use. Differences are the ham (deukari) and belly (samkeubsal).

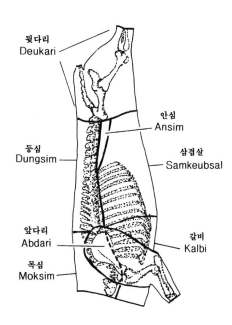

Figure 125. Korea – pork cuts.

(Source: Joo, 1999).

KOREYKA

In → Russia, this cut is the back ribs of a pork carcass but the whole back (thoracic and lumbar vertebrae) of a lamb carcass.

KORTHUGGEN SADEL

A short saddle of lamb in → Sweden.

KORV

Swedish for sausage.

KOTELET

Posterior rib chops or roast on a → Netherlands lamb carcass.

KOTELETT

In →German pork cutting , this is an equivalent primal cut to the loin of pork in North America, but the term also may be used for a single chop or cutlet. In German lamb cutting, the postero-dorsal part of the rib-cage also may be called the Kotelett.

KOTELETT-KAM

In → Norway, the kotelett-kam or midtkoteletter is a prime-rib roast of a beef carcass or the rib and loin of a pork or lamb carcass.

KÓTILETTUR

Icelandic for a cutlet.

KOTLETT

The kotlett may be a whole loin of a → Swedish pork carcass. The bone-in loin of pork is the kotlett med ben, but becomes the crown cutlets or kronkotlett after boning out. For Swedish lamb chops, a simple cutlet (kotlett, enkel) is from one side, while the double cutlet (kotlett, dubbel) is a bilateral chop through the saddle. For Swedish veal, with the *psoas major* and *iliacus* removed as a whole filet (filé), the loin is the kotlettrad utan filé.

KRABBA

Swedish for → crab.

KRABBE

Danish for → crab.

KRAKOWSKA

A lean, smoked → sausage made from leg of pork.

KRK

The neck of a beef lamb or veal carcass in → Slovakia.

KRKOVIČKA

The neck of a pork carcass in → Slovakia.

KRONKOTLETT

In → Swedish pork cutting, the bone-in loin of pork (kotlett med ben) becomes the crown cutlets or kronkotlett after boning out. Slices of *longissimus dorsi* are called kronkotlett.

KRUIS

The rump containing the ilium of a → South African beef carcass.

KRUSPELSPITZ

An →Austrian beef cut containing the anterior thoracic vertebrae.

KRVAVICA

Blood sausage in → Serbia & Montenegro.

KTAFF

Forelimb of a beef carcass in → Morocco.

KTAFF GSSIR

Forelimb of a camel carcass in → Morocco.

KUDU

The kudu, *Tragelaphus strepsiceros*, is a large African antelope reaching over 300 kg live weight. The → South African cutting pattern is shown.

(Source: Dorst and Dandelot, 1970)

KUGEL

In → German beef cutting, the *quadriceps femoris* group of muscles proximal to the patella in the hindlimb are likened to a ball, as in the boule d'aloyau of French beef cutting. The German Kugel generally includes *tensor fascia lata* as well. A Kugel cut may be taken from the anterior part of the ham in German pork cutting.

KULMAPAISTI

In → Finnish beef cutting, the *quadriceps femoris* group of muscles creates an angular cut of meat called the kulmapaisti.

KUNNI

Muscles of the sacrum and first five coccygeal vertebrae of beef, veal and camel in → Pakistan.

KYCKLING

Swedish for → chicken.

KYLJYS

A chop or cutlet of meat in → Finland.

KYLKI

In → Finnish meat cutting, the kylki is the plate or flank of a beef carcass, or a belly of pork.

LACTATE DEHYDROGENASE (LDH)

LDH is an enzyme that adds hydrogen to pyruvate to produce lactic acid. This is required for the continuation of post-mortem → glycolysis, which can only occur under anaerobic conditions once a meat animal has been slaughtered. LDH exists in a number of forms or isoenzymes. If LDH-1 is prevalent, it facilitates aerobic metabolism, where possible, because it is inhibited by pyruvate and lactate. LDH-5 is not inhibited by high levels of lactate and pyruvate, and it facilitates anaerobic metabolism. LDH-1 is typical of cardiac muscle while LDH-5 is typical of skeletal muscles, particularly those adapted for anaerobic conditions during contraction. In skeletal muscles, the ratio of LDH-1 to LDH-5 corresponds to the dominant activity pattern of a muscle. Thus, red muscles capable of sustained activity and which only use aerobic metabolism have high levels of LDH-1, whereas white muscles have high levels of LDH-5.

LÆRI

The hindlimb of a beef, pork, lamb or horse carcass in → Iceland.

LÆRISSNEIÐAR

Slices through the hindlimb of a carcass in → Iceland.

LÆRISTUNGA

Semitendinosus of a beef carcass in → Iceland.

LÆRSKANKI

The foreshank of a carcass in → Iceland.

LAGARTILLO

Distal forelimb meat on a beef carcass in → Ecuador.

LAGARTO

A "lizard-like" cut of beef in → Portugal that includes *biceps brachii* and the extensors of the foreshank. But in → Brazil, the lagarto is the *semitendinosus* of beef or veal while in → Chile it is the *biceps brachii*.

LÄGG

In Sweden, the shank of a → reindeer carcass.

LAKS

Danish for salmon, with the plural being laksørred, → Salmoniformes.

LALOK

The jowl of a pork carcass in → Slovakia.

LAMB

Species of sheep

There are many different wild species and domestic breeds of sheep in five main groups; (1), the moufflon from Mediterranean countries; (2), urial from southern Russia; (3), argali from the Himalayas; (4), bighorn from Canada and eastern Russia; and (5), domestic sheep, *Ovis aries*. Sheep were domesticated at an early stage in the transition from nomad to settled farmer. Goats probably were domesticated before sheep, but the domestication of sheep precedes that of cattle and pigs. Numerous characteristics have been changed by domestication. Many wild types of sheep have a wool-hair mixture and, in hot climates, certain species are almost naked. Wool bearing sheep probably were derived from animals originating in cold or mountain conditions. Domestic sheep show a range from very short to very long tails, but all wild types have short tails. Some sheep deposit fat in their tails. The lop-eared characteristic is not found in wild sheep and was produced very early during domestication. A convex nose is a striking feature of many breeds of sheep and is associated with a decrease in length of the jaws, which is a common feature in many other domesticated animals such as the pig and dog. Wild sheep often wield an array of elaborately shaped horns. During domestication the number has been reduced to a single pair, or horns have been lost altogether (polled). Animals kept in arid, rocky conditions derive an advantage from long legs, while smaller sheep are better for winter housing in colder climates.

(Sources: Zeuner, 1963; Ucko and Dimbleby, 1969)

LAMBA-

In → Iceland, a prefix used to denote cuts of lamb, as in the straight rack of lamb or lambakóróna.

LAMMFLEISCH

German for lamb meat.

LAMMKÖTT

Swedish for lamb.

LAMPREY

Lampreys are the survivors of an ancient group of fishes, the Agnatha, which predate the evolution of jaws. The mouth of the lamprey is a sucking disk with numerous sharp little teeth with which it gouges out a chunk, or sucks blood from another fish. Lamprey muscle is an acquired taste, the overeating of which is reputed (by Robert Fabyan, Sheriff of London in 1493) to have killed King Henry I of England while in Normandy. Although this is factually dubious, it does not detract from lampreys as being edible, especially *Petromyzon marinus*, the sea lamprey. However, the blood and mucus of *Lampetra fluviatilis*, the river lampern, may be poisonous.

So beware! Was Fabyan right, did the king's cook serve up lamperns instead of lampreys?

(Source: Pivnička and Černý, 1987)

LAMSBORST

Breast of lamb in the →Netherlands.

LAMSBOUT

Leg of lamb in the → Netherlands.

LAMSRUG

Antero-dorsal region of the thorax in a → Netherlands lamb carcass.

LAMSZADEL

Loin of lamb in the → Netherlands.

LANDJAGER

Originally carried by hunters and hikers, the Landjager (land hunter) sausage contains lean beef and pork fat in a large casing. The sausage is flattened before being heat-set by smoking so that it is easy to pack and maintain.

LANGA

Icelandic for → ling.

LANGE DE BOEUF

Equivalent to côte découverte of a → Belgian beef carcass.

LANGHRYGGUR

In → Iceland, a long primal cut including most of the thoracic and all the lumbar vertebrae.

LANGOUSTE

A spiny lobster, *Palinurus*, → Lobster.

LAPA

In → Finnish meat cutting the lapa is a shoulder cut. In beef, it contains the scapula, humerus and associated muscles lifted from the anterior ribcage. In pork, it is taken level with the humerus. In → Greece, the lapa is a cut originating from the abdominal muscles of the flank.

LAPIN

French for → rabbit.

LAPOCKA

The shoulder of a →Hungarian beef carcass.

LARD DE DOS

Back-fat on a Belgian pork carcass.

LARD DE JOUE

The jowl of a pork carcass in → Belgium.

LARD DE POITRINE

Lard maigre or belly of pork in →Belgium.

LARD GRAS

Back-fat on a → Belgian pork carcass.

LARD MAIGRE

Belly of pork in → Belgium.

LARD MOLÉ

Equivalent to lard maigre or belly of pork in → Belgium.

LÅR MED LÄGG

In → Swedish beef cutting, the whole hindlimb from the pubis posteriorly is called the lår med lägg. Lår med lägg utan filé is the whole hindlimb of a Swedish veal carcass, but with the *iliopsoas* (filé) removed. It is subdivided like a beef carcass into a medial innanlår, a lateral ytterlår, an anterior fransyskan, and a rostbiff through the ilium.

ŁATA

The abdominal flank of a → Polish beef or veal carcass.

LATISSIMUS DORSI

A major muscle with a flat, triangular shape, located laterally on the mammalian ribcage (Figure 126). Its origin is on the lumbodorsal fascia of much of the thoracic and lumbar region, plus the latter ribs. It insertions include the humerus and coracoid of the scapula.

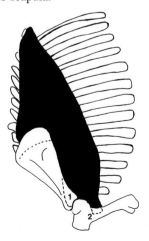

Figure 126. *Lateral view of the extent of beef* latissimus dorsi *(solid black), showing its insertion (broken line) on the scapula (1) and teres tuberosity of the humerus (2).*

LAX

Swedish and Icelandic for salmon, → Salmoniformes.

LAY

The shoulder region of a beef or buffalo carcass in → Thailand.

LECHÓN

Spanish for suckling pig.

LEDJA

Posterior thoracic cut of beef in → Serbia & Montenegro.

LEG OF LAMB

In → North America, a whole leg of lamb composed of the sirloin plus leg, is removed by cutting perpendicularly through the vertebral column near the anterior part of the ilium (Figure 127). The muscles appearing on the anterior face of a typical leg of lamb with the sirloin removed depend on the angle and position of the main cut.

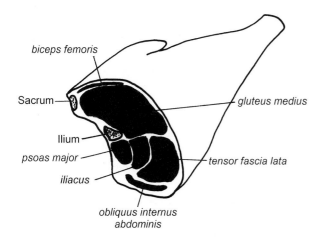

Figure 127. Main muscles of a leg of lamb with the sirloin removed, but there is considerable variability, depending on how much sirloin is removed.

There is considerable variability in the way that legs of lamb are used in North America. A whole leg roast might include parts 2 to 5 of Figure 128, and sometimes might be boned and rolled to make a long boneless roast. A short-cut, sirloin-off leg roast is composed of parts 2 to 4, or parts 4 and 5 may be left together to make a sirloin roast. Parts 3 and 4 are sometimes used as a center roast, or are cut into center slices of leg with a muscle grouping very similar to that seen in the full round of beef.

LEG OF MUTTON CUT

Despite its name, the leg of mutton cut is a cut of beef found around London, → England. It originates from the rib region ventral to the scapula.

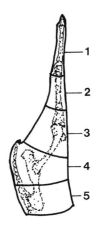

Figure 128. Retail cuts from a leg of lamb in North America. There is very little meat on the cannon bone of the lower shank (1), whereas extensors and flexors build up the meat around the tibia in the main part of the shank (2). Center slices may be taken through the center of the leg (3). The distal part of the sirloin (4) may be left on a leg roast, while the proximal part (5) may be used for sirloin chops of lamb.

LEG OF PORK

The whole hindlimb of the pork carcass is typically taken as a primal cut and may be called the ham even if it is → fresh meat. Sometimes the term leg is kept for fresh pork, with the cured leg then becoming the ham. The main subdivisions of the primal cut in North America are the rump (proximal), center, and shank (distal). The muscle groups are the same as those shown for the ham (Figure 96).

LEG TOPS

A dorsal part of the → leg of mutton cut of beef found in the London area.

LEMOS

The neck of a carcass in → Greece.

LEMPEN

The flank of a → Swiss beef carcass.

LENDE

→ German for the loin. Also Dutch for the loin in the → Netherlands, and sometimes used to denote primal cuts in the lumbar region, especially lamb and veal. Also the lumbar region of a → South African beef carcass is called the lende.

LENGUA

Tongue in Spanish.

LEPRE

Italian for → hare.

LEPROTTO

Young → hare or leveret in Italy.

LEPTOCEPHALUS

A larval eel, → Anguilliformes.

LEVATORES COSTARUM

A series of very small muscles located dorsally in the intercostal spaces (Figure 129).

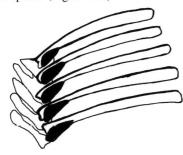

Figure 129. Some of the levatores costarum *(solid black) in the posterior ribs of a beef forequarter.*

LEVERET

A young → hare.

LIEBRE

Spanish for →hare.

LIESSTUK

The *tensor fascia lata* of a beef or veal carcass in the → Netherlands, alternatively known as the ezeltje.

LIÈVRE

French for → hare.

LIFTER STEAK

In → North American beef cutting, a slice across the *infraspinatus.*

LIGAMENTUM NUCHAE

A large ligament supporting the head and attached to the vertebral column. It is located dorsally in the carcass, in the midline between left and right sides. It is composed mainly of elastin, which is heat stable, so that it must be removed from retail cuts of meat, otherwise it remains as a thick, yellow back strap with mechanical properties similar to those of very strong rubber.

LILLA FRANSYSKAN

A cut through the shaft of the ilium on a → Swedish beef carcass.

LINEA ALBA

Tendinous connective tissue in the midline of the mammalian abdomen.

LING

The ling of the North Atlantic, *Molva molva*, → Gadiformes, is a long fish lacking the tubular nostrils or barbels found in other members of the cod family. The Spanish ling, *Molva macrophthalmia*, occurs in the Mediterranean.

LINGCOD

From California to Alaska, *Ophiodon elongatus* sometimes has blue-green muscles, → Scorpaeniformes. It is not a real cod.

LINGUIÇA

Fresh or smoked Portuguese → sausage.

LISTA

The stripe or lista is a long cut level with the midlength of the ribs on an → Italian beef carcass.

LIVER SAUSAGE

An emulsion → sausage, often containing at least some skeletal muscle.

LLAMA

The llama (*Lama glama*) is a multipurpose domesticated camelid of the Andes. Its meat use is mainly domestic, but some types are raised primarily for meat production in southern Bolivia. Llama meat is similar in taste to mutton. A llama produces about 12 to 15 kg of → charqui.

(Source: Calle Escobar, 1984; Iniguez *et al.*, 1998)

LOBSTER

Lobsters are decapod crustaceans of the suborder Macrura-Reptantia (→ Crustacea). Around the world, there are over 150 species of Crustacea that pass as lobsters of one type or another. The clawed lobsters (Nephropidae) have greatly enlarged claws on the first pair of legs so that they provide large masses of muscle from both the claws and the abdomen or lobster tail. The spiny lobsters (Palinuridae) only have simple unclawed tips on all five pairs of walking legs and their paired antennae are long and whip-like with a very stout base. Both the clawed lobsters and the spiny lobsters have stalked eyes, but the clawed lobsters have a spike-like rostrum pointing forwards from between their eyes while the spiny lobsters have a horn over each eye. The slipper lobsters (Scyllaridae) have simple walking legs and lack any rostrum between, or horns above their sunken eyes. The most noticeable feature of slipper lobsters, however, is that their carapace is flattened from above. Their antennae are also flattened.

The common clawed lobster in England, *Homarus gammarus* or *vulgaris*, occurs along the Atlantic coast of Europe down to the Mediterranean. It is very similar to *Homarus americanus* (Figure 130), which is caught in baited wood-frame traps or pots along the Northwest Atlantic continental shelf from Labrador down to North Carolina. Both species of lobster are scavengers that frequent rocky bottoms. A typical length is from 18 to 30 cm with a weight from 0.2 to 1 kg, although they may grow much larger. Their colour ranges from blue to red-brown. The left-right asymmetry that develops between the lighter cutter claw and the heavier crusher claw is accompanied by changes in → histochemical types of myofibres. The cutter has 81% fast and 19% slow muscle, while the crusher is all slow.

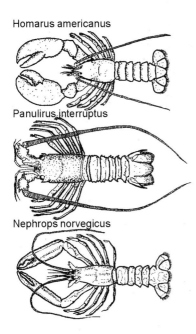

Figure 130. *Lobster (*Homarus americanus*), spiny lobster (*Panulirus interruptus*) and Norway lobster (*Nephrops norvegicus*).*

Nephrops norvegicus (Figure 130) is called by names such as the Norway lobster or the Dublin Bay Prawn, although the myosystem derived from this species is best known on a menu by its Italian name - scampi. *N. norvegicus* occurs in the Northeast Atlantic from Iceland down to Morocco, as well as through the Mediterranean to the Adriatic. It inhabits muddy sea bottoms and is commonly obtained as a by-catch from trawlers. Its colour ranges from pink to orange, and its claws often have red and white bands. The maximum length of the carapace and abdomen is about 24 cm.

Palinurus elephas or *vulgaris* is the European spiny lobster or crawfish caught from shallow rocky bottoms in baited pots or trammel (tangle) nets from the Southwest British Isles, along the coast of Northwest Africa, as far into the Mediterranean as the Adriatic. The crawfish is brown with pale markings, and only the tail is taken and immediately frozen. In France, as langouste, it may be valued more highly than lobster.

Panulirus interruptus and *P. argus* are the two species of spiny or rock lobsters fished commercially in the United States (note the confusing similarity in name between *Palinurus* and *Panulirus*). *Panulirus interruptus* is caught off South Carolina, and *P. argus* is caught along the east coast of Florida.

(Sources: Costello and Lang, 1979)

LOHAT EL KTAFF

Anterior rib cut on a beef carcass in → Morocco.

LOIN

In → North America, loins tend to be rather large cuts. Whole pork loins include the vertebral column and axial bones and muscles from the region of the scapula back to the pelvis, and may be subdivided as cuts from the blade loin roast or rib portion, through the center loin, back to the sirloin region. In North America, the backribs originate from the primal loin of pork, and are a series of ribs plus intercostal muscles where they have been detached to keep the ribs short in cutting rib chops. Elsewhere, loins range from large, as in → English lamb, to small, as in English beef.

LOMBARDIA

A brandy flavoured → salami.

LOMBATA

A primal cut containing lumbar and posterior thoracic vertebrae in an → Italian beef carcass. On an Italian lamb carcass it is just the lumbar region. In Italian, the lombi are the ilia.

LOMBINHO

The filet or *psoas* muscles of a → Portuguese beef carcass.

LOMBO

Posterior thoracic and lumbar vertebrae of a → Brazilian beef or lamb carcass.

LOMITO

Beef *iliopsoas* in → Costa Rica.

LOMO

The loin of a beef or pork carcass in → Spain, of a beef, pork or lamb carcass in → Mexico , or a beef *longissimus dorsi* in → Costa Rica and → Cuba. In → Ecuador, beef lomo extends all along the thoracic and lumbar vertebrae.

LOMO ALTO

The high loin or prime rib region of a → Spanish beef carcass.

LOMO DE PALETA

Beef *teres major* and *triceps brachii* in → Costa Rica.

LOMO LISO

In → Chile, lomo liso or smooth loin in beef extends from rib 10 to the ilium.

LOMO RAYADO

Beef *spinalis dorsi* in → Costa Rica.

LOMO VETADO

Beef rib steaks from the mid-thoracic region in → Chile.

LONDON BROIL

A poorly defined North American retail cut of beef (see fanciful names). Sometimes it is the tip from a → diamond round or hip, boned and sliced to contain *rectus femoris* and one of the *vastus* muscles, but also it might be prepared from the flank to contain *pectoralis* muscle.

LONDON LAMB

In → Iceland, this is the local name for lightly smoked lamb.

LONDON ROAST

Topside in British → Proten beef.

LONG DIGITAL EXTENSOR

Part of the *extensor digitorum longus* of the mammalian hindlimb.

LONGE

In Quebec, this may be loin of beef, pork or lamb. In → Luxembourg it is a cut near the femur in beef.

LONGISSIMI CAPITIS ET ATLANTIS

Longissimus capitis plus *longissimus atlantis*.

LONGISSIMI THORACIS ET LUMBORUM

Longissimus dorsi.

LONGISSIMUS ATLANTIS

A small, flat muscle attached to the atlas in the mammalian neck.

LONGISSIMUS CAPITIS

A small, tendinous muscle in the mammalian neck, originating from transverse processes of the fifth cervical to the seventh thoracic vertebrae (Figure 131).

Figure 131. *Lateral view of beef* longissimus capitis *(solid black).*

LONGISSIMUS CERVICIS

An anterior extension of *longissimus dorsi* in the mammalian neck, originating on transverse processes of the first to seventh thoracic vertebrae, and inserting on transverse processes of the third to seventh cervical vertebrae (Figure 132).

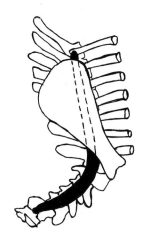

Figure 132. *Lateral view of beef* longissimus cervicis *(solid black and broken lines).*

LONGISSIMUS COSTARUM

Iliocostalis.

LONGISSIMUS DORSI

Although *longissimi thoracis et lumborum* may be a preferable name from an anatomical viewpoint, the name *longissimus dorsi* is very well entrenched in the literature of the animal and food sciences. In the *Journal of Animal Science*, the name may appear shortened even further to the *longissimus* muscle. The *longissimus dorsi*, almost universally abbreviated to LD, is a very large compound muscle comprising > 6% of the total muscle weight. Anteriorly, it first becomes noticeable beneath the scapula, then runs back along the carcass (Figure 133). Most of the LD terminates on the anterior face of the ilium. Thus, the LD forms the major eye of meat dorsal to the ribs in all of the commercially valuable rib steaks and roasts, as well the dorsal eye of meat in all lumbar steaks.

Figure 133. *Lateral view of beef* longissimus dorsi *(solid black and dotted lines).*

LONGISSIMUS LUMBORUM

Lumbar part of *longissimus dorsi.*

LONGISSIMUS THORACIS

Thoracic part of *longissimus dorsi.*

LONG PLATE OF BEEF

In → North American beef cutting, a long plate is the brisket plus the plate.

LONGUS CAPITIS

A long, lateral muscle in the mammalian neck, widest at its most anterior position. Its origin is on the ventral parts of transverse processes of the third to seventh cervical vertebrae. Its insertion to the skull is severed when the carcass is decapitated.

LONGUS COLLI

This compound muscle covers the ventral surfaces of the vertebrae in the cervical region, and as far back as the fifth thoracic vertebra in sheep, and seventh thoracic vertebra in beef (Figure 134). In pork, muscles from left and right sides

may not meet, thus allowing the vertebrae to be seen in a hanging carcass.

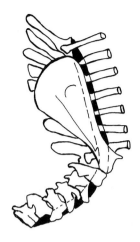

Figure 134. *Lateral view of beef* longus colli *(solid black and broken lines).*

ŁOPATKA

A shoulder cut from a → Polish pork or veal carcass.

LOPATOCHNAYA CHAST

The shoulder of a carcass in → Russia.

LOUZA

Quadriceps femoris muscles on a beef carcass in → Morocco.

LÚÐA

Icelandic for halibut (→ Pleuronectiformes)

LUMPFISH

Bottom-dwelling fishes of the North Atlantic in the order → Scorpaeniformes, such as *Cyclopterus lumpus*. United with a circular flap of skin, the pelvic fins form a sucker by which the fish adheres to rocks or floating objects. Lumpfish roe is the main product, but the myosystem is quite edible and sometimes is smoked. Buckland thought the taste similar to that of turtle.

(Sources: Bompas, 1886; Gavaris, 1985)

LUNCHEON MEAT

Typically a meat product with a → binder enabling it to be sliced as a convenience food.

LUNDI

Lundi is Icelandic for the puffin (*Fratercula arctica*), which is a relatively large (30 cm), muscular bird which walks, flies, swims and dives very well. Breeding in vast colonies, it is a traditional game bird in Iceland, with reyktur lundi being the smoked product. Puffins are in the family Alcidae, which also includes the guillemots, another traditional Icelandic game bird (svartfugl).

(Source: Bertin, 1967; IAIS)

LUNDIR

A filet of *iliopsoas* in → Iceland.

LUNGENBRATEN

Psoas and *iliacus* muscles of an → Austrian beef carcass.

LUXEMBOURG - BEEF CUTS

Unlike the longe in Quebec, which is the loin, in Luxembourg (Figure 135) the longe is in the hock region, perhaps where animals were tethered or restrained by the hock (longe = tether). Part of the longe is known as the quasi or chump. The posterior muscles of the hindlimb, possibly just the eye of the round or *semitendinosus*, may be known as the filet d'Anvers. The noix, or nut, is restricted to the patella region, unlike the more widespread noix cuts of France. The rumsteak contains the gluteal muscles. The lumbar and posterior thoracic steaks are the roastbeef, as in → Germany and → Switzerland, whereas the abominal flank muscles are named flanchet, as in → France.

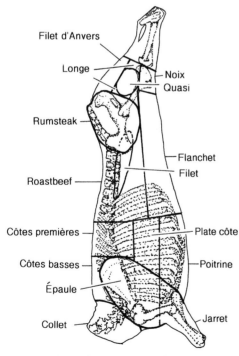

Figure 135. Luxembourg – beef cuts.

Removal of the shoulder or épaule leaves the prime rib (côtes premières) and deep rib (côtes basses). The more ventral remainder of the ribs are included in a ladder of rib sections (plate côte) and the brisket (poitrine), as in France. The neck is the collet and the shank is the jarret.

(Source: OEEC, 1961)

LUXEMBOURG - PORK CUTS

The sacral or tail region of the ham is removed as the queue, leaving the pubic and femoral parts as the jambon, plus a ham knuckle or jambonneau de derrière, and the trotter or pied (Figure 136). The lumbar region or loin is called the filet, and between this and the queue, the cut containing the wing of the ilium is called the entredeux, literally, between the two. In French, another connotation of entre is in the layering of fat and lean, either as in streaky bacon (entrelardé) or in the culinary layering of a dish with fat (entrelarder). In heavy pigs, the backfat is called simply the lard, while the lesser fat layer on the abdominal wall is the thin lard, or lard maigre. The ventral region of the belly is called the breast or poitrine. In the angle between the ham and belly is the griffe, which literally means a claw, notch or point of attachment. Perhaps this is the point from which side bacon was hung during curing?

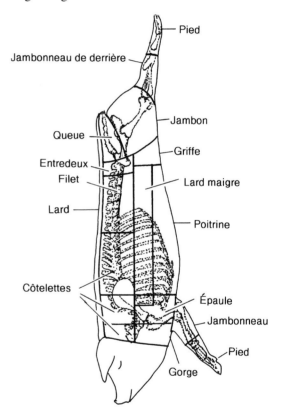

Figure 136. Luxembourg – pork cuts.

The vertebral region from the neck to the midthoracic region is called côtelettes or cutlets. The shoulder (épaule) is removed with the knuckle (jambonneau) and trotter (pied), to leave the throat or gorge.

(Source: OEEC, 1961)

LYMPH

The lymphatic system, in parallel with the venous system, drains fluids from many types of tissues, including skeletal muscles, where drainage is via surface connective tissues.

Lymph fluid flows to lymph nodes and then back into the venous circulation. Lymph nodes detect and respond to infections in the tissues drained, which is why they are examined by meat inspectors to verify animal health. Some of the variability in fluid losses from meat could originate from the lymph status of a muscle at slaughter.

(Source: Havas *et al.*, 1997).

MACCHLI

Fore- or hindshanks of a lamb or mutton carcass in → Pakistan.

MACKEREL

An Atlantic bony fish with silver belly and blue and black bars along its back, *Scomber* → Perciformes.

MACREUSE

In → French beef cutting, a series of macreuse cuts may be made from the forelimb to include the *triceps brachii*, deltoids and *subscapularis*.

MADURAS

A hybrid between zebu cattle and → banteng, named after the Indonesian island of Madura. They are small, compact animals with well developed forequarters. Bulls reach only 375 kg live weight. They have a relatively large rib-eye area and a high yield of lean meat. Subcutaneous fat is minimal, reaching 15 mm at maximum.

(Source: NRC, 1983)

MAFRACH

Camel flank in → Morocco.

MAGERES MEISEL

Supraspinatus on an → Austrian beef carcass.

MAHRAD

Thoracic vertebral region of a camel carcass in → Morocco.

MAIGRE

Meaning thin and lean, in → French beef cutting the maigres may be the macreuse slices through the *triceps brachii*.

MAKREL

Danish for mackerel, → Perciformes.

MAKRILL

Swedish for mackerel, → Perciformes.

MALAYA

Damn! Either a Latin-American expletive or the *cutaneus* muscle from a beef carcass in → Chile.

MALIGNANT HYPERPYREXIA

→ Porcine stress syndrome

MALIGNANT HYPERTHERMIA

→ Porcine stress syndrome

MANATEE

The manatees (*Trichechus* species) and dugong (*Dugong dugon*) of the order Sirenia are threatened by their slow rate of breeding and disappearing habitat, but also by the high value placed on their meat. When roasted it has the taste of pork with a flavour of veal. The rate of autolysis is very slow which makes pickling possible in a tropical climate. However, the intermuscular fat may be greenish in colour with a fishy flavour.

(Sources: Buckland, 1894; Burton, 1962)

MANHATTAN ROAST OR STEAK

A → North American retail cut of beef from the top round, typically containing *semimembranosus* or *adductor*.

MANJI

Hindshank of a beef carcass in → Morocco.

MAN KHANG

Pork carcass back fat in → Thailand.

MANO

The forefoot of a → Spanish pork carcass.

MANTIS SHRIMP

Crustaceans with an incomplete carapace, such as *Squilla* (→ Crustacea).

MARBLING

Fine lines and specks of intramuscular fat. Many butchers and cooks believe that marbling enhances meat taste and tenderness (myself included). Many meat scientists have tried to find quantitative data in support of this traditional belief but have failed. It is important to remember that failure to find evidence of an experimental effect does not prove that the effect is absent, especially when dealing with things that are hard to measure with precision, accuracy, sensitivity and repeatability. The visual appearance of marbling (to both the human eye and to machine vision) may change with the direction in which the meat is sliced and on the degree of optical contrast between fat and lean. Similarly, the protocols used for experimental meat cooking are often far removed from the techniques of gourmet cooking. Most meat scientists happily freeze their meat samples in a haphazard manner, store them for months, cook them in strange ways, core them rather than slice them, and serve them lukewarm on paper plates. No wonder the results are random.

MARIPOSA

In → Cuban beef cutting, a butterfly-shaped cut of pelvic muscles including *obturatorius internus* and *externus*.

MARVIPIBEBOV

In → Danish beef cutting, the marvipibebov or marrow-bone shoulder contains the distal part of the humerus.

MASNA POTRBUSINA

Posterior part of the abdominal muscles on a pork carcass in → Serbia & Montenegro.

MATAMBRE

The ventral region of an → Argentinean beef carcass, its name indicating a culinary use as rolled beef.

M-BRIDGE

Part of the → M-line.

MEAT LOAF

Chunks or particles of meat are heat set in a cereal matrix to make a bread-like product that usually can be sliced. Vegetable ingredients typically include tomato, garlic and/or onion.

MEATY RIBS

A cut of → Proten beef in Canada.

MECHANICALLY-DEBONED MEAT

Meat remaining on bones after they have been removed from the carcass may be retrieved by grinding the bones and then subjecting them to great pressure behind a metal screen or sieve. The residual fragments of meat are pushed through the screen and collected as mechanically deboned meat. Thus, mechanical deboning enables the recovery of meat from bones such as vertebrae that are difficult to clean manually. However, bone particles about 0.1 mm across as well as fat and marrow droplets may also be extruded through the screen.

MEDALLONES

Lumbar chops through a saddle of lamb in → Mexico.

MEDIAL DIGITAL EXTENSOR

Part of the *extensor digitorum longus* of the mammalian hindlimb.

MEGRIM

A flatfish, *Lepidorhombus*, → Pleuronectiformes.

MELLEMBOV

In → Danish beef cutting, the mellembov or intermediate

shoulder passes across the anterior ribs and the proximal part of the humerus.

MELLEMSKÆRT

In → Danish beef cutting, the neck is called the mellemskært or intermediate cut.

MELLOMMØRBRAD

In a → Norwegian beef carcass the mellommørbrad is posterior to the mørbrad, through the shaft of the ilium.

MENHADEN

A low-grade, herring-like fish, *Brevoortia*, → *Clupeiformes*.

MERLAN

The *sartorius* muscle when part of the tranche in → French beef cutting.

METLEH

Rib of beef in → Jordan.

METMYOGLOBIN

→ Myoglobin.

METTWURST

Soft, cooked → sausage for spreading in sandwiches.

MEXICO - BEEF CUTS

The primal cuts of beef in Mexico are shown in Figure 137. Distally, the leg or pierna is sliced transversely to the tibia, but proximally is separated into medial and lateral muscle groups which are boned and rolled. Quite likely, following the natural way for a hip of beef to be lowered onto a butcher's block (i.e., with its lateral surface downwards), the medial cut is the cara (facing upwards) while the contra is lateral (against the block). The gluteal muscles and all of the pelvis are included in the cadera, which is sliced into North American style sirloin steaks. The lomo or loin is a major cut of beef extending from the anterior face of the ilium to the midlength of the scapula. The *psoas* muscles are removed as the filete, from which are sliced filete mignon or turnedó steaks. The *longissimus dorsi* together with the flap of abdominal muscles may be boned and rolled or, alternatively, sliced transversely (without the flap of abdominal muscles) to produce a series of steaks on the typical North American pattern: entrecot steaks with ribs, T-bone steaks in the mid-lumbar region, and a few Porterhouse steaks just anterior to the ilium.

The espaldilla y costillar (shoulder blade and rib cutlets) is sliced transversely to the vertebral axis to give, from anterior to posterior, the pescuezo (neck), tapa de espaldilla (top of the blade), costilla (rib cutlet) and chuletas. The *supraspinatus* may be removed as the pulpa. The chambarete (foreshank)

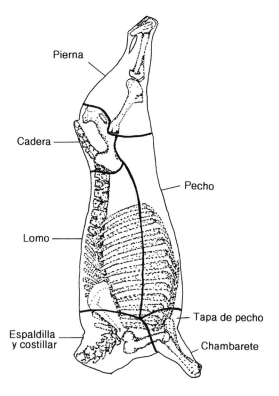

Figure 137. *Mexico – beef cuts.*

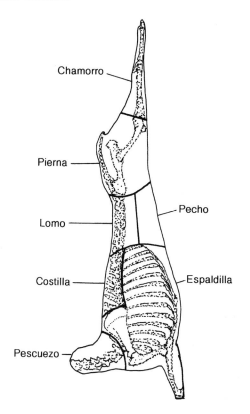

Figure 138. *Mexico – lamb cuts.*

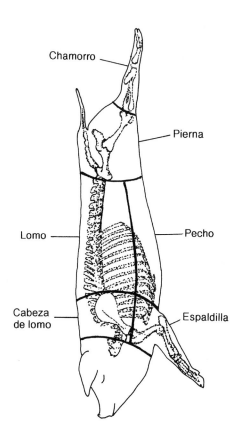

Figure 139. *Mexico – pork cuts.*

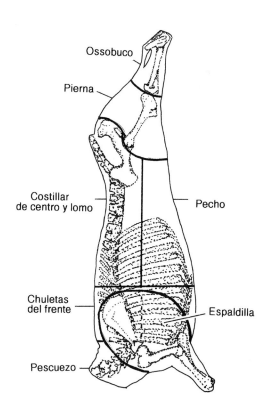

Figure 140. *Mexico – veal cuts.*

is separated from the underlying tapa de pecho (top of the chest) containing the anterior part of the sternum. The remainder of the pecho may be boned and rolled (pecho enrollado) or sold sliced as agujas.

(Source: Lourdes Pérez, 1999)

MEXICO - LAMB CUTS

Variations in cutting the hindlimb include separation of the chamorro and pierna, as shown in Figure 138, or cutting chops through the ilium and gluteal muscles (chuletas de pierna) to leave the muscles around the femur and tibia as a shank-end leg called the pierna. The lumbar chops from the lomo are called T-bone. The costilla, or dorsal thoracic region, is separated as a roast (costillar) or cut into chops (chuletas del centro). The espaldilla is a large primal cut that includes the scapula. The scapula may be lifted from the ribcage to make a smaller shoulder cut also called the espaldilla. The underlying ribs are cut into chops called chuletas de aguja. Neck of lamb is the pescuezo. The abdominal muscles are rolled as pecho enrollada.

(Source: Lourdes Pérez, 1999)

MEXICO - PORK CUTS

The hindlimb is split into a proximal pierna and distal chamorro, with the chamorro varying in size from the relatively small tibial cut shown in Figure 139 to a larger cut taken through the femur. Chuletas, like North American sirloin steaks, may be cut through the ilium of the pierna. The lomo or loin is split into a rib portion (lomo con costilla), loin chops (chuletas de lomo), and *psoas* (filete).

The head end of the loin, cabeza de lomo, is split into chops, chuletas de cabeza de lomo. The distal portion of the forelimb, espaldilla, may be cut to include the jowl, as espaldilla y codillo (blade and elbow). Note that little if any of the scapula or espaldilla occurs within the cut of pork called the espaldilla. The pecho is split into rib chops called costillitas.

(Source: Lourdes Pérez, 1999)

MEXICO - VEAL CUTS

Separation of the forequarter from the hindquarter may be along a straight line just posterior to the scapula (as in Figure 140), or an L-shaped forequarter may include the abdominal muscles of the pecho (breast). The shoulder or espaldilla is lifted from the anterior ribcage to leave the pescuezo (neck) and chuletas del frente (front chops). The primal cut called the costillar de centro y lomo (ribs and centre loin) is sectioned into anterior chuletas del centro (centre chops) and posterior chuletas de riñonada (kidney chops). The primary division of the hindlimb is into the pierna and ossobuco. The pierna may be used to prepare escalopas de bola.

(Source: Lourdes Pérez, 1999)

M-FILAMENT

Part of the → M-line.

MIAMI ROLL

In North America, a boneless, rolled cut of beef prepared from the brisket.

MIDDLE NECK

In → English lamb cutting, the scapula and associated muscles are removed from the underlying thoracic vertebrae and ribs, which remain as the middle neck. English veal carcasses may be cut in a similar manner.

MIDDLE RIBS OF BEEF

In → English beef cutting, anterior to a 4-rib forerib may be located a 4-rib middle ribs cut. However, the middle ribs only contains three ribs if the forerib is Scotch-cut with five ribs. The dorsal part of the middle ribs contains the scapula and may be separated as a back ribs cut. The ventral part of the middle ribs may be separated as the top ribs cut. Thus, the top ribs are ventral to the back ribs. Landmark muscles are shown in Figure 141.

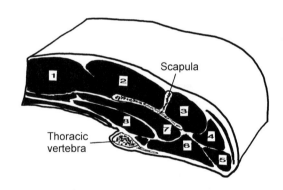

Figure 141. *English middle ribs of beef with some landmark muscles;* triceps brachii *(1),* infraspinatus *(2),* supraspinatus *(3),* rhomboideus *(4),* splenius *(5),* complexus *(6),* serratus ventralis *and* subscapularis *(7),* *and* longissimus dorsi *(8).*

MIÐLÆRI

Medial muscles of the hindlimb in → Iceland.

MIDTKOTELETTER

Equivalent to the loin or kotelett-kam of a → Norwegian pork carcass.

MIDTRIBBE

A → Norwegian pork cut including the full dorso-ventral depth of the posterior thoracic ribs.

MIELGA

The brisket of a → reindeer in Saami language.

MIELGANJAARCAA

The flank of a → reindeer in Saami language.

MIESO ZŁOPATKI

The shoulder of a → Polish beef carcass.

MILIEU DE FILET

In → French pork cutting, milieu de filet is a loin of pork including the lumbar vertebrae.

MITHAN

The mithan or gayal (*Bos frontalis*) is a semi-domesticated bovine in Asia, possibly derived from the wild gaur (*Bos gaurus*). An average bull reaches 1.5 m at the shoulder and weighs about 540 kg. The slaughter is a ritual and the meat is only eaten ceremonially.

(Source: NRC, 1983; Clutton-Brock, 1987).

MITTELLBRUST

A → German beef cut containing the middle part of the sternum.

MITTLERES- UND DÜNNES KÜGERL

Costal cartilage region of an → Austrian beef carcass.

MJAÐMASNEIÐAR

The gluteal region of a carcass in → Iceland, such as steaks through the *gluteus medius*.

MLAJ

Camel forelimb in → Morocco.

M-LINES

At the midlength of the thick → myofilament (which is composed mainly of myosin molecules) there is a surface thickening caused by proteins such as glycogen debranching enzyme, creatine kinase and myomesin. When looking at electron micrographs, these thickenings, when all in register across the myofibril, create a transverse M-line about 75 nm wide across the myofibril (Figure 142).

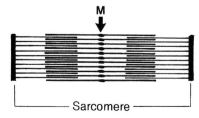

Figure 142. The M-line of the sarcomere caused by alignment of thickenings on the thick myofilaments.

But, at high magnification, M-line structure is very complex (Figure 143). Cross-bridges linking adjacent thick myofilaments may be seen and, if the preservation of myofibrils for electron microscopy is enhanced by the use of low-temperature techniques, the M-line may appear as several lines, all perpendicular to the long axis of the sarcomere and located near its midlength. The appearance of these M-lines may differ between different physiological types of myofibres.

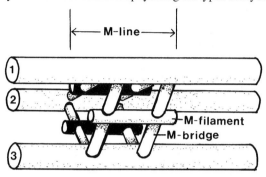

Figure 143. The location of M-filaments and M-bridges at the M-line relative to three thick myofilaments (1, 2 and 3).

A particularly thick part of the M-line is caused by short lengths of thin (5 to 6 nm) M-filaments running parallel to the thick myofilaments. The lengths of the M-filaments correspond to the width of the M-line. The M-filaments are joined to the thick myofilament by M-bridges about 4 nm thick. There are usually at least three arrays of M bridges in the thickness of the M-line, but there may be several extra sets of arrays in some types of myofibres. In transverse sections through the M-line, each thick myofilament is connected with its six neighbouring thick myofilaments by a triangular pattern of M-bridges. At the midlength of each M-bridge, the cross-sectioned M-filament may appear as an enlarged electron dense area about 5 to 6 nm across.

The structure of the M-line does not change much with sarcomere lengths from 1.4 to 1.8 μm, or even if the sarcomere is stretched to about 3 μm. But, in contracted sarcomeres it may be difficult to see the M-line because the thin myofilaments pass through and overlap in the M-line region. In stretched muscle, the M bridges may appear curved because the thick myofilaments to which they are attached are now much closer together. In sarcomeres that have been overstretched to between 3.9 and 4.0 μm, the M-line is lost but reappears when the sarcomere is allowed to resume a normal length, although the arrangement of the M filaments is disturbed.

(Sources: Wallimann *et al.*, 1975; Luther and Squire, 1978; Mani *et al.*, 1980; Grove *et al.*, 1987; Carlsson and Thornell, 1987; Edman *et al.*, 1988)

MOCK DUCK

A creative cut of lamb made from the foreshank, utilising the bend of the foreshank to create the mock duck's neck and head. The body, wings and tail are made from the *pectoralis* muscles of the breast, sewn into shape. A cranberry may be used to make the eyes.

(Source: NLMB, 1937).

MOKSIM

The neck and anterior chuck of a → Korean beef carcass or the neck and anterior rib region of a pork carcass.

MOLLET

In → French beef cutting, the "soft" may be the *quadriceps femoris* group of muscles forming the tranche grasse.

MONGOLIA

The main meat sources for the nomads of the Mongolian Gobi are sheep, goat, cattle, horse and camel.

(Source: Seishi, 1999)

MONK FISH

Squatina squatina, also called the angel shark, has a body shape intermediate between the sharks and rays (→ Chondrichthyes).

MOOSE

Explaining the transatlantic confusion between elk and moose is not easy and is largely guesswork. The Germans of the middle-ages were familiar with *Alces alces* (the moose in North America, but the elk in Europe). After the retreat of the glaciers, *A. alces* was widespread across Europe and, by Roman times, was still present in the Black Forest of Germany (according to a reliable source - Julius Caesar). The Swedish for *A. alces* is älg, similar to the High German word, *elch*, which slipped into the English language to indicate a large beast with antlers. But the English soon forgot how big an elk really is. When they reached North America and got trampled by the unbelievably large *A. alces*, they needed a superlative name. So they dropped elk (which would have been correct), and used the local name which, in the Narrangsett language, was *moos*. In Quebec, the French altered the Basque word for a deer, *orenac*, which evolved through *originac* to the present name, l'original. Some zoological sources call the North American moose *A. americana*, as distinct from the European elk *A. alces*.

Regardless of linguistics and taxonomy, moose produce a lot of meat, especially from bulls whose live weights range up to about 600 kg. The carcass has a short neck and long limbs. Moose meat is similar to beef, but is a little drier and darker than beef. Moose meat can be regarded as → venison, like the meat from other members of the deer family, but usually it is named separately.

(Sources: Lydekker, 1894; Burton, 1962; Rue, 1981)

MØRBRAD

Cuts through the ilium in → Norwegian beef, pork, lamb or veal carcasses.

MØRBRADSTYKKE

A → Danish cut of pork from the sacral region.

MORCILLO

Distal extensors and flexors from the forelimb or hindlimb of a → Spanish beef carcass. Morcillo also is used for Spanish blood sausage, morcilla.

(Source: Santos *et al.*, 2003)

MOROCCO – BEEF CUTS

As shown in Figure 144, the hindquarter of beef (Elajl) contains the Manji (hindshank), Fakhda (posterior muscles), Louza (*quadriceps femoris* group), and Diala (caudal region). The *psoas* muscles are the Fili. The steaks from posterior to anterior are the El Katna (lumbar), Eladlaa (posterior thoracic) and Lohat El Ktaff (anterior thoracic). The forelimb mass forms the Ktaff, leaving the Dala (ribs), Sadria (brisket), Bouuite (flank) and Angra (neck).

(Source: Daoudi and Benkerroum, 2003)

MOROCCO – CAMEL CUTS

The hump is removed from the camel carcass (Skitat El Ibil) as the Droua (Figure 145). The Gaima is the hindlimb. The forelimb is split into the proximal Ktaff Gssir and the distal Gandouz or Mlaj. The Mahrad is the thoracic vertebral region while the lumbar region is the Falka or Katna. The Mafrach or Janb is the flank and the Saedania or Gargra is the rib region. The neck is the Ragba or, as in beef, the Angra.

(Source: Daoudi and Benkerroum, 2003)

MOROCCO – CHICKEN AND TURKEY CUTS

Cuts of the chicken or turkey carcass (Figure 146) include the Ounk (neck), Sdar (breast), Fkhad (proximal leg) and Sak (distal leg). The Kfass contains a triangular region of thoracic vertebrae and ribs. The wing is divided into three parts from proximal to distal, the Blanquette, Jniah and Tarf Jnah. The Oussous is the tail.

(Source: Daoudi and Benkerroum, 2003)

MOROCCO – LAMB AND MUTTON CUTS

As shown in Figure 147, the shank (El Kaeba) and tail (Diala) are removed from the leg (Fkhad). The sternal breast (El Sadr) is removed from the abdominal flank (Boussuite). And the loin (El Salsoul) is separated from the rib (El Dloue).

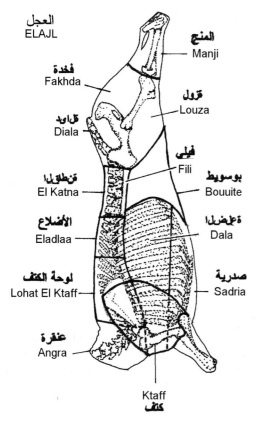

العجل
ELAJL

المنج
Manji

فخدة
Fakhda

قزول
Louza

ديالق
Diala

فيلي
Fili

القاطنة
El Katna

بوسويط
Bouuite

الأضلاع
Eladlaa

الضلعة
Dala

لوحة الكتف
Lohat El Ktaff

صدرية
Sadria

عنقرة
Angra

Ktaff
كتف

Figure 144. *Beef cuts in Morocco.*

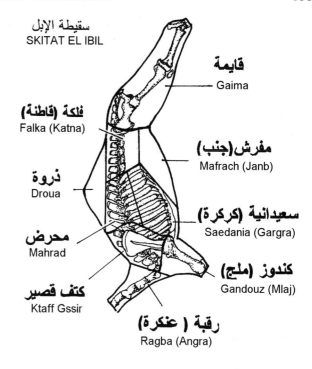

سقيطة الإبل
SKITAT EL IBIL

قايمة
Gaima

فلكة (قاطنة)
Falka (Katna)

مفرش(جنب)
Mafrach (Janb)

ذروة
Droua

سعيدانية (كركرة)
Saedania (Gargra)

محرض
Mahrad

كندوز (ملج)
Gandouz (Mlaj)

كتف قصير
Ktaff Gssir

رقبة (عنكرة)
Ragba (Angra)

Figure 145. *Camel cuts in Morocco.*

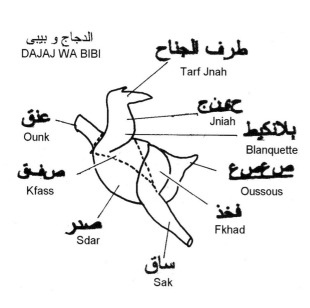

الدجاج و بيبى
DAJAJ WA BIBI

طرف الجناح
Tarf Jnah

جناح
Jniah

عنق
Ounk

بلانكيط
Blanquette

كفاس
Kfass

عصعص
Oussous

صدر
Sdar

فخذ
Fkhad

ساق
Sak

Figure 146. *Chicken and turkey cuts in Morocco.*

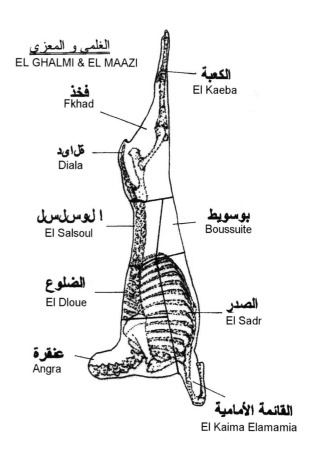

الغلمى و المعزي
EL GHALMI & EL MAAZI

الكعبة
El Kaeba

فخذ
Fkhad

ديالق
Diala

الوسلسل
El Salsoul

بوسويط
Boussuite

الضلوع
El Dloue

الصدر
El Sadr

عنقرة
Angra

القائمة الأمامية
El Kaima Elamamia

Figure 147. *Lamb and mutton cuts in Morocco.*

The Angra is the neck and the El Kaima Elamamia is the foreshank.

(Source: Daoudi and Benkerroum, 2003)

MORTADELLA

A smoked, dry → sausage packed in a beef bladder → casing and distinguished by cubes of pork fat.

MOSTEK

Polish for sternum, this may be the sternal region of a → Polish beef or veal carcass, or a breast of lamb.

MOUSE BUTTOCK

An ancient beef cut from → England containing the *gastrocnemius*.

MOUVANT

The mouvant is part of the tranche grasse in → French beef cutting and is composed of *vastus internus* and *vastus intermedius*.

MOYENNE CÔTE

In → Belgium, this may be the same as the belle côte at the thoraco-lumbar junction.

MOZEH AMAMIEH

Foreshank of beef in → Jordan.

MOZEH KHALFIEH

Hindshank of a beef carcass in → Jordan.

MUGILIFORMES

Although these fishes are perch-like in nature, they differ in having their pelvic fins in an abdominal position (Figure 148). The family Atherinidae contains some relatively primitive marine and freshwater fish without a lateral line. The mouth is small with weak teeth. Individual fishes usually are quite small (≈ 10 to 15 cm), but they may occur in vast schools, such as: *Atherinopsis*, the California jacksmelt, and *Atherina*, the Atlantic silverside. *Chirostoma* is a freshwater food fish of Mexico.

Figure 148. Chirostoma.

MUIS-VOOR

In the → Netherlands, a slice before the "ball" of the shoulder of a beef carcass.

MULLET

Fish of various species of *Liza* and *Mugil*, → Mugiliformes,

MULTIFIDI CERVICIS

A compound muscle flexing the mammalian neck by pulling the dorsal vertebral spines towards the transverse processes of vertebrae. It may be found anywhere from the third cervical to the second thoracic vertebra (Figure 149).

Figure 149. *Lateral view of beef* multifidus cervicis *(solid black and broken line).*

MULTIFIDUS DORSI

A compound muscle (which may be written *multifidi dorsi*) flexing the mammalian vertebral column, and located laterally to the spinous processes of vertebrae from the neck posteriorly to the sacrum (Figure 150). In rib or loin steaks or chops, sections of the *multifidus dorsi* are located between the spinous processes of the vertebrae and the main eye of *longissimus dorsi* meat.

Figure 150. *Lateral view of beef* multifidus dorsi *(solid black and broken line).*

MUSCOLO

Respectively, muscolo anteriore and muscolo posteriore are the distal extensor and flexor muscles of the fore- and hindlimb on an → Italian beef carcass.

MÚSCULO DO DIANTEIRO

Flexors and extensors of the foreshank of a → Brazilian beef carcass, cut from the paleta.

MÚSCULO DURO

Around the tibia on a → Brazilian beef carcass.

MÚSCULO MOLE

The *gastrocnemius* and associated muscles of a → Brazilian beef carcass.

MUSKELLUNGE

A fish in the pike family (→ Salmoniformes).

MUSK OX

The musk ox, *Ovibos moschatus*, has been hunted for thousands of years and now has a restricted range in Arctic regions. Re-introduced from Greenland to Alaska and Nunavut, hunting may be sustainable if there is no predation by wolves. The musk ox has potential for domestication, but most efforts have not got very far. Perhaps the only possibility for sustainable meat production is as a byproduct of quiviut production - the animal's fine underwool which is extremely valuable. The meat tastes good, but is drier and darker than beef.

(Source: Rue, 1981; White *et al.*, 1989)

MUTTON

Mutton is meat derived from sheep. The premium market is mostly concerned with lamb from younger sheep.

MYOBLAST

A cell capable of contributing to the formation of a → myofibre. Myoblasts have two elongated cytoplasmic extensions capable of searching for other myoblasts with which to fuse. Myoblasts no longer divide by mitosis and their cytoplasm stains strongly with basic dyes because of a high content of RNA involved in synthesising new muscle proteins.

MYOCOMMA

A partition or septum of connective tissue separating adjacent myomeres (→ myomere) in the musculature of a fish. The plural is myocommata.

MYOFIBRES

Striated myofibres (muscle fibres) are the basic cellular units of living muscle and of food myosystems (Figure 151). Myofibres are unusual cells because they are multinucleate (have many cell nuclei) and extremely long (commonly several centimetres). Myofibre diameters are microscopic (usually less than one tenth of a millimetre). Myofibre diameters increase during the growth of a muscle, but they also increase temporarily when a myofibre contracts. Thus, when measuring myofibre diameters, care must be taken to avoid or to correct for differences in the degree of muscle contraction. Myofibres contain contractile organelles (→ myofibrils) surrounded by cytoplasm (→ sarcoplasm).

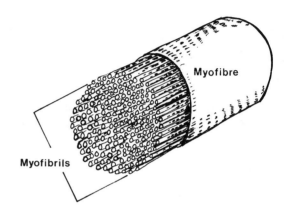

Figure 151. *Myofibrils protruding from the end of a short length of myofibre, as seen in many comminuted meat products under the microscope.*

MYOFIBRIL

Myofibrils are contractile organelles packed tightly within skeletal myofibres of meat and food myosystems. Myofibrils are → transversely striated.

MYOFILAMENT

With an electron microscope, the → transverse striations along → myofibrils can be seen to originate from the regular longitudinal arrangement of sets of thick myofilaments (10 to 12 nm in diameter and dominated by → myosin molecules) and thin myofilaments (5 to 7 nm in diameter and dominated by → actin molecules). When stretched, both thick and thin myofilaments may be extensible. In a transverse section cut through overlapping thick and thin myofilaments (as at the edge of the A-band), each thick myofilament is surrounded by six thin myofilaments (Figure 152), although this hexagonal lattice may change to a tetragonal lattice when sarcomeres are stretched. When a muscle contracts, the thick myofilaments slide relative to the thin myofilaments so that the I-band gets shorter. The length of the A-band generally remains constant during muscle contraction in vertebrates.

(Sources: Schiereck *et al.*, 1992; Wakabayashi *et al.*, 1994; Blange *et al.*, 1997)

MYOFILAMENT LATTICE

Negative electrostatic repulsion between → myofilaments maintains a regular lateral separation between myofilaments, so that a lattice pattern is produced (Figure 152). As the pH decreases in meat post-mortem, the repulsion is reduced, the myofilaments move closer together, and water is lost from the myofilament lattice. The fluid moves through the sarcoplasm and may accumulate between myofibres and between fasciculi. When the meat is cut, the fluid is lost (sometimes at great economic cost).

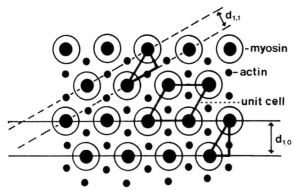

Figure 152. A diagrammatic plan of a transverse section through a very small part of the myofilament lattice showing x-ray diffraction designations ($d_{1,1}$ and $d_{1,0}$).

The preparation of meat samples for transmission electron microscopy involves embedding and heat curing in resin. This causes the myofilament lattice to shrink, so that measurements of filament spacing are too low. X-ray diffraction is a better method. An x-ray beam is directed through a thin (0.5 to 1 mm) sample of meat and a diffraction pattern perpendicular to the x-ray beam and the sample axis is recorded on film. The two main diffraction lines, designated $d_{1,0}$ and $d_{1,1}$, originate from the alignment of myofilaments as shown in Figure 152. With a known distance between the sample and the plane in which the diffraction pattern was obtained, Bragg's equation is used to find the lateral separation of myofilaments. The main problem with the method is that, since the sample must be maintained in a buffer solution as it is being measured, the pH of the buffer has a major effect on myofilament spacing. Thus, the pH must be matched to the pH anticipated in the meat if the sample had not been removed. Shorter exposure times using high intensity x-rays may solve this problem.

(Source: Diesbourg *et al.*, 1988)

MYOGLOBIN

In a healthy live animal, myoglobin is an iron-containing red pigment dissolved in the → sarcoplasm between the → myofibrils within skeletal and cardiac myofibres. It can hold or help transport oxygen brought to the surfaces of myofibres by capillaries. Myoglobin concentration is high in myofibres that store or transport large amounts of oxygen (as in fatigue-resistant myofibres using blood-borne sources of energy such

as fatty acids), but myoglobin concentration is low where myofibres mainly use stored energy (as in fast-contracting myofibres using internally stored glycogen granules without an immediate need for oxygen from the blood). This accounts for the great range in the redness of meat, from dark red beef to white chicken breast. Myoglobin may colour any fluids released from myofibres, but may have been diluted with water from within the → myofilament lattice. Thus, meat juice may be pale yellow-pink in colour. The colour of myoglobin changes with its state. It is very dark and almost purple without oxygen (as in the interior of a large chunk of beef at the instant it is first cut). It is bright red when oxygenated (oxymyoglobin - as in beef presented for sale to the general public). It is brown when oxidised (metmyoglobin - as in well-aged beef for the connoisseur), and it may be bright pink after curing the meat with nitrite (dinitrosylhaemochrome - as in ham). Oxygenated means the myoglobin is merely carrying the oxygen, while oxidised means the myoglobin has been chemically altered by the oxygen.

MYOMERE

About half the total body mass of many fishes is composed of skeletal muscle, depending on the locomotory ability of the species. Very active fishes such as tunas have nearly 70% muscle. In most fishes, the major locomotory muscle mass is the axial musculature, while the muscles of the paired fins control direction and stabilization. However, in species such as the skates and rays, the fin musculature has a dominant role in propulsion. A myomere is a gross, structural segment of the axial musculature of a fish (Figure 153).

Figure 153. An isolated myomere from a teleost. The head of the fish is to the left, the tail to the right.

Fishes swim by a series of rhythmic lateral undulations, thrusting the tail against the resistance of the water behind them. Movement originates from myomere contraction in waves along each side of the fish, with left and right sides being out of phase. The number of myomeres is matched to the number of vertebrae, and the segmental nature of both the vertebrae and the myomeres is derived from the embryonic arrangement of blocks of somitic mesoderm. Superficially, each myomere appears to have the shape of a letter W on its side, but the internal structure is more complex. The shape of the myomere is derived embryologically from an elaboration of a basic V shape, still found in primitive chordates such as amphioxus. Within each myomere, the myofibres tend to be arranged in an anterior to posterior direction. Although the most superficial myofibres may appear to be parallel to the long axis of the fish, the deeper myofibres are usually arranged in a more complex manner so that they subtend acute angles with respect to the long axis.

On each face of the myomere, the myofibres are anchored in connective tissue. The partitions or septa of connective tissue that separate the adjacent myomeres are called the myocommata (singular = myocomma). The myocommata extend inwards (medially) into the body of the fish and are attached to the vertebral column. The vertebral column provides a laterally flexible but incompressible axis that converts dominantly antero-posterior myofibre contraction to lateral movement. The ribs are situated within these segmental myocommata and, in some fishes, there may be many extra rib-like intermuscular bones. A major function of the ribs is to prevent excessive distortion of the shape of the myomere during contraction. The amount of myocommata collagen is related to the way in which the fish swims. Fish with greater flexibility tend to have more collagen.

The zigzag pattern of the myomeres and myocommata enhances the efficiency of propulsion, and there is naturally some variation between different types of fish in relation to their mode of swimming. The myomere is a three-dimensional structure and, in its depth (into the body of the fish), it has a further dimension of geometrical complexity. In horizontal section, the myomere may have a parallelogram shape, but often the shape is more complex and curvilinear with the most acute angle usually being the deepest and most anterior (that is, at the center top of the "W" shape). In many fishes the points of the W-shaped myomeres are almost conical so that there are three anterior cones and two posterior cones.

A further complexity is created by the horizontal septum of connective tissue that often divides the body of the fish into dorsal and ventral parts. This usually occurs just below the tip of the forward-facing V of the myomere. The dorsal ribs (uppermost set of ribs) occur at the level of the horizontal septum. The muscle mass above the horizontal septum is called the epaxial musculature while that located below this level is called the hypaxial musculature. An important commercial problem, gaping, occurs if the myomeres becomes separated.

(Sources: Alexander, 1969; Thiemig and Oelker, 1998)

MYOMESIN

A protein in the → M-line.

MYOSIN

Myosin is a protein with two globular heads at one end of a long rod (Figure 154). The rod-like parts of myosin molecules are bound together to form a thick → myofilament so that

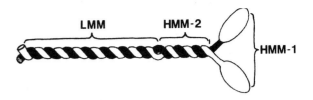

Figure 154. *A myosin molecule composed of light meromyosin (LMM) and two fractions of heavy meromyosin (HMM).*

the myosin molecule heads protrude along the length of the myofilament (Figure 155). There are several hundred myosin molecules along a thick myofilament (the length of a thick myofilament is about 1.5 µm). Myosin accounts for about 55 to 60% of the protein content of meat. The heads of myosin molecules are the motors that cause thick myofilaments to slide past thin myofilaments when a living muscle contracts. In meat (which, of course, is no longer contractile), many of the myosin molecule heads are locked onto the thin myofilaments.

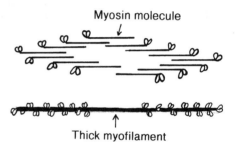

Figure 155. *The general idea of how individual myosin molecules (top) can be packed together by their rods to make a thick myofilament (bottom).*

Myosin is composed of two heavy chains and four light chains. Two of the light chains have a regulatory function and the other two are essential for ATPase activity. A digestive enzyme, trypsin, may be used experimentally to split the myosin molecule into two large fragments - light meromyosin (LMM, much of the rod-like part of the molecule) and heavy meromyosin (HMM, the rest of the rod plus the two heads). When light meromyosins are placed together under conditions similar to those inside a myofibre, they self-assemble into a bundle that resembles the backbone of a thick myofilament. Another enzyme, papain, may be used experimentally to split heavy meromyosin into two subfragments. Subfragment HMM-1 of heavy meromyosin is composed of the two active heads of the myosin molecule, and these retain their ability to bind to actin and hydrolyse ATP. The other subfragment (HMM-2) is the rest of the rod. Papain is an enzyme obtained from tropical melons and may be used commercially to tenderize meat.

The exact way in which myosin molecules are arranged in a thick myofilament is difficult to determine, but involves a threefold radial symmetry with nine subfilaments at any plane of sectioning. Half of the myosin molecules in a thick myofilament face towards one Z-line of their sarcomere, while the remainder face towards the opposite Z-line. Although the heads protrude at regular intervals along the thick myofilament, there is a short headless zone at the midlength of the thick myofilament where the light meromyosins of opposite sides overlap by their tails.

(Sources: Ebashi and Nonomura, 1973; Ashton *et al.*, 1992).

MYOTENDON JUNCTION

The myotendon junction of a muscle is where it connects to a tendon. Sometimes this is at an easily seen point, but often it is spread irregularly over a sheet of connective tissue. The term is also used for the junction between a → myofibre and one or more collagen fibres.

At the end of a myofibre, at its myotendon junction, the plasma membrane is highly infolded to allow all the → myofibrils within the myofibre to be attached. Mechanical forces are transmitted through the → basement membrane around the myofibre by junctional microfibrils. Vinculin is abundant at myotendinous junctions. Meta-vinculin is similar to vinculin in distribution but has a slightly higher molecular weight. Regions where vinculin is concentrated on the inner face of the plasma membrane are matched by extra → endomysial fibres on the outside of the myofibre.

MYOTUBE

Early in prenatal development, → myoblasts fuse together. Eventually they will form new → myofibres but, at an intermediate stage of development, the nuclei are located within a hollow tube composed of myofibrils. Maturation of a myotube to become a myofibre occurs when the nuclei move to their final position, mostly under the plasma membrane in mammals.

NABORST

The posterior part of a brisket on a → Netherlands beef carcass.

NACHBRUST

The Nachbrust or after breast is a → German or → Swiss beef cut including the costal cartilages.

NACHE

A rump of beef, which gave its name to the → aitch bone.

NACKEN

In → German, this refers to the nape of the neck, but also it is a specific cut of pork from that region.

NAGE

Alternate spelling of → nache.

NAKKE

In → Norway, the neck of a beef, pork, lamb or veal carcass.

NAKKEKAM

→ Danish for the crest or nape of the neck, and used for a cut of pork from that region.

NALGA

The buttock - a lateral cut from the hindlimb of an → Argentinean beef carcass.

NANDU

South American name for the → rhea.

NAUTAKJÖT

Beef in → Iceland.

NAUTARIF

In → Iceland, a cut of beef composed of a ladder of mid-rib sections.

NEBULIN

Nebulin is an insoluble protein comprising about 3% of the total → myofibrillar protein. It has a high molecular weight (around 800 kD) and binds to → actin at multiple sites so that the thin → myofilament seen by electron microscopy now may be regarded as a composite filament containing not only actin, tropomyosin and troponin, but nebulin as well. Nebulin may regulate the length of the thin filament as it assembles. Nebulin is rapidly degraded by proteolysis post-mortem, particularly by calcium-activated protease.

(Sources: Labeit *et al.*, 1991; Wright *et al.*, 1993)

NECK OF LAMB

Neck of lamb (Figure 156) typically is sliced or chopped for stewing. The muscle grouping is quite complex and changes along the neck.

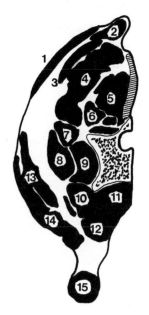

Figure 156. Neck of lamb, a slice taken through the fourth cervical vertebra with some landmark muscles; trapezius (1), rhomboideus (2), splenius (3), semispinalis capitis (4), spinales dorsi (5), multifidus dorsi (6), longus capitis (7), longissimus atlantis (8), longissimus cervicis (9), intertransversarius longus (10), longus colli (11), longus capitis (12), omotransversarius (13), brachiocephalicus (14) and sternocephalicus (15). The ligamentum nuchae is shown by cross hatching.

NEFRAMIA

The lumbar region of a → Greek lamb carcass.

NEK

The neck of a lamb carcass in the → Netherlands or of a → South African beef carcass.

NERKÓWKA

Relating to the kidneys, the cut of this name is the lumbar region of a → Polish veal carcass.

NETHERLANDS - BEEF CUTS

As shown in Figure 157, in the hindquarter (de runderachtervoet), the muscles of the round adjacent to the femur are divided into a medial bovenbil (*semimembranosus*, *adductor* and *pectineus*), a lateral platte bil (*semitendinosus* and *biceps femoris*) and an anterior spierstuk (*quadriceps femoris* group of muscles). Removal of major muscles creates cuts such as the vast deel (*semimembranosus* plus *adductor*), staarstuck (*biceps femoris*) and achtermuis (*semitendinosus*). The dikke lende or thick loin is dominated by the gluteal muscles, while the *tensor fascia lata* is the liesstuk

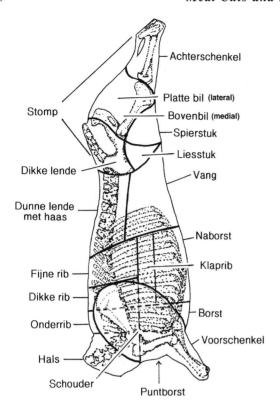

Figure 157. Netherlands – beef cuts.

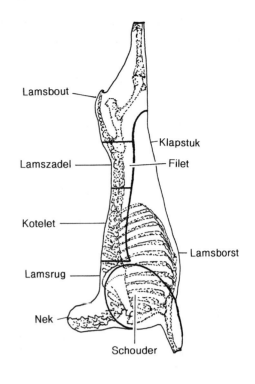

Figure 158. Netherlands – lamb cuts.

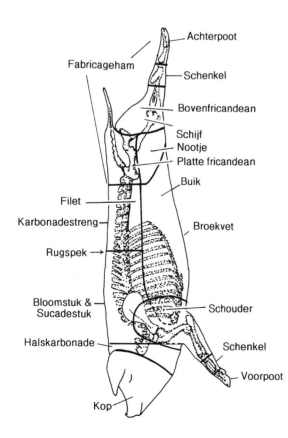

Figure 159. Netherlands –pork cuts.

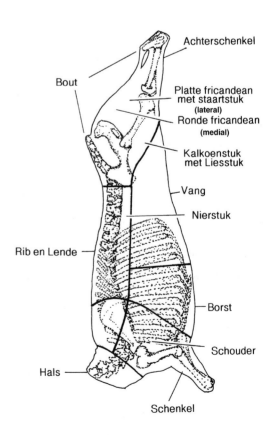

Figure 160. Netherlands – veal cuts.

(alternatively called the ezeltje). The remaining cut of the primal hip or stomp is the hindshank or achterschenkel. The flank is called the vang.

Along the vertebral axis from hindquarter to forequarter (de rundervoorvoet), are located the thin loin (dunne lende met haas), prime rib (fijne rib), thick rib (dikke rib), under rib (onderrib) and neck (hals). The shoulder (schouder) is removed and boned, with the major muscles becoming the bloemstuk (*triceps brachii*), sucadestuk (*infraspinatus*), schoudermuis (*supraspinatus*), schouderlap (*subscapularis*) and boeglapje (*biceps brachii*). The borst or breast is divided into two parts, from anterior to posterior, the puntborst and naborst, to leave a ladder of ribs, the klaprib. The foreshank is the voorschenkel.

(Sources: SVO; Toorop, 1998).

NETHERLANDS - LAMB CUTS

The lamsbout includes the leg up to the shaft of the ilium (Figure 158). The vertebral column from posterior to anterior includes the lamszadel, kotelet, lamsrug and nek. The thoracic part of the breast is called the lamsborst while the posterior part is the klapstuk.

(Sources: Schön, 1958; SVO; Toorop, 1998).

NETHERLANDS - PORK CUTS

There appear to be many ways to divide the whole manufacturing ham or fabricageham into retails cuts. Schön (1958) indicated separation into the hind trotter (achterpoot), shank (schenkel), a medial bovenfricandean, a lateral schijf, the nootje in the *quadriceps femoris* region, and the platte fricandean containing sacrum and pelvis (Figure 159).

Removal of the back-fat (rugspek) gives a short loin or karbonadestreng (a karbonade is a chop) plus filet. The belly is the buik, but the ventral region is termed the broekvet, which has a fascinating literal translation as "greasy knickers". Removal of the shoulder (schouder), shank (schenkel) and front trotter (voorpoot), leaves the dorsal thoracic region containing cuts such as the bloomstuk and sucadestuk. The halskarbonade is the neck chop, and the head is the kop.

(Sources: Schön, 1958; SVO; Toorop, 1998).

NETHERLANDS - VEAL CUTS

The bout or leg of veal is divided into the achterschenkel or shank, a medial ronde fricandean, and a lateral platte fricandean, with the *quadriceps femoris* group contributing to the kalkoenstuk met liesstuk (Figure 160). The filet or nierstuk is removed from the rib en lende (rib and loin), leaving the abdominal muscles as the vang. The schouder composed of forelimb skeleton and muscles is removed to leave the underlying hals (neck) and breast (borst).

(Sources: Schön, 1958; SVO; Toorop, 1998).

NEUROMUSCULAR JUNCTION

The point at which a → myofibre is innervated by a terminal axon derived from extensive branching of the main axon of a motor neuron. If the neuromuscular junction is compact it may be called a motor end plate.

NEW YORK ROUND

A primal cut of beef in North America, essentially a hip of beef minus its sirloin tip (Figure 161).

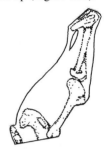

Figure 161. New York round of beef.

NEW YORK SHOULDER OF PORK

In Canada, a shoulder of pork may be called a New York shoulder or épaule, genre New York.

NEW ZEALAND - BEEF CUTS

Beef cutting in New Zealand (Figure 162) shares some similarities to that in → Australia, although the hindquarter is much shorter, with the separation at the last rib. As in Australia, the rump and sirloin both refer to the same region of meat cut through the ilium, like a North American sirloin Figure 162 shows a rump D cut, but if it extended farther ventrally it could have been labelled as a sirloin butt or a long-cut rump. Alternatively, the rump D cut may be left with the loin attached as a rump and loin.

O.P. (oven prepared) ribs are prepared from a 12-rib forequarter, cutting between ribs 5 and 6. The shoulder clod is variable in shape, either rounded or square-cut, and includes both scapula and humerus before boning. The major muscle is *triceps brachii*. The New Zealand shoulder clod of beef resembles a Canadian shoulder clod of veal (pointe d'épaule) which is a boned and rolled joint prepared from the large shoulder muscles after removal of the scapula. The crop shown in Figure 162 can be extended more anteriorly into the chuck. Some cuts not used for primals, such as thin flank, are used for boxed beef.

(Source: NZMPB, 1985)

NEW ZEALAND - LAMB CUTS

The leg may be cut with the chump on (long cut) or off (short cut), or the chump may be left on the loin (long loin) (Figure 163). The rack (also known as best end neck, as in England)

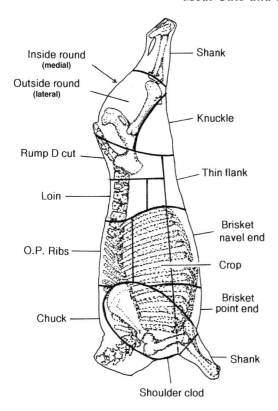

Figure 162. New Zealand – beef cuts.

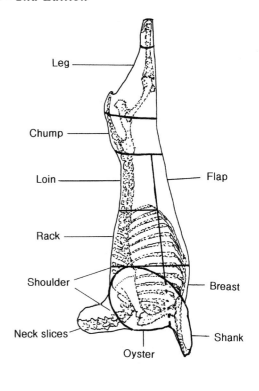

Figure 163. New Zealand – lamb cuts.

typically contains seven ribs, with the flap removed. The shoulder may be square cut or a rounded oyster cut. A saddle is formed by the chump and loins from both sides of an unsplit carcass.

(Source: NZMPB, 1985)

NIAGARA STEAKS

North American retail broiling steaks cut from the primal chuck or rib, originating near the dorsal part of the scapula (removed when boneless).

NIERVET

Renal fat in the Netherlands.

NIGERIA – BEEF CUTS

The forequarter is separated from the hindquarter leaving two ribs on the hindquarter. Only three small cuts are taken – thin flank, brisket and neck (Figure 164). Local names in the three main languages are as follows.

English	Yoruba	Ibo	Hausa
Hindquarter	Itan	Apata Ukwu	Chinya
Forequarter	Apa	Aka	Kafargaba
Thin Flank	Abonu	Afor	Chiki
Brisket	Ige	Obi	Kirji
Neck	Orun	Olue	Wuya

(Source: Abiola, 2003)

NIGERIA – LAMB CUTS

The cutting pattern and local names are the same as those for beef (Figure 165).

(Source: Abiola, 2003)

NIGERIA – PORK CUTS

The cutting pattern and local names are the same as those for beef (Figure 166). The head is called the ori in Yoruba, the isi in Ibo, and the kai in Hausa.

(Source: Abiola, 2003)

NINE HOLES

A → Scottish cut of beef in the postero-ventral region of the ribcage.

NISKA

A neck of beef in → Finland.

NITRITE

Sodium nitrite is used in meat → curing to inhibit the growth and division of bacteria such as *Clostridium perfringens* and *C. botulinum*. *C.perfringens* is a common cause of food poisoning. *C. botulinum* causes a rare but often fatal food poisoning called botulism. Uncontrolled bacterial production of nitrite from nitrate is undesirable, so that sodium nitrate is excluded from meat products and sodium nitrite is added directly in small quantities (typically < 125 parts per million).

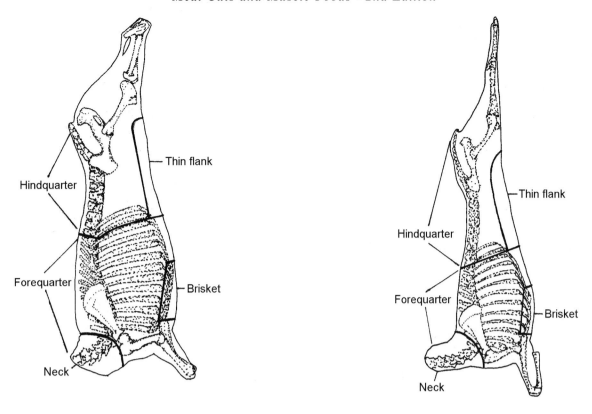

Figure 164. *Beef cuts in Nigeria.*

Figure 165. *Lamb cuts in Nigeria.*

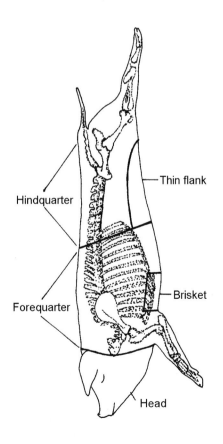

Figure 166. *Nigeria – pork cuts.*

Nitrite levels are tightly regulated because nitrite contributes to the formation of carcinogenic nitrosamines by adding a nitroso group (-N=0) to nitrogen atoms present in the structure of the meat. Nitrite in cured meat converts myoglobin to a heat-stable bright pink pigment often called dinitrosylhaemochrome.

NÍZKA ROŠTENKA

Lumbar and posterior rib region of a beef carcass in → Slovakia.

N-LINE

The N-line across → myofibrils was discovered by light microscopy in the 1880s, but it is difficult to see in electron micrographs. Quite likely, the old techniques used to visualise the N-line caused the deposition of silver on the N-line, thus making it easier to see. With the light microscope, two N-lines may be seen in each → sarcomere. One N-line runs across the → I-band at each end of the sarcomere, so that the N-line is parallel to the Z-line. When seen in electron micrographs, the N-line is about 100 to 150 nm wide and appears to be formed by lines of electron-dense thickenings along the thin myofilaments. Each N-line may be subdivided into two smaller parallel lines termed N_1 and N_2 lines. The N_1-line is narrow (about 50 nm) and is located at a fixed distance of about 100 to 200 nm from the Z-line. The N_2-line is wide (up to 150 nm) and exhibits a very curious change in position when the sarcomere is stretched. As the I-band gets wider, the N_2-line maintains its position along the I-band relative to both the Z-line and the edge of the A-band. When the sarcomere is stretched, fluorescent antibodies for nebulin follow the N_2-line as it moves along the thin myofilaments to maintain its relative position in the I-band. The way the N-line moves when the sarcomere is stretched indicates that the N_2-line is connected to elastic titin filaments in the I-band.

(Sources: Wang and Williamson, 1980; Locker and Wild, 1984)

NOCE

An → Italian beef cut from the anterior of the hindlimb based on the *quadriceps femoris* group of muscles.

NOGI

The trotters of a → Polish pork carcass.

NOIX

A small cut of beef around the patella in → Luxembourg.

NOIX DE PORC

In → Belgium, this may be the pointe au filet, a cut of pork containing sacral and posterior lumbar vertebrae.

NOIX PLANNURE

The pièce levée of a → Belgian beef carcass.

NOOTJE

A cut from the *quadriceps femoris* region of a → Netherlands pork carcass.

NORTH AMERICA - BEEF CUTS

Canada generally follows the same pattern of meat cutting as in the USA but, being a bilingual country, has extra French names for meat cuts (given here in parentheses). Separation of the forequarter from the hindquarter leaves only the last of the 13 ribs on the hindquarter (Figure 167). On a hanging side, the separation is started seven vertebral centra down from the sacral-lumbar junction plus almost the length of a half a centrum. A saw-cut is made perpendicularly through the vertebral column at this point, then continued with a knife cut separating the forequarter from the hindquarter by cutting through the intercostal and abdominal muscles, following the curvature of the twelfth rib. The forequarter may be dropped onto a table or held suspended by its own hook from a hoist.

In the USA, the chuck (French, block) is separated from the primal rib with a perpendicular cut through the vertebral column and the intercostal muscles between ribs 5 and 6. Thus, a chuck containing five ribs is seen in most meat cutting charts (such as ICMIST, 1973; and Fabbricante and Sultan, 1978). However, at least in Canada, it is not difficult to find butchers who are cutting four-rib chucks to gain on the rib roasts. The rib (côte) is separated from the plate (poitrine) by an anterior to posterior cut. The chuck is separated from the brisket (pointe de poitrine) by a cut perpendicular to the ribs. If the shank (jarret avant) is sliced, the shank knuckle slices are proximal.

Excess fat is trimmed from the hindquarter, around the pubis and over the posterior part of the abdominal muscles. Anterior to the *rectus femoris*, at a point where the *tensor fascia lata* reaches its most distal extent, a separation is started to end on rib 12, about 20 cm from the vertebral column. This detaches the flank (flanc), from the primal hip, sirloin and loin. The round (ronde) is separated from the rump (croupe) about 1 cm distal to the ischium, terminating the separation just after passing through the head of the femur. The rump is separated from the sirloin (surlonge) from between sacral vertebrae 4 and 5 to a point just ventral to the acetabulum of the pelvis. The sirloin is separated from the short loin (longe) with a cut perpendicular to the vertebral column, passing between lumbar vertebrae 5 and 6.

The rib cut is separated into rib steaks or standing rib roasts by cutting perpendicularly to the vertebral column. Rib-eye (faux-filet) or delmonico steaks contain sections of *spinalis dorsi* as well as *longissimus dorsi* muscles. The chuck is sliced in planes parallel to rib 5 to make blade (palette) steaks or blade pot roasts. Arm steaks, arm pot roasts or cross cut ribs

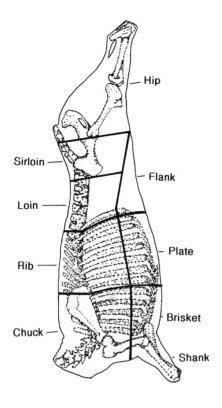

Figure 167. *North America – beef cuts.*

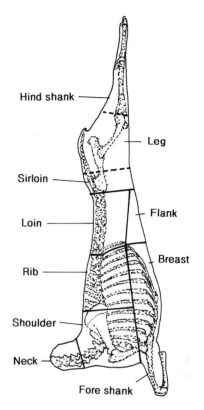

Figure 168. *North America – lamb cuts.*

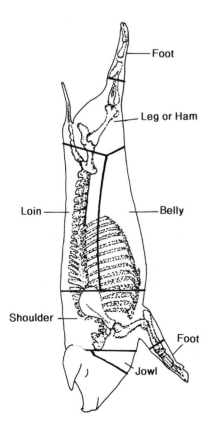

Figure 169. *North America - pork cuts.*

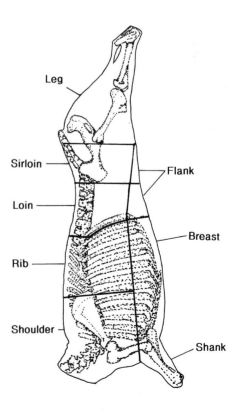

Figure 170. *North America – veal cuts.*

(côte croisées) are sliced perpendicularly to the humerus. Brisket is sold in chunks for braising or moist cooking. The shank is cut into thick slices perpendicular to the radius and ulna. The plate may be divided into cubes of rib bone and muscle to give short ribs (bout de côtes). Muscles located ventro-laterally to the rib cage are usually rolled, tied, and cut into cylindrical cuts of plate. Abdominal muscles may be isolated from the flank as flank steaks.

The short loin steaks are sliced perpendicularly to the vertebral column. The most anterior steaks are the wing or club steaks (côte d'aloyau), and nearly all their meat is derived from the *longissimus dorsi*. Next are the T-bone steaks (aloyau) which gain extra meat from the *psoas major* towards the posterior of the loin. The last steaks are two or three porterhouse steaks with large sections of *longissimus dorsi* and *psoas major*. In the porterhouse region at the posterior end of the short loin, the vertebrae may be removed to create New York strip-loin steaks (*longissimus dorsi*) and tenderloin or filet steaks (*psoas major* and *minor*).

Sirloin steaks cut perpendicularly to the shaft of the ilium are named by the shape of the sectioned ilium from anterior to posterior: (1), pin bone sirloin steaks with an oval section of the anterior projection of the ilium; (2), flat bone or double bone sirloin steaks with a flat section of the wing of the ilium where it joins with the wing of the sacrum; (3), round bone sirloin steaks with a round section of the slender shaft of the ilium; and (4), wedge bone sirloin steaks with a triangular cross section of the ilium near to the acetabulum. The rump is the ischial region in North America and this may be sliced into rump steaks or used for roasts such as standing rump or boneless rump. Farther down the hindlimb, the round may be used for full cut round steaks perpendicular to the femur, or cut into chunks parallel to the femur to create the inside or top round (mostly *semimembranosus* and *adductor*) and the outside or bottom round (mostly *semitendinosus* and *biceps femoris*). Sometimes *semitendinosus* is removed and sliced as eye of the round. The sirloin tip (pointe de surlonge) from the round includes the muscles which pull on the patella.

(Sources: Reynolds, 1963; NLMB, 1973; Swift Canadian, 1973; Agriculture Canada, 1974; Romans and Ziegler, 1974; Fabricantte and Sultan, 1978; Canadian Meat Council, 1980, 1988)

NORTH AMERICA - LAMB CUTS

The main North American cuts of lamb (l'agneau in Quebec) are shown in Figure 168. The whole pin bone leg composed of sirloin (gigot surlonge) plus leg (gigot jarret) is removed by cutting perpendicularly through the vertebral column at a point level with the anterior face of the ilium. If the anterior part of the ilium remains in the posterior loin chump, this produces a chump chop with a very high bone content, even though both anterior and posterior faces look quite meaty.

In the lamb carcass, the loin includes part of the abdominal wall. The loin (longe) is removed by a cut that passes between ribs 12 and 13 and which then continues perpendicularly

through the vertebral column. Usually the whole breast (poitrine) and the shank (jarret) are removed with a single cut from the anterior of the sternum to the ventral part of rib 12. Alternatively, the dominant cut may be made between ribs 5 and 6, to separate the rib from the shoulder, and to divide the breast into anterior and posterior sections. The rib is often called the rack (carré). In Figure 168, note how the metacarpal cannon bone is fixed back.

The leg may be divided a number of ways, either into leg chops or steaks cut perpendicularly to the femur, or into large or small roasting cuts. Like many other decisions made by the butcher, seasonal preferences are taken into account. Steaks and chops are popular in the summer while large roasts are more popular in the winter. Similarly, the sirloin either may be cut into sirloin chops or left as a roast. The flap of abdominal muscle on the loin is removed, and is added to the breast meat. The loin is sliced into loin chops or left whole as a roast. The rib or rack of lamb may be subdivided into rib chops, or left whole as a rib roast. The rack makes an excellent → crown roast when the vertebral column is trimmed and bent back on itself.

There are a number of ways in which to divide the shoulder. It may be made into blade chops, or left largely intact as a square shoulder roast. Parts of the shoulder may be boned and rolled to make saratoga chops. The neck usually is sliced perpendicularly to the vertebral column. The foreshank is removed intact, and the remaining breast is subdivided in an arbitrary manner. Much of the fat on the breast may be removed, and the remaining lean may be rolled or cut into riblets conforming to local preferences. Sometimes left and right sides of the lamb carcass are kept together, and the major cuts are made crosswise to produce a foresaddle and a hindsaddle.

(Sources: Reynolds, 1963; NLMB, 1973; Swift Canadian, 1973; Agriculture Canada, 1974; Romans and Ziegler, 1974; Fabricantte and Sultan, 1975; Canadian Meat Council, 1980, 1988)

NORTH AMERICA - PORK CUTS

In preparing the primal cuts of the pork carcass, the hindfoot (pied arrière in Quebec) is removed by cutting through the tuber calcis (Figure 169). Then the front foot is removed just distally to the ulna and radius. The leg of pork (jambon) is removed with a cut from between sacral vertebrae 2 and 3 towards the *tensor fascia lata*, and may be separated into a distal shank end (jambonneau) and proximal butt end (jambon croupe). Alternatively, the line of cutting may change direction so that most of the *tensor fascia lata* remains on the leg. The shoulder butt (soc) and picnic (épaule picnic) are removed together as a shoulder, cutting perpendicularly to the vertebral column and starting between thoracic vertebrae 2 and 3. The butt is separated from the picnic, skimming past the ventral region of the cervical vertebrae at a tangent. This keeps the top of the picnic relatively square. The jowl (bajoue) is removed from the picnic with a cut following the crease lines in the skin. The remainder of the side of pork is split into the loin (longe) and belly (flanc) with a cut following the curvature of the vertebral column. One end of the curve is just ventral to the ilium, the other end is just ventral to the blade of the scapula.

The loin is divided into chops: from anterior to posterior, rib chops, center loin chops, then tenderloin chops. Alternatively, the thoracic, lumbar and iliac regions may be left intact as large roasts: rib end roast, center loin roast and tenderloin end roast. The *psoas* muscles may be removed from the lumbar region as pork tenderloin, and the *longissimus dorsi* and adjacent small muscles may be removed to make a boned and rolled loin roast. A crown roast can be made by twisting thoracic vertebral column into a circle with ribs radiating outwards like the points of a crown. This facilitates rapid carving and distribution of portions.

In the USA, the *longissimus dorsi* may be cured and smoked to make Canadian Style bacon. In Canada, it used for peameal bacon and back bacon. The rib cage and its immediately adjacent muscles are removed from the belly to make spareribs. The remaining muscles of the abdomen, together with those that insert on the ribcage, constitute a side of pork. Side of pork may be cured and smoked to make slab bacon. The picnic may be sliced to make picnic shoulder chops through the humerus, or it can be partly subdivided to make picnic shoulder roasts. Picnic shoulder roasts may be boned and rolled, or smoked and cured in a variety of ways. The butt, or Boston butt, is usually divided into a number of blade steaks cutting through the scapula from dorsal to ventral. The more anterior part then forms a butt roast. The leg may be subdivided to create, from proximal to distal, the butt end roast and the shank end roast. Alternatively, the leg may be cured and smoked to make ham. The feet, the hocks, the knuckles and the tail can be baked or cooked in liquid.

(Sources: Reynolds, 1963; NLMB, 1973; Swift Canadian, 1973; Agriculture Canada, 1974; Romans and Ziegler, 1974; Canadian Pork Council, 1975; Fabricantte and Sultan, 1975; Canadian Meat Council, 1980, 1988)

NORTH AMERICA - VEAL CUTS

Veal carcasses are smaller than beef carcasses, and a simplified form of cutting like that of lamb may be used (Figure 170). Thus, in North America, a small veal carcass might be cut into foresaddles and hindsaddles, like lamb, while a larger veal carcass might be cut like a miniature beef carcass.

The leg (cuisseau) may be cut into whole round steaks or roasts, plus a heel roast. Select veal cutlets (escalopes choisies) may prepared by flattening any of the largest leg muscles, which implies that the less expensive (delicated, escalopes attendries) cutlets have been produced by folding and knitting of lesser muscles whose connective tissues have been removed. The sirloin may be cut into steaks or used as a whole roast, as may the loin (with or without kidney). The rib or rack (carré) may be used as a whole roast, cut into

steaks, or trimmed to make a crown roast like that of lamb. The breast and shank may treated the same way as that of lamb. The shoulder (épaule) may be used as a whole roast, cut into shoulder steaks, or boned and rolled.

(Sources: Reynolds, 1963; NLMB, 1973; Swift Canadian, 1973; Agriculture Canada, 1974; Romans and Ziegler, 1974; Fabricantte and Sultan, 1975; Canadian Meat Council, 1980, 1988)

NORTHERN IRELAND - BEEF CUTS

The pattern documented by Pomeroy and Williams (1976) and shown in Figure 171 contains a number of interesting features. Naming of several of the cuts, such as the ventral mid-rib region as the plate, follows nearby terminology in Northeast → England and → Ireland, but including the ilium and associated muscles in a primal cut called the chump is somewhat unusual because this name is more typically used for lamb, pork or veal. The *quadriceps femoris* group was not named by Pomeroy and Williams (1976).

Current specifications given by the Livestock & Meat Commission differ considerably from the above, although the last three ribs remain on the hindquarter as previously (Figure 172). The hindquarter is broken into topside, silverside, thick flank and rump in a typical British pattern. But the chuck, clod and bladeneck is more unique. The salmon cut is the *semitendinosus* from the silverside, with the remainder and bulk of the *biceps femoris* being the silverside flat. *Gastrocnemius* is removed as the heel muscle, from which the *flexor digitorum superficialis* may be removed as the heel muscle core. The fillet is dominated by *psoas major* with the *iliacus* being termed the chain (fillet, chain on, includes *psoas major* plus *iliacus*). French terminology is adopted for the → bavettes or flank steaks. The rib cut may include four or five ribs. The fingerbone rib is an → entrecote steak. Around the scapula, the inside feather blade is *subscapularis*, the outside feather blade is *infraspinatus*, and the chuck tender is *supraspinatus*.

(Source: Pomeroy and Williams, 1976; LMC, 2003)

NORTHERN IRELAND – LAMB CUTS

There is considerable variation in the leg of lamb, which may be a leg pair from an unsplit carcass, or unilateral with or without chump and/or shank (Figure 173). Likewise the loin may be a single side or a bilateral saddle. The loin may be separated into an anterior best end with ribs and a posterior short loin with lumbar vertebrae.

(Source: LMC, 2003)

NORWAY - BEEF CUTS

As shown in Figure 174, the major division of the beef round in Norway is into a lateral banquet cut or Bankekjøtt containing *biceps femoris* and *semitendinosus*, and a medial flat-beef cut or flatbiff containing *pectineus, adductor* and

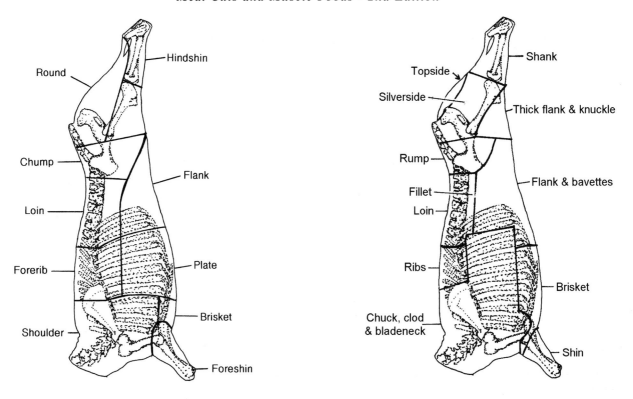

Figure 171. Northern Ireland - beef cuts.   *Figure 172.* Northern Ireland – LMC beef cuts.

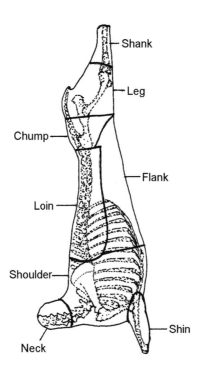

Figure 173. Northern Ireland – LMC lamb cuts.

semimembranosus. However, transverse slices through the round also may be taken, from distal to proximal: lårstyyke, lårskive, and issebenstek. Both methods culminate in an intermediate sirloin or mellommørbrad through the shaft of the ilium and a sirloin or mørbrad containing the blade of the ilium. The lumbar region or filet-kam may be separated into an outer filet of *longissimus dorsi* (ytrefilet) and an inner filet of *psoas* muscles (indrefilet). The prime ribs of the posterior thoracic region (back or kam) may be separated into a posterior kotelett-kam and an anterior entrecote-kam. The høyrygg or high backbone contains the anterior thoracic vertebrae. The shoulder region is sliced transversely into two or three shoulder (bog) slices, with or without the scapulo-humeral ball joint (bog m/kuleben and u/kuleben, respectively). Apart from the neck (nakke), foreshank (for-knoke) and hindshank (bak-knoke), the remainder of the carcass is divided into a brisket (bryst), plate (bibringe) and flank (slagside).

(Sources: OEEC, 1961; Braathen, 1996; Opplysningskontoret for kyøtt)

NORWAY - LAMB CUTS

The main roast is the stek or leg, which may be cut long or short, with or without the mørbrad (Figure 175). The kidney piece or nyrestykke is subdivided into a loin (kotelett-kam) and flank (side). The shoulder or bog is removed to contain the scapula, humerus, ulna and radius, leaving the anterior ribs, and cervical and anterior thoracic vertebrae as a breast with neck (bryst m/nakke).

(Sources: OEEC, 1961; Braathen, 1996; Opplysningskontoret for kyøtt)

NORWAY LOBSTER

A clawed → lobster, *Nephrops norvegicus.*

NORWAY - PORK CUTS

The kotelett-kam, midtkoteletter or back cutlets of a Norwegian pork carcass is from rib 5 to the anterior face of the ilium, leaving the ventral parts of posterior ribs in the flatribbe (Figure 176). The buklist, ister or belly may be left on the flatribbe to form a sideflesk. But instead of the elongate antero-posterior kotelett-kam and sideflesk, the carcass also may be divided transversely to give a deep midtribbe with thoracic vertebrae and the full length of the ribs. The mørbrad or sirloin is taken through the shaft of the ilium, leaving a whole, round ham or skinke, hindshank or bak-knoke, and trotter or bak-labb.

The neck or nakke extends from the atlas back to include the scapula and underlying thoracic vertebrae, and may be subdivided into outer and inner parts (utside and innside, respectively). The bog is the shoulder cut, the fore-knoke is the foreshank, and the for-labb is the trotter.

(Sources: OEEC, 1961; Braathen, 1996; Opplysningskontoret for kyøtt)

NORWAY - VEAL CUTS

Veal cutting (Figure 177) is similar to lamb cutting in Norway, apart from the removal of the foreshank (for-knoke) and hindshank (bak-knoke). The loin is called the kam rather than the kotelett-kam, and the bryst and nakke are separated rather then being left together.

(Sources: OEEC, 1961; Braathen, 1996; Opplysningskontoret for kyøtt)

NÖTKÖTT

Swedish for beef.

NOUA KAI OURA

A cut taken laterally to the femur in → Greek beef.

NUA LOOK MA PRAW

The ischial region of a beef or buffalo carcass in → Thailand.

NUSS

The *quadriceps femoris* muscles of an → Austrian beef, pork, lamb or veal carcass. As in → French cuts called noix, the nuss is German for a nut and probably relates to the patella.

NYAK

The neck of a carcass in → Hungary.

NYRESTYKKE

On a → Norwegian lamb or veal carcass, the kidney piece or nyrestykke is a primal cut though the full dorso-ventral depth of the posterior thoracic and lumbar region.

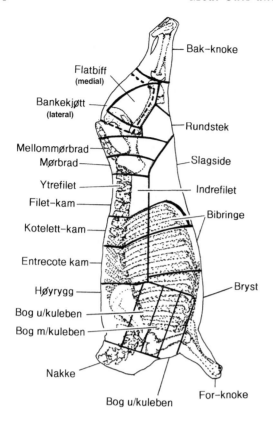

Bak-knoke

Flatbiff
(medial)

Bankekjøtt
(lateral)

Rundstek

Mellommørbrad

Slagside

Mørbrad

Ytrefilet

Indrefilet

Filet-kam

Kotelett-kam

Bibringe

Entrecote kam

Høyrygg

Bryst

Bog u/kuleben

Bog m/kuleben

Nakke

For-knoke

Bog u/kuleben

Figure 174. Norway – beef cuts.

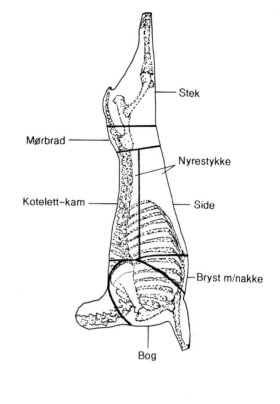

Stek

Mørbrad

Nyrestykke

Kotelett-kam

Side

Bryst m/nakke

Bog

Figure 175. Norway – lamb cuts.

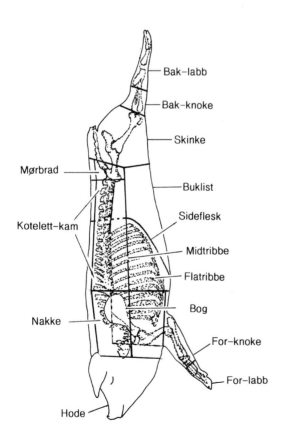

Bak-labb

Bak-knoke

Skinke

Mørbrad

Buklist

Kotelett-kam

Sideflesk

Midtribbe

Flatribbe

Nakke

Bog

For-knoke

For-labb

Hode

Figure 176. Norway – pork cuts.

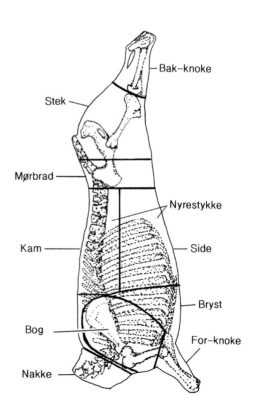

Bak-knoke

Stek

Mørbrad

Nyrestykke

Kam

Side

Bog

Bryst

Nakke

For-knoke

Figure 177. Norway – veal cuts.

OBERSCHALE

In → German beef cutting, the Oberschale or top dish is in the medial part of the round, comparable to an English → topside. A similar cut may be taken from the lateral part of the ham in → German pork cutting.

OBI

Ibo for the brisket in → Nigeria.

OBLIQUELY STRIATED MYOFIBRES

It is important to remember that mammals and birds have highly developed myofibres with transverse striations caused by myofilaments being packed tightly and neatly into sarcomeres for maximum efficiency of contraction. In animals for which rapid locomotion is less important for survival, the arrangement of myofilaments may be far looser and irregular, or may follow a different packing pattern. For example, the tentacles of squid and the adductor muscles of bivalve molluscs may have obliquely striated myofibres, with structures analogous to sarcomeres spiralling along the length of the myofibre.

(Source: Kier, 1985)

OBLIQUUS CAPITIS CAUDALIS

A muscle located dorso-laterally to the atlas and axis in the mammalian neck. It is well developed in beef and lamb, but poorly developed in pork.

OBLIQUUS CAPITIS CRANIALIS

A short, thick muscle located laterally to the atlas and axis in the mammalian neck.

OBLIQUUS EXTERNUS ABDOMINIS

A massive sheet of muscle originating from ribs 5 to 13 and thinning down to form one of the major sheets of connective tissue in the mammalian abdomen.

OBLIQUUS INTERNUS ABDOMINIS

A thick muscle of the mammalian abdomen, stretched between the ilium, lumbar vertebrae and last rib, and connecting with the linea alba.

OBTURATORIUS

A medium-sized, fan-shaped muscle on the floor of the pelvis, passing through the obturator foramen to insert on the trochanteric fossa of the femur (Figure 178). In lamb and pork the *obturatorius* may be divided into internal and external layers (*obturatorius internus* and *externus*).

OCA

Italian for → goose.

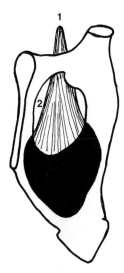

Figure 178. A boned-out beef aitch bone (pubis on the left of the page, and ilium at the bottom) showing how the obturatorius *originates (solid black) from the dorsal or inner face of the pelvis, to insert on the femur (1) after passing through the obturator foramen (2).*

OCEAN PERCH

A spiny bony fish, but not a true perch, → Scorpaeniformes.

OCTOPUS

Eight-armed cephalopod molluscs such as *Octopus vulgaris* in Europe and other species around the world have long been used for culinary purposes. Small, young octopus are preferred. The myosystem is similar to that of → squid.

OGON

→ Polish for tail.

OIE

French for → goose.

OK

In → Thailand, the brisket of a beef or buffalo carcass.

OKSEKØD

Danish for beef.

OLIVES OF BEEF

In England, olives of beef are thin sections of → toprump that have been flattened.

OLU

Ibo for the neck in → Nigeria.

OMOHYOIDEUS

Cervicohyoideus.

OMOTRACHÉLIEU

Omotransversarius.

OMOTRANSVERSARIUS

A strap-like muscle in the mammalian neck, originating from the shoulder fascia and inserting on the atlas (Figure 179). Only in pork is it normally possible to make a convincing separation of this muscle from the overlying *brachio-cephalicus*.

Figure 179. Lateral view of beef omotransversarius (solid black). A tapering insertion on the wing of the atlas has been freed (1), and a general indication of the origin on the shoulder fascia near the acromion process is given (2). In beef, separation of this muscle from the overlying brachiocephalicus *is difficult.*

OMPROSTHIO KOTSI

The foreshank of a carcass in → Greece.

ONDERRIB

A beef cut medial to the scapular blade of a → Netherlands beef carcass.

ONGLET

French for the muscular part of the diaphragm.

OPISTHIO KOTSI

In → Greece, the hindshank of a carcass.

OPOSSUM

The opossum, *Didelphis virginiana*, is the only wild marsupial of North America (its range extends into South America as well). It looks and breeds like a large rat, and is extensively hunted for its winter fur and tasty meat.

O.P. RIBS

Standing for oven prepared ribs of beef in → New Zealand.

ORANGE ROUGHY

The orange roughy, *Hoplostethus atlanticus*, is orange-red in colour with a large bony head, large eyes, and rich deposits of oil, especially in a waxy layer beneath the skin (Figure 180). It is harvested at great depths (800 to 1500 m) where the seawater is cold and dark. It occurs in the Atlantic, Indian and West Pacific Oceans as far as New Zealand, where it is the basis of an important domestic and export fishery. Management of the fishery is vital because the orange roughy only grows very slowly, taking well over a century to reach full size (3.6 kg at 50 cm length). A typical commercial size is 30 to 40 cm in length weighing 0.9 to 1.9 kg. The myosystem is firm, white and delicately flavoured.

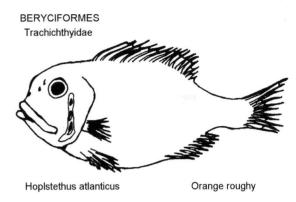

BERYCIFORMES
Trachichthyidae

Hoplstethus atlanticus   Orange roughy

Figure 180. Hoplostethus.

ORI

Yoruba for the head in → Nigeria.

ORIGINAL

L'original (sometimes l'originac) is → moose in Quebec.

ORUN

Ibo for the neck in → Nigeria.

ORYX

The East African oryx or beisa, *Oryx beisa*, is a desert-adapted African antelope of interest for sustainable meat production in relatively arid areas. It produces a carcass of about 100 kg, dressing at about 58%. About 50% of the carcass is hindquarter meat. Relative to other African game, oryx meat has a low myoglobin concentration so that the meat appears quite pale. *Oryx gazella* is the gemsbok, another large antelope reaching live weights around 200 kg. The → South African cutting pattern for gemsbok is shown.

(Sources: Ledger, 1968; Dorst and Dandelot, 1970; Eltringham, 1984; Onyango et al., 1998)

OSOBUCO

In → Chile, osobuco de mano is forelimb shank and osobuco de pierna is hindlimb shank.

OSSO BUCCO

Transverse sections across a shank of veal, either forelimb or hindlimb, depending on the country. The basic idea is that the immature bone has a high content of easily gelatinised collagen which contributes to the dish when cooked.

OSSO DI COPPA

Neck cut of an → Italian pork carcass.

OSSO PIATTO

Italian for blade-bone or scapula.

OSTEICHTHYES

The higher bony fish or Osteichthyes are divided into two major subclasses depending on whether their fins contain fleshy lobes (Sarcopterygii) or are supported entirely by rod-like rays (Actinopterygii). The Sarcopterygii might be regarded as the descendants of the fish that gave rise to the first amphibious land vertebrates with muscular limbs. The only surviving fishes of this group are the lung fishes, which can breathe atmospheric oxygen when caught in a drought, plus the most famous of the living fossils - the Coelacanth.Thus, all the remaining familiar edible fish with a bony skeleton are in the subclass Actinopterygii and, of these, the Teleostei reign supreme. However, there are a couple of exceptions. The non-Teleost Actinopterygii include sturgeons of the order Chondrostei, and two North American fresh-water fish, the gar-pike and the bowfin, which both belong to the Holostei. Sturgeons of the genus *Acipenser* are bottom-living fish reaching 9 m in length. The sturgeon fisheries of Russia and Europe are economically important for sturgeon eggs or caviar, but the myosystem is also a major food item.

(Sources: Jordan, 1923; Hardy, 1959; Jones, 1967; Royce, 1972; Simms and Quin, 1973; Browning, 1974; Scott and Messieh, 1976; Lagler, 1977; Wade and Fadel, 1997)

OSTRICH

The ostrich, *Struthio camelus*, is gaining in popularity as a source of red meat - starting from South Africa, moving to Israel and Australia, and finally to North America and Europe. The main advantage of ostrich meat seems to be its tenderness and low level of intramuscular fat (0.5%), but its main disadvantage for further processing is its high pH (> pH 5.9 is common). Suspension of the carcass from the leg produces a shape such as that shown in Figure 181. Typical primal cuts are the thigh, rib cage and drum (drumstick or leg).

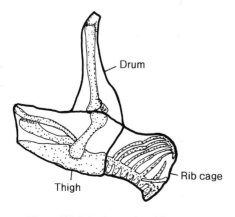

***Figure 181.** Primal cuts of ostrich carcass.*

Typically, the meat is removed from the bone and surface connective tissues are stripped away (de-silvering). The meat is retailed by its major muscle groups (similar to those of the → emu), but there is considerable variation in naming the muscles between different countries. Ostrich associations in various countries are attempting to standardise the terminology by numbering the muscles. In South Africa, most muscles are simply known as steaks and only a few are specially named. For example, Bergie filet is *iliotibialis* and tournedo filet is *ambiens*. In Australia, the long muscles tend to be called filets of one sort or another, while the chunky muscles are designated as rumps. In North America, the muscles are designated as strips of one type or another, but the typical poultry term for the oyster muscle is used for the *iliofemoralis*. Other fairly consistent trends are: (1), separation of medial and lateral parts of the *gastrocnemius* into inside and outside leg (respectively); (2), calling the *iliofibularis* muscle the fan; and (3) using the *obturatorius medialis* as a loin, tenderloin or back tender.

(Source: Sales and Oliver-Lyons, 1996; Paleari *et al.*, 1998).

OUNK

Neck of poultry in → Morocco.

OUSSOUS

The tail of a poultry carcass in → Morocco.

OUTSIDE

In → Australian and → New Zealand beef cutting, the outside is the *biceps femoris* plus *semitendinosus*.

OUTSIDE FLAT

In Australian beef cutting, an outside flat is part of the → outside, restricted to *biceps femoris*.

OUTSIDE ROUND

In → North American beef cutting, the outside or bottom round is taken from the lateral face of the hindlimb and contains *semitendinosus* (eye of the round) and *biceps femoris*.

OXYMYOGLOBIN

→ Myoglobin

OYSTER

In → England, an oyster of veal is the scapula, humerus and associated muscles separated from the underlying thoracic vertebrae and ribs. In → New Zealand, an oyster may be a rounded shoulder cut of lamb.

PÁ

The shoulder of a → Portuguese beef, pork, lamb or veal carcass. In → Brazil, the pá of a beef carcass is part of the paleta. It includes the scapula, humerus and associated muscles.

PÄÄ

The head of a pork carcass in → Finland.

PAAHTOPAISTI

In → Finnish beef and pork cutting, paahtopaisti is a cut taken through the gluteal muscles overlying the ilium.

PACHWINA

Tensor fascia lata and a strip of antero-ventral muscle and fat from a → Polish pork carcass.

PADDYWAX GRISTLE

An English name for the flat strap of → ligamentum nuchae (yellow elastic connective tissue) attached to the neural spines of vertebrae. The ligamentum nuchae is removed from rib roasts to facilitate carving or subdivision into steaks. Straps of paddywax gristle can be slapped noisily against a butcher's block, while the butcher sings "knick-knack paddy-whack" to special customers.

PAIDAKIA

The dorsal thoracic region of a → Greek lamb carcass.

PAILLASSE

Equivalent to the flanc of a → Belgian beef carcass.

PAINE

→ Panne or kidney fat from a Belgian pork carcass.

PAISTI

Finnish for a retail cut of meat or roast.

PAKISTAN – BEEF, CAMEL AND VEAL CUTS

In Urdu, the official language of Pakistan, there are names for various parts of the carcass, transliterated in Figure 182. The bong is the shank and associated muscles of the ulna and radius, or the tibia. The guttor is the thigh containing the femur and associated muscles, whereas the raan is the whole hind limb (guttor + bong). The kunni is based on the sacral vertebrae and five coccygeal vertebrae. The burk contains the gluteal muscles and the ilium, while the batti is contiguous and contains all the dorsal abdominal muscles. The palla contains the lateral and ventral muscles of the flank. The puth is the loin and the chaanp contains the rib cuts. The seena is equivalent to the brisket and contains the sternum and costal cartilages. The dasti includes the proximal forelimb bones from the scapula to the ulna and radius, together with associated muscles. The garden is the neck.

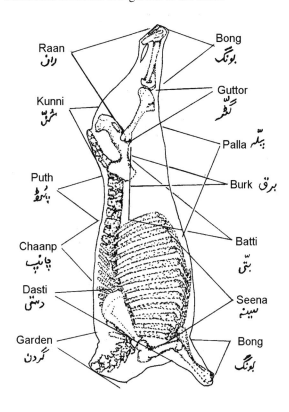

Figure 182. *Parts of a beef carcass in Pakistan.*

(Source: Anjum, 2003)

PAKISTAN – LAMB AND MUTTON CUTS

The names for lamb or mutton cuts are similar to those of beef, but with less subdivision. Thus, the primal cuts are raan (hindlimb), puth (loin), chaanp (posterior thoracic rib cuts), dasti (forelimb), garden (neck), palla (abdominal muscles of the flank) and seena (sternum and ventral rib region). The fore- and hindshanks are called macchli instead of bong.

(Source: Anjum, 2003).

PALANCA

Posterior abdominal muscles in → Chilean beef cutting.

PALERON

A primal cut of → French beef including most of the forelimb. It is broken into the macreuse, charolaise, jumeaux and derrière de paleron.

PALE, SOFT, EXUDATIVE (PSE)

The normal conversion of muscle to meat is marked by the development of → rigor mortis and a decrease in → pH caused by anaerobic → glycolysis. If the rate of pH decline is very rapid, so that proteins are still at body temperature, or if glycolysis is extended to reach a very low ultimate pH, then meat may become pale, soft and exudative. Paleness is caused by a high degree of light scattering so that incident light is reflected back to the observer with little penetration into the meat. Most of the extra scattering originates from denatured sarcoplasmic proteins, with a small contribution from increased scattering by → myofibrils. The softness is caused by fluid filled spaces between myofibrils and myofibres, sometimes exacerbated by detachment of → myosin molecule heads from thick → myofilaments. The exudation is caused by the release of fluid from the → myofilament lattice, because myofilaments move closer together as the pH decreases towards the isoelectric point at which negative electrostatic repulsion is minimal. PSE is a particular problem in pork, but it may occur in any species and is a growing problem in white poultry meat. PSE pork may be caused by any factor that accelerates post-mortem glycolysis, either the → porcine stress syndrome or the stressing of normal pigs.

(Source: Bendall and Swatland, 1988)

PALETA

The scapular region of a pork carcass in → Spain or → Brazil. In → Cuba, the paleta is a boned-out set of shoulder muscles including *infraspinatus* (punta de paleta), *supraspinatus* (boliche francés), plus *biceps* and *triceps brachii* (yema de paleta). In Brazilian beef cutting, the paleta is a primal cut containing the limb bones and associated muscles and is subdivided into the pá, which contains the scapula and humerus together with their surrounding muscles. Further separation gives the raquete (*infraspinatus*), peixinho (*supraspinatus*), and the coração da paleta (the heart of the paleta, which is the *triceps brachii*).

PALETTE

In Quebec, the beef chuck is sliced in planes parallel to rib 5 to make palette or blade steaks. In → French lamb cutting, the palette is the scapular part of the épaule.

PALLA

For beef, veal and camel in → Pakistan the palla includes the lateral and ventral abdominal muscles of the flank. For lamb or mutton, the palla includes all the muscles from dorsal to ventral.

PALOMILLA

In Cuban beef cutting, the palomilla (Figure 183) is composed of the *gluteus* muscles which, when boned out and spread out, may have a v-shape somewhat like a moth?

PALOMITA

In → Argentina, the beef palomita is in the region of the *tensor fascia lata*. The name indicates approval or approbation, but is strangely close to the Cuban → palomilla.

Figure 183. *The palomilla in Cuban beef cutting.*

PANCETA

The ventral part of the belly on a → Spanish pork carcass.

PANCETTA

The ventral part of the belly on an → Italian pork carcass.

PANCIA

The flank of an → Italian beef carcass.

PANNE

Kidney fat from a Belgian pork carcass.

PANNEAU

In → French beef cutting, the panneau is a long primal from the ribs back to the abdominal musculature. It is subdivided into the bavette d'aloyau, and the plat-de-côtes couvert and découvert.

PANNICULUS CARNOSUS

Cutaneous muscle.

PAPADA

The jowl of a pork carcass in → Spain or → Brazil.

PAPILLON

A papillon is the collet plus épaule in → French lamb cutting.

PARADISE ROAST

See fanciful names.

PARIFILEESELKÄ

In → Finnish beef cutting, the parifileeselkä is a grilling filet from the posterior lumbar region.

PARTRIDGE

The partridges are grouped together with the pheasants in the Family Phasianidae in the Order → Galliformes. The partridge is a relatively small bird with a rounded shape and without extended tail feathers. Cocks are larger than hens, but the sexes otherwise may be difficult to distinguish. In Europe, the three most important partridges are the red-legged partridge (*Alectoris rufa,* Figure 184), the rock partridge or chukor (*A. graeca*) and the common or grey partridge (*Perdix perdix*). The common partridge was introduced into the upper Mississippi valley in the USA where it may be called the Hungarian partridge. Otherwise a "partridge" in the USA is a grouse. Partridges are suitable for roasting and their breast muscle is dominated by fast-contracting anaerobic myofibres (→ histochemical types of myofibres).

Figure 184. *Red-legged partridge.*

(Source: Vesey-Fitzgerald, 1949; Bertin, 1967; Pyornila *et al.,* 1998)

PASCHINA

In → Russia, the flank of a beef or lamb carcass or the belly of a pork carcass.

PASTRAMI

Cured and smoked beef muscles (→ curing), typically *pectoralis* or abdominal muscles.

PATA

The foot of a pork carcass in → Brazil.

PATAGONIAN TOOTHFISH

Dissostichus eleginoides → Perciformes.

PATINHO

The *quadriceps femoris* muscles of a → Brazilian beef or veal carcass.

PATIO CHOP

See fanciful names.

PATO

Spanish for → duck.

PATTE DE PORC

In → Belgium, a pork trotter from fore or hindlimb.

PAVO

Spanish for → turkey.

PAXILLIN

An adhesion protein found at the → myotendon junction.

PEA MEAL BACON

Typically containing an eye of *longissimus dorsi*, pea meal bacon is cured (→ curing) but not smoked, and is given a thick coating of bright yellow corn meal.

PECETO

An → Argentinean beef cut located posteriorly in the hindlimb.

PECHO

Anterior breast meat of a → Spanish beef carcass or beef *pectoralis profundus* and *superficialis* in → Costa Rica. In → Mexico, the pecho enrollado is boned and rolled beef brisket or veal breast. In Mexico, the pecho is also a primal cut of pork.

PECHO DE POLLO

Spanish for chicken breast.

PEČIENKA

The loin of a veal carcass in → Slovakia.

PECTINEUS

A muscle deep in the mammalian hindlimb. It originates on the anterior edge of the pubis and on the prepubic tendon, and inserts on the posterior face of the femur (Figure 185). It is located anteriorly to the *adductor*, from which it may be difficult to separate in lean animals. In the pork ham, the *pectineus* is quite well developed.

Figure 185. *Lateral view of beef* pectineus *(solid back), most of which is posterior to the femur (1) but difficult to separate from the* adductor *in lean carcasses.*

PECTORALIS

In birds, the *pectoralis* is the large flight muscle located outermost in the breast muscle mass (→ chicken).

PECTORALIS PROFUNDUS

In mammals, the deep pectoral muscle is the larger of the two pectoral muscles (≈ 3.7% total muscle weight) and has abundant connective tissue. It originates on the sternum, from the fourth sternebra to the xiphoid cartilage, and inserts onto both the scapula and humerus (Figure 186). Two subdivisions of the muscle may be evident in pork, with three in lamb.

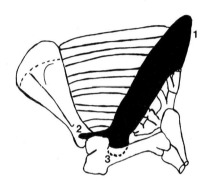

Figure 186. Lateral view of beef pectoralis profundus *(solid black and broken line) showing its origins on the sternum (1) and insertions on the scapula (2) and humerus (3).*

PECTORALIS SUPERFICIALIS

In mammals, the superficial pectoral originates on the first four sternebrae and is inserted on the distal, posterior face of the humerus (Figure 187). It is less than half the weight of the *pectoralis profundus*, and is subdivided into two parts in beef, almost three parts in pork, and only one main part in lamb.

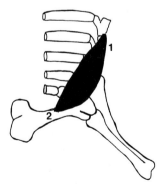

Figure 187. Lateral view of beef pectoralis superficialis *(solid black) showing its origins on the sternum (1) and insertion on the humerus (2).*

PEES

Dutch for tendon or gristle, but it may also be a cut of beef such as the heel of the round leading to the Achilles tendon.

PEITO

The brisket of a → Portuguese beef or veal carcass. In beef, the peito may be divided into three parts from anterior to posterior - maça do peito, peito alto, and prego do peito. In lamb, the peito is taken with a flap of abdominal muscle as the peito e fralda. The peito is also the name used for a beef brisket in → Brazil.

PEIXINHO

The *supraspinatus* in a → Brazilian paleta of beef.

PELÉ D'ÉPAULE

A shoulder cut of beef in → Belgium.

PELÉ DE PLATINE

→ Pelé d'épaule.

PENTHOUSE STEAKS

In North America, a retail cut of beef which, essentially, is a slice of eye of the round (*semitendinosus*). Also, the name was used for silverside in British → Proten beef.

PEPPERONI

A dry → sausage.

PERCH

A spiny-finned freshwater fish, → Perciformes.

PERCIFORMES

This order of bony fish contains a large number of food fishes in several families. Their most notable feature is generally the occurrence of spines in the first of two dorsal fins. The pelvic fins are located far forward close to the pectoral fins, sometimes even in front of the pectoral fins.

The family Serranidae includes the sea basses and groupers. They have 24 vertebrae and a maximum of three anal spines. Their scales are ctenoid (with a fringe-like edge, especially posteriorly), like most other percoid fishes. *Roccus saxatilis*, the North American striped bass is a large fish (sometimes > 40 kg) that spawns in fresh water. *Roccus chrysops*, the white bass of the Great Lakes is smaller. *Paralabrax clathratus* is the kelp bass of Southern California.

Perches of the family Percidae have two anal spines. The two dorsal fins may be separate or fused. The family includes the freshwater perches of the northern hemisphere. A single genus, *Perca* (Figure 188), includes most of the important sport perches of North America and Europe. The pelvic fins are close together. *Stizostedion* includes the wall-eyed pike or pike-perch, while *Lucioperca* includes the commercial pike-perches of Europe and Russia.

The family Carangidae includes the horse-mackerels, jacks, scads and pompanos, comprising about 200 species of strong, carnivorous fish, many of which are important commercial food fishes. *Trachurus*, the horse mackerel, is a low grade fish that is usually smoked. *Trachinotus* is the pompano of the tropical Western Atlantic.

Family Lutjanidae includes the brightly coloured snappers of tropical and subtropical seas, such as *Lutjanus*, the red snapper or pargo (Figure 188). Family Anarhichadidae are the wolffish distinguished by their rather large teeth. They have no pelvic fins and the body has an elongated shape with a continuous dorsal fin. The wolffish *Anarrhichas* is sometimes called the ocean catfish.

PERCIFORMES
PERCIDAE

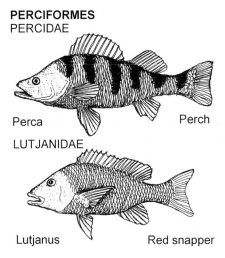

Perca   Perch

LUTJANIDAE

Lutjanus   Red snapper

Figure 188. Perch-like fishes.

PERCIFORMES
SCOMBRIDAE

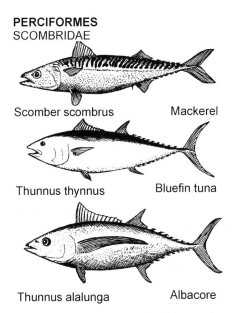

Scomber scombrus   Mackerel

Thunnus thynnus   Bluefin tuna

Thunnus alalunga   Albacore

Figure 189. Mackerel, tuna and albacore.

Families Scombridae and Thunnidae include the important mackerel and tuna (Figure 189) which, together with other well known fishes such as the swordfish (*Xiphius*) may be included in the suborder Scombroidea. *Scomber scombrus*, the Atlantic mackerel, is a migratory pelagic fish which lacks a swim bladder and can change its swimming depth very rapidly. It is a plankton feeder and occurs in large schools, usually caught with drift or gill nets. The myosystem is fatty and delicious, but liable to become bitter during storage. *Auxis* is the frigate mackerel. *Euthynnus* includes the little tunas with the alternative name *Katsuwonus*. *E. pelamis* is the extremely important skipjack which can swim very rapidly for periods of time and is metabolically adapted for dealing with the lactate produced by its muscles (→ glycolysis). *Sarda* includes the bonitos. *Thunnus* includes the important large tunas that support major fisheries. The bluefin tuna, *T. thynnus* is the largest and reaches 800 kg. The smaller albacore, *T. alalunga*, is highly valued for canning because of its white myosystem. *T. albacares* is the yellowfin tuna, while *T. obesus* is the bigeye tuna. Being active predators, tuna are energetic fish, using a heat exchange mechanism to maintain their musculature at up to 10°C higher than the surrounding water. Mercury accumulation is a problem in the myostems of older, larger tuna. Bluefin tuna, highly valued for sushi dishes, is evaluated by freshness, oil content and colour. To maintain freshness, carcasses are chilled in brine to a core temperature < 5°C, which may require 20 to 30 hours. After washing and inspection, carcasses are shipped in plastic-lined boxes packed with ice, often en route to Tokyo.

Scombroid fish such as tuna, mackerel, bonito and saury, plus other related fishes, have high levels of histidine in their myosystems. Histidine decarboxylase from otherwise unnoticed mesophilic (> 15°C) bacterial spoilage may produce high levels of histamine. Ingestion of histamine may produce nausea, vomiting, etc., but also localized inflammatory responses such as irritation and hypotension. There is a fluorescence assay for histamines. Most scombroids are rejected at > 10 mg% histamine, while the level for anchovies is 20 mg%.

Family Nototheniidae includes the Patagonian toothfish, *Dissostichus eleginoides*, another recent victim of illegal fishing in South American waters. This large fish (up to 2 m in length) occurs in deep waters (300 to 1,500 m). Like other long-lived fish (up to 50 years) it is slow breeding and not readily replaced.

Also in the diverse order Perciformes are the families Cichlidae and Xiphiidae. Cichlids of the family Cichlidae are freshwater tropical fish found in Africa, India, Southwest Asia and South America. Some species are used in aquaculture. The family Xiphiidae contains only one species, the well known swordfish, *Xiphius gladius*. Its sword is an elongation of its upper jaw, and its skin is smooth and lacks scales. The swordfish is a large deepsea fish found on both sides of the Atlantic, and the myosystem may be sold fresh, as frozen steaks or canned.

Also in the Perciformes are fishes such as the sand lance, *Ammodytes*, which has a well established fishery in the North Sea. This eel-like fish, also distributed across the North

Atlantic, reaches a length of 37 cm and can burrow in sand intertidally.

(Sources: Eitenmiller *et al.*, 1982; Ahrens, 1985; Clay and Hurlbut, 1988; Perry *et al.*, 1985; Scott, 1993; DFO, 1993)

PERDREAU

French for a young → partridge.

PERDRIX

In France, a perdrix is a → partridge (*Perdix perdix*).

PERIMYSIUM

Fibrous connective tissue binding several skeletal → myofibres into a bundle or → fasciculus. Thus, the perimysium is around the fasciculus. Most of the connective tissue fibres of the perimysium are composed of Type I → collagen.

(Source: Schmitt *et al.*, 1979)

PERLHUHN

German for → guinea fowl.

PERNA

The leg of a pork, lamb or veal carcass in → Portugal. In → Brazil, the perna is a leg of lamb.

PERNICE ROSSA

Italian for → partridge.

PERNIL

Leg of pork in → Spain and → Brazil.

PERONEUS LONGUS

The *peroneus* or *peronaeus longus* is a small, lateral muscle in the mammalian hindlimb. It originates mainly from the lateral condyle of the tibia. In a pork carcass, the *peroneus longus* can be removed rapidly after slaughter for various experiments, without appreciable damage to the ham. The main part of the muscle is white, but the medial face is red. The *peroneus longus* lacks intrafascicularly terminating myofibres.

PERONEUS TERTIUS

The *peroneus* or *peronaeus tertius* is the most medial and superficial muscle of the extensor group in the distal part of the mammalian hindlimb. In beef and lamb it is very thin, but it is much larger in pork. It originates on the femur and has tendons inserting on the first and second tarsals and third metatarsal.

PESCOÇO

In → Brazil, the neck of a beef, pork, lamb or veal carcass.

PESCUEZO

The neck of a → Spanish or → Mexican lamb or veal carcass. In → Costa Rica, the beef pescuezo includes the *atlantis, brachiocephalicus, splenius* and *trapezius* muscles.

PETIT CROISÉ

Posterior part of the brisket on a → Belgian beef carcass.

PETITE CROISURE

→ Petit croisé.

PETITES CÔTES

A cut from the anterior part of the sternum in a → Belgian pork carcass.

PETITE TÊTE D'ALOYAU

Sacral region of a → Belgian beef carcass.

PETIT SALÉ

Lard maigre or belly of pork in → Belgium.

PETTO

The brisket of an → Italian beef carcass, or the → dewlap of a live animal.

PETTO O PANCIA

Breast of lamb in → Italy.

pH

The acidity of meat may be measured by the amount of sodium hydroxide required to neutralize a sample of comminuted meat containing a phenolphthalein indicator, but pH measurements made with glass or solid-state electrodes are more convenient. Except for muscles that contain little or no glycogen at the time of slaughter, the pH decreases postmortem because of anaerobic → glycolysis. But meat contains buffers (ATP and phosphate) that may capture hydrogen ions, so that the pH may give a poor measure of the total acidity. Carnosine and anserine in meat also contribute to buffering capacity and these dipeptides may vary between red and white muscle.

The pH measured by an electrode pushed into meat may not represent the pH in the bulk of the sample. For example, lactate formation may be slower on the surface of meat exposed to atmospheric oxygen. Glass pH electrodes may give erroneous readings when they are pushed into a piece of meat. Very often the pH reading varies from one minute to another and it may never settle down to give a constant reading, particularly if the meat has a high pH, is relatively dry and has insufficient fluid to make good electrical connections.

To avoid these problems, pH may be measured after a meat sample has been macerated in water. But this creates two new problems: (1), acceleration of glycolysis by cellular disruption, and (2), the dilution of buffers in the meat. Thus, sodium iodoacetate (5 mM) is needed to prevent further lactate formation, and the ionic strength of the meat sample must be maintained by using 150 mM potassium chloride instead of water. The pH of meat also is affected by temperature. It decreases by about 0.15 when the meat is warmed from 20 to 38°C, and it increases by about 0.2 when the meat is cooled from 20 to 0°C. The pH of frozen meat changes as it is thawed.

If the pH of meat drops to the isoelectric point at which its proteins bear no net charge, solubility and water binding capacity are minimal. The isoelectric point of muscle proteins is near to a pH of 5.5. The bulk of commercial meat gets very close to the isoelectric points of its various proteins, and this accounts for the ease with which meat will release water by evaporation or as drip loss from the → myofilament lattice.

(Source: Bendall, 1973; Bendall and Swatland, 1988)

PHALAN

The aloyau of a → Belgian beef carcass.

PHEASANT

The pheasants are grouped together with the partridges in the Family Phasianidae in the Order → Galliformes. The pheasants have beautiful long tail feathers which, traditionally, are used to decorate the eviscerated carcass, so that they protrude from the eviscerated body cavity. They should be removed before cooking!

There are numerous species of pheasant around the world, but the most important is the ring-necked pheasant, *Phasianus colchicus,* which now is feral in many rural areas or reared commercially in both Europe and North America (Figure 190). Originally, pheasants were introduced into Europe from Asia by the ancient Greeks, then spread westwards by the Romans. Farmed birds may be available in supermarkets, after being processed and packaged like chickens. To develop their full taste potential, however, pheasants are best hung intact (guts in, feathers on) in cool conditions for several weeks until they ripen. This makes them difficult to pluck and eviscerate because the skin rips easily, but it is well worth the labour. Watch out for sharp slivers from shattered bones. Exposed muscle areas can be wrapped with pork fat. Surface larding is important to baste the breast as the pheasant is roasted. The dressed weight of a larger pheasant may be over 1 kg. Small, young males which have not yet developed long, sharp spurs produce the best combination of yield and tenderness.

Old pheasants may be too tough to roast and are best casseroled. When my parents had pheasant for Sunday lunch I could smell the fabulous aroma at the end of the lane as I hurried home from the pub.

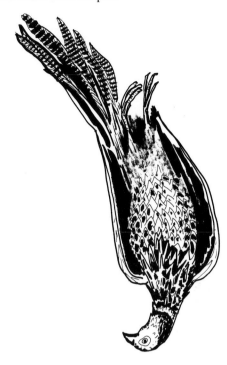

Figure 190. Ring-necked pheasant.

PHOSPHATE

Phosphates are used in meat processing because, like salt, they cause swelling of the → myofilament lattice and an increase in → water-holding capacity.

(Source: Bendall, 1954)

PICCIONE

Italian for → pigeon.

PICHONCILLO

Spanish for fledgling →pigeon.

PICKEREL

Stizostedion vitreum, a major north American sport fish with a large and highly reflecting eye, → Perciformes.

PICNIC

In → North American pork cutting, the picnic is part of the shoulder of pork. The posterior face is cut perpendicularly to the vertebral column, often between thoracic vertebrae 2 and 3. The butt is separated from the picnic, cutting past the ventral region of the cervical vertebrae at a tangent. This keeps the top of the picnic relatively square. The jowl is removed from the picnic following the crease lines in the skin of the neck. The picnic is often boned and rolled, then subdivided into small roasts

PICNICBOG

The *triceps brachii* of a bog from a → Swedish pork carcass.

PIE

The hindfoot of a → Spanish pork carcass.

PIÈCE À L'ABATTU

The pièce levée of a → Belgian beef carcass.

PIÈCE LEVÉE

A distal cut from the beef hindlimb in → Belgium.

PIÈCE RONDE

Tranche grasse of a → Belgian beef carcass.

PIED

In Quebec, the hindfoot of a pork carcass is the pied arrière, while the forefoot is the pied avant.

PIERNA

Spanish for leg, as in the hindlimb of a → Mexican beef, pork, lamb or veal carcass; or lamb and pork in → Argentina.

PIERNA CON BOLICHE

In → Cuban beef cutting, this is a boneless cut containing the following hindlimb muscles: *biceps femoris* (pierna), *semitendinosus* (boliche blanco) and *gastrocnemius* (sapo).

PIG

→ Pork.

PIGEON

The Wood-pigeon or Ring-dove, *Columba palumbus*, is the largest of the European pigeons, ranging into North Africa and Asia. It lives and nests in wooded areas, but is regarded as a pest rather than game because of its vast numbers and grain-feeding habits. It is, however, difficult to shoot because of its excellent vision and general wariness. *C. livia*, the Rock-dove or common pigeon, was domesticated and housed in traditional, free-access dovecotes to provide a source of fresh meat during the winter. The modern equivalent farm building has nesting boxes and flight areas. *C. livia* was introduced into North America and is now feral. A fledgling pigeon is called a squab, but the term is used in the meat trade to denote any plump, tender, young pigeon. Mature wood-pigeons are generally too tough to roast, and are best

used in a casserole. Squabs can be roasted. Mourning doves, *Zenaida macroura*, are hunted for their meat in the USA.

(Source: Vesey-Fitzgerald, 1949; Bertin, 1967; Petchey, 1991; Burger *et al.*, 1997)

PIKE

An elongated, predatory freshwater fish, *Esox*, → Salmoniformes.

PILCHARD

A large sardine, such as *Sardinops*, → Clupeiformes.

PIMIENTO LOAF

A → meat loaf flavoured with sweet peppers.

PIN BONE

The blade and shaft of the ilium, especially in a sirloin steak.

PIN BONE SIRLOIN STEAK

In North America, a steak of beef showing part of the shaft or blade of the ilium, and transitional between a porterhouse and a sirloin steak. The pin bone sirloin has a very high bone content.

PINTADE

French for → guinea fowl.

PINWHEELS OF LAMB

In North America, rolled strips of breast of lamb may be called pinwheels.

PIRIFORMIS

In lamb, the piriformis is a caudal lobe of the gluteus accessorius. It originates from the lateral surface of the sacrosciatic ligament and inserts onto the greater trochanter of the femur.

PIROSKI

A cooked pork sausage containing nitrite, starch and milk products.

(Source: Pexara *et al.*, 2002).

PIS-DE-BOEUF

In → French beef cutting, the pis-de-boeuf is a primal cut in the sternal and abdominal region. It includes the poitrine, tendron and flanchet.

PISTOLA

This is a pistol-shaped hindquarter of beef, with the vertebral column forming the barrel. The designation is used for →

Danish beef exports, as well as for lamb and horse in → Iceland (pistóla).

PITKÄ PAISTI

In → Finnish beef cutting, the whole round is called a long roast or pitkä paisti, and is divided into a medial part or sisäpaisti (mostly *semimembranosus, adductor* and *pectineus*), and a lateral part or ulkopaisti (mostly *semitendinosus* and *biceps femoris*).

PLAICE

A medium-sized flatfish, → Pleuronectiformes.

PLANTARIS

Flexor digitorum superficialis of the hindlimb.

PLASMA MEMBRANE

The lipid membrane around all animal cells.

PLAT-DE-CÔTES

In → French beef cutting the plat-de-côtes couvert and découvert are, respectively, posterior and anterior rib cuts.

PLAT DE TRANCHE GRASSE

The flat (plat) is part of the tranche grasse in → French beef cutting and is composed of *vastus externus*.

PLATE

A primal cut of beef containing the ventral portions of the posterior ribs in → North America, in Northeast → England, for → Proten beef in Canada, in → Ireland and → Northern Ireland.

PLATEADA

Rib of beef cut in → Chile, probably named for a silvery connective tissue surface.

PLATE CÔTES

A ladder of beef rib sections in → Belgium or → Luxembourg.

PLATTE

Abdominal muscles of an → Austrian beef carcass.

PLATTE BIL

In → Netherlands beef cutting, the platte bil is located laterally to the femur and includes *semitendinosus* and *biceps femoris*.

PLATTE FRICANDEAN

A lateral part of a → Netherlands pork or veal carcass.

PLECE

The shoulder of a beef, pork, lamb or veal carcass in → Slovakia.

PLECKA

The shoulder of a beef, pork, lamb or veal carcass in → Serbia & Montenegro.

PLEIN FILET

The filet d'Anvers of a → Belgian beef carcass.

PLEURONECTIFORMES

Flatfishes in this order have many perch-like features (→ Perciformes) but they undergo a metamorphosis so that the adults are asymmetrically flattened. The underside against the substrate loses its pigmentation and the eye of that side moves around to the upper side. Being bottom dwellers, flatfish have no swim bladder. The naming of flatfish, like that of shrimps and prawns, is rather confusing. Sole are relatively thin fish of several families. Dab, flounder and plaice are medium-sized fish. Turbot are any of the largest species of families Pleuronectidae and Bothidae, while halibut are extremely large species of Pleuronectidae. As far as myosystems go, it is important to remember that flatfish fillets are easily spoiled by overcooking if they are thin.

The family Pleuronectidae are the righteye flounders. They lie on their left side and have both eyes on the right side of the body. The operculum has a free edge and the pelvic fins are symmetrical. This includes the plaice *Pleuronectes*, and the Atlantic halibuts, *Hippoglossus*, caught from great depths and weighing up to several hundred kilograms (Figure 191). Atlantic halibut (*H. hippoglossus*) is a valuable fish with a firm white myosystem and a distinctive delicate flavor. *H. stenolepis* is found in the North Pacific. Other species, however, may be quite small such as *H. platessoides*, the rough dab, so named because of its rough scales. *Limanda ferruginea* is the yellowtail flounder of the Northwest Atlantic, sold as a filleted, frozen product. The abundant American plaice of the Northwest Atlantic is *Hippoglossoides platessoides*, while *Pseudopleuronectes americanus* is the inshore winter flounder and *Glyptocephalus cynoglossus* is the witch flounder, with its smooth scales and copious mucus making it very slippery.

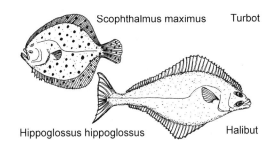

Figure 191. Turbot and halibut.

The family Bothidae includes the lefteye flounders such as *Paralichthyes*, the California halibut. *Scophthalmus maximus*, the European turbot (Figure 191), has a diamond-shaped body without scales, but the upper side is covered with blunt tubercles. *S. rhombus*, the brill, has an oval body covered with smooth scales. The myosystem may have a relatively high triglyceride content and a relatively firm texture. It is often sold cured or smoked. Other members of the family include *Lepidorhombus*, the megrim; *Arnoglossus*, the scale fish; and *Zeugopterus*, the topknot.

The family Soleidae includes soles with an elongated shape and both eyes on the right side of the body. Teeth are only present on the jaw of the blind side. *Solea*, is the sole, while *Buglossidium* is the solenette.

(Sources: Bowering, 1990; Zwanenburg, 1990; Pitt, 1989, 1990, 1993)

PODGARDLE

The jowl of a → Polish pork carcass.

PODPLECIE

Under-the-shoulder cut from a veal carcass in → Slovakia.

POINTE AU FILET

In → Belgium, a cut of pork containing sacral and posterior lumbar vertebrae.

POINTE D'ALOYAU

The petite tête d'aloyau of a → Belgian beef carcass.

POINTE DE FILET

In → French pork cutting, pointe de filet is the cut including the ilium and sacrum.

POINTE DE POITRINE

Brisket of beef in Quebec or, similarly, the petit croisé in → Belgium.

POIRE

The *pectineus* muscle, part of the tranche in → French beef cutting.

POITRINE

In → France, Quebec and → Luxembourg, the poitrine may be a brisket of beef, or breast of lamb or veal, or belly of pork.

POJADOURO

In → Portuguese beef cutting, the pojadouro contains the muscles medial and anterior to the femur.

POLAND - BEEF CUTS

Separation of fore and hindquarters follows closely around the last rib (Figure 192). After removal of the hindshank or pręga tylna, the major division of the round or udziec is into the following retail cuts. The *semitendinosus* is the ligawa (biała pieczeń), the *biceps femoris* is the zrazowa zewnętrzna, the gluteal muscles are the krzyżowa, the *quadriceps femoris* muscles are the skrzydło, and *adductor* and *semimembranosus* dominate the zrazowa wewnętrzna. The *psoas* muscles comprise the filet or polędwica which may be removed from the loin or rostbef, leaving the abdominal muscles as the łata.

The scapula, humerus and associated muscles are detached as the mieso złopatki from the underlying distal ribs (szponder), but leaving shapely standing roasts from the posterior thoracic (antrykot) and anterior thoracic regions (rozbratel). The anterior cervical region or szyja is separated from the posterior cervical region or karkówka. The whole of the sternum is contained in the mostek. The foreshank is the pręga przednia.

(Source: Pospiech, 1998)

POLAND - LAMB CUTS

The leg of lamb is called the udziec (Figure 193), from which is detached the tail (ogon) and hindshank (goleń tylna). The saddle or loin is the comber. The anterior and posterior thoracic roasts are the górka and antrykot, respectively. The breast or mostek is separated from a wedge-shaped cut, the goleń przednia, that includes the anterior sternum, distal humerus, ulna and radius.

(Source: Pospiech, 1998)

POLAND - PORK CUTS

As shown in Figure 194, removal of the ham (syznka) with a tibial shank (golonka tylna) and a trotter (nogi) fits the pattern for many other countries. But what is quite unusual is the curvilinear division of the abdominal musculature into an anterior, slab-like boczek and a posterior, L-shaped pachwina. Thus, in Figure 194, the spare-rib rack or żeberka is medial to the fleshy slab of boczek which includes muscles lateral to the ribcage. The shoulder or łopatka is lifted from the underlying ribcage, leaving the karkówka dorsally in the anterior thoracic region, plus a jowl of podgardle. The foreshank is divided into a proximal golonka przednia and a distal trotter or nogi. The głowa is the head. The biodrówka contains posterior lumbar chops with areas of both *longissimus dorsi* and *psoas*, whereas the remainder of the lumbar and thoracic chops with an eye of *longissimus dorsi* are in the schab. Backfat is removed as słonina.

(Source: Pospiech, 1998)

POLAND - VEAL CUTS

As in many other countries, Polish veal cutting is a miniature version of beef cutting, with the removal of a round (udziec) and a shoulder (łopatka) detached from the ribcage (Figure

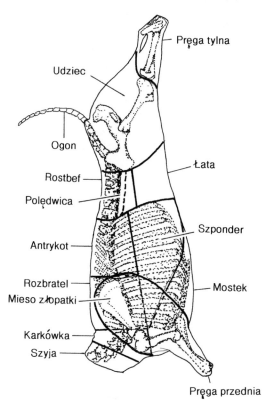

Pręga tylna
Udziec
Ogon
Rostbef
Polędwica
Łata
Antrykot
Szponder
Rozbratel
Mieso złopatki
Mostek
Karkówka
Szyja
Pręga przednia

Figure 192. *Poland – beef cuts.*

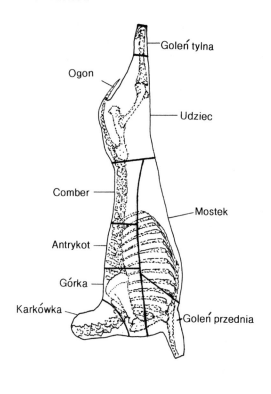

Goleń tylna
Ogon
Udziec
Comber
Mostek
Antrykot
Górka
Karkówka
Goleń przednia

Figure 193. *Poland – lamb cuts.*

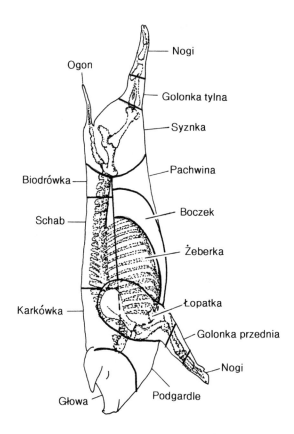

Ogon
Nogi
Golonka tylna
Syznka
Pachwina
Biodrówka
Boczek
Schab
Żeberka
Karkówka
Łopatka
Golonka przednia
Nogi
Głowa
Podgardle

Figure 194. *Poland – pork cuts.*

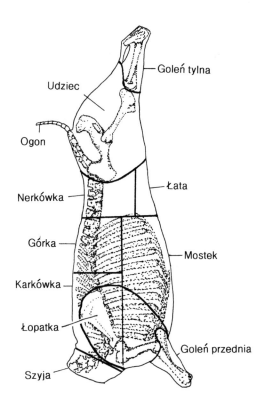

Goleń tylna
Udziec
Ogon
Łata
Nerkówka
Górka
Mostek
Karkówka
Łopatka
Goleń przednia
Szyja

Figure 195. *Poland – veal cuts.*

195). The mostek is much larger, relatively speaking, in veal than in beef because it includes the ribs which are removed as szponder in beef. The vertebral meat is divided into karkówka, górka and nerkówka, from anterior thoracic through to lumbar regions. Fore- and hindshanks are removed as golén przednia and golén tylna, respectively.

(Source: Pospiech, 1998)

POLĘDWICA

The *psoas* muscles of a → Polish beef carcass.

POLIÈAN

A Czech dry-fermented sausage containing beef and pork.

(Source: Komprda et al., 2001).

POLLACK

Pollack or pollock, a gadoid fish of the North Atlantic such as *Pollachis*, → Gadiformes.

POLLO

Italian and Spanish for → chicken, but pollo ganso is the *semitendinosus* of the hindlimb in → Chilean beef cutting, while pollo barriga is diaphargm.

POLPA DI COSCIA

The whole hindlimb of an → Italian beef carcass.

POLPA DI SPALLA

An → Italian beef cut containing the *biceps brachii* and *pectoralis profundus*.

POMPANO

A warm-sea bony fish of the family Carangidae, → Perciformes.

PONDEROSA HIP OF BEEF

A large Canadian joint of beef suitable for restaurant or banquet carving (Figure 196). The pelvis and rump are removed from a hip of beef. The tibia is cut just proximal to the tibio-tarsal joint, and the flexors and extensors are trimmed away to leave the bare tibia protruding as a handle for the carver. All loose tissue is trimmed away to enable the carver easy access to the major muscles of the round.

PONTIKI

The *biceps brachii* of a → Greek beef carcass.

PONY

Middleribs plus steakmeat of beef in → England.

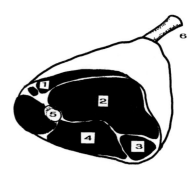

Figure 196. *Canadian Ponderosa hip of beef, with some landmark muscles and bones;* gracilis *(1),* pectineus, adductor *and* semimembranosus *(2),* semitendinosus *(3),* biceps femoris *(4),* femur *(5), and* tibia *(6).*

POPLITEUS

A small triangular muscle behind the joint between the femur and tibia in mammals.

POP-UP STEAK

See fanciful names.

PORBEAGLE

A shark (→ Chondrichthyes), *Lamna nasus*, the myosystem of which is prized as "sea veal".

(Source: Pivnička and Černý, 1987)

PORCINE STRESS SYNDROME (PSS)

The porcine stress syndrome is caused by a mutation in the calcium ion release channel of the sarcoplasmic reticulum (→ ryanodine receptor). Thus, affected pigs tend to overcontract their muscles and are very prone to overheating (malignant hyperthermia). Typically, PSS pigs are unable to cope with the stress of being moved to an abattoir. They may become lethargic, develop hyperthermia, or start to take rapid, shallow breaths through an open mouth (dyspnea). Their skin may appear very pale with a slight blue venous colouration (cyanosis), and their muscles may tremble, or become rigid or weak. If a PSS pig survives as far as the abattoir, it may produce severe PSE pork (if muscles still contain high glycogen levels at slaughter) or severe DFD pork (if glycogen levels are severely depleted at slaughter). Sometimes a PSS pig may be intermediate between these extremes and may produce pork with a normal appearance.

PORCUPINE

The porcupine of North America (*Erethizon dorsatum*) may have been a source of meat in pioneer days, but now is seldom eaten. The meat has been described as having a flavour like turpentine. The African brush-tailed porcupine (*Atherurus africanus*), however, is a favourite source of meat in Gabon, Nigeria, Cameroon and the Congo and captive breeding has been proposed to supplement natural sources.

(Sources: Rue, 1981; Jori *et al.*, 1998)

PORK

Species of pigs

Fossil pig skeletons have been found in geological deposits dating back to the Pliocene period in Europe and Asia. Domestic pigs of Europe and North America appear to be a mixture of two original species of wild pig: *Sus scrofa*, the wild boar of Europe found north of the Alps, and *S. vittatus*, the wild pig now only found wild in the Malay Peninsula. Wild pigs of the same genus (*Sus*) but of different species to domestic pigs are found in India and Ceylon (*S. cristatus*). The domestic pigs now found in China are usually considered to be *S. vittatus*. Whether or not *S. scrofa* and *S. vittatus* should be considered as separate species is a difficult question because transitional races are now widespread, thus demonstrating the obvious point that the hybrids are fertile. The scientific distinction between *S. scrofa* and *S. vittatus* is based on the shape of the lacrimal bone in the skull (located round the orbit of the eye and supporting the tear duct from the eye to the nose). Several different subspecies of wild swine are recognised: *Sus scrofa scrofa*, Europe; *S. s. meridionalis*, Mediterranean; *S. s. barbarus*, North Africa; *S. s. attila*, Eastern Asia; and *S. s. palustris*, found in the archaeological excavations of Swiss Neolithic lake dwellings.

Early evidence of pork consumption

In the bone heaps around the eating areas of prehistoric peoples are found the remains of three types of pigs: bones of wild pigs obtained from hunting, bones of large pigs probably put out to forage, and bones of small pigs probably kept in confined or covered areas. Remains of domesticated pigs are not found before Neolithic times (the agricultural revolution when man became a settled farmer) and, since pigs are difficult to control (they do not easily form herds like the ruminants), the nomadic farmers of earlier times probably did not have any pigs. Tribal conflict between settled farmers and warlike nomads may explain why domestic pigs, the invention of the settled farmer, were first prohibited by some religions. Another factor is the existence of parasites such as the pork tapeworm and trichinella.

Because of their rooting habits when foraging, pigs probably produced a dramatic change in the local ecology by reducing woodland undergrowth and allowing grass to grow. Before the invention of ploughing, pigs may have been driven over seeded ground to embed the seeds. Pigs may be used to hunt for underground mushrooms (truffles) or to retrieve game, and these habits might have been important to primitive farmers.

Breeds of pigs

In medieval times, herded pigs had a long snout and long legs. Around the year 1800, Chinese pigs were introduced into Europe and combined with *Sus scrofa*. This resulted in a dramatic phenotypic change as pigs became thick-set in shape, smaller in size and laid down fat earlier in life.

Early development of pig breeds was influenced by factors such as ease of taming, socially structured behaviour, large numbers of offspring at relatively short intervals, early rapid growth and maturation, and longevity. In the 1800s, the ability of pigs to store large amounts of fat was considered a desirable feature because, before the widespread use of fossil fuel energy for industrial machines, ordinary people expended large amounts of energy in their daily work. The high caloric content of fat and the high fat content of pork once provided important food energy. Nowadays, however, there is intensive selection against fatness and in favour of lean muscle development. For the gourmet, however, nothing comes close to fat pork from an old-fashioned pig, especially if it is properly → conditioned.

(Sources: Zeuner, 1963; Ucko and Dimbleby, 1969)

PORKLETS

In North America, cubed scraps of meat, typically from a fresh leg of pork.

ÞORSKUR

Icelandic for cod (→ Gadiformes).

PORTERHOUSE STEAK

In Canada, Mexico and USA, the T-bone steak nearest to the sirloin is large and well endowed with muscle (Figure 197), and is called a porterhouse steak, supposedly after an ale-house or porter-house in New York. It must have been a great place.

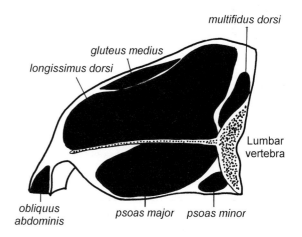

Figure 197. North American beef porterhouse steak.

PORTUGAL - BEEF CUTS

In Portugal, the major separation of the chã or round of beef around the femur is into a medial pojadouro and a lateral chã de fora (outside round). The edges of both cuts are quite far anterior so as to include the *vastus* muscles (Figure 198). Anteriorly this leaves the small rabadilha. Oxtail is rabada, but the rabadilha is the knuckle cut containing the patella. The hindshank is the chambão posterior or chambão da perna. The alcatra is a wide cut, right across the carcass from the sacrum, through the gluteal region to the *tensor fascia lata*.

Lombo is the general region of the loin, while the *psoas* is the lombinho and the remainder of the loin containing the *longissimus dorsi* is the vazia. The cuts in the thoracic region of the vertebral column are the acém redondo (round) and acém comprido (long). The main cut including the scapula is named the pá or shovel, which is reminiscent of the Spanish and German (tapa de paleta in Cuba, and Schaufelstück in Germany). Perhaps this is a folk memory of the use of the scapula as a digging tool in ancient times. The brisket is the peito (breast). The aba das costelas is a ladder of mid-rib sections, while the aba descarregada contains the abdominal muscles. The neck is the cachaço. The *biceps brachii* and extensor muscles of the shank are included in the lagarto, the literal translation of which is the lizard.

(Source: Simões, 1999)

PORTUGAL - LAMB CUTS

The basic plan for lamb cuts (Figure 199) is similar to that of pork (Figure 200), from the leg (perna) forwards to the loin chops (costeletas do lombo), rib chops (costeletas com pé), anterior rib chops (costeletas do fundo), shoulder (pá) and neck (cachaço). The breast and flap of abdominal muscle are removed as a single cut (peito e fralda).

(Source: Simões, 1999).

PORTUGAL - PORK CUTS

From the leg or perna is taken quite a large shank called the chispe, which contains tibia, fibula, tarsals, metatarsals and phalanges (Figure 200). The tail is the rabo. The lombada or thoraco-lumbar length of the vertebral column is divided into loin chops (costeletas do lombo) and chops with shoulder (costeletas com pé). The cervical region may be called the nape of the neck (cachaço) or deep chops (costelas do funda). The jowl or faceira is separated from the remainder of the head or cabeça. The shoulder or pá contains the scapula and humerus, while all the more distal bones and muscles are contained in the chispe of the forelimb (chispe da mão, literally, of the hand). The ventral part of the carcass is divided into the entrecosto in the thoracic region and the entremeada containing the abdominal muscles. If the entrecosto is cut as a ladder of rib sections, then the larger cut containing abdominal muscles plus costal cartilages and sternum is called the banha or fat.

(Source: Simões, 1999).

PORTUGAL - VEAL CUTS

The layout and naming of veal cuts follows the pattern for lamb (Figure 201), except that the foreshank is named the shank of the hand (chambão da mão).

(Source: Simões, 1999).

POSTA DE PALETA

The "handle of the shovel" in → Chilean beef cutting contains *triceps brachii*.

POSTA NEGRA

The medial muscle mass of the hindlimb in → Chilean beef cutting. These muscles are usually very dark, almost black.

POSTA ROSADA

The rosy pink cut of beef in → Chile is the *quadriceps* group.

POTRBUSINA

Abdominal muscles in → Serbia & Montenegro.

POULAIN

Meat from a colt in France, → horse meat.

POULE

French for a boiling fowl, an old → chicken.

POUSSIN

French for a young → chicken.

POYASNICHNAYA CHAST

The loin of a → Russian beef carcass.

PRASKY

Cured pork sausage in a beef weasand.

PRAWN

Prawns are decapod crustaceans of the suborder Macrura-Natantia (→ Crustacea). The distinction between a prawn and a shrimp depends on the geographical context (→ shrimp).

PREDNÁ NOŽIČKA

The trotter of the forelimb on a → Slovakian pork carcass.

PREDNÁ NOŽINA

The foreshank of a carcass in → Slovakia.

PREDNÉ KOLENO

A cut containing the ulna and radius on a → Slovakian pork carcass.

PRĘGA PRZEDNIA

The foreshank of a → Polish beef carcass.

PRĘGA TYLNA

The hindshank of a → Polish beef carcass.

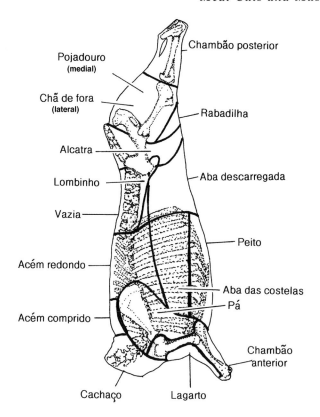

Figure 198. *Portugal – beef cuts.*

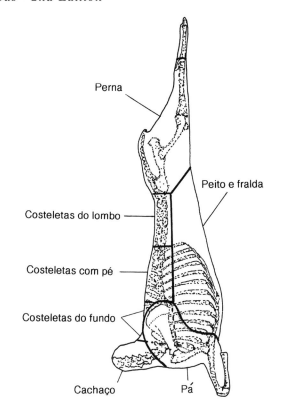

Figure 199. *Portugal – lamb cuts.*

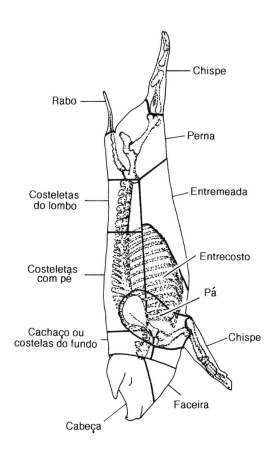

Figure 200. *Portugal – pork cuts.*

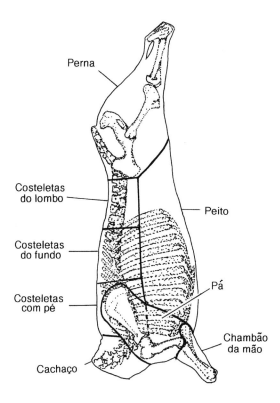

Figure 201. *Portugal – veal cuts.*

PREMIÈRES CÔTES

Rib of beef, entrecôte avant, in → Belgium.

PRIMARIB

The prime rib cut of a → South African beef carcass.

PRIME RIB OF BEEF

Some years ago in North America, calling something a prime rib roast would have been a double statement, the equivalent of saying a rib roast from a carcass graded as Prime. Back then the proper name for the cut of meat itself, regardless of carcass grade, was standing rib. However, prime rib is an appealing name, and now has become a cut of meat in its own right.

PRONATOR TERES

A poorly developed muscle on the medial surface of the elbow. In pork it is fusiform, in beef very thin and, in lamb, mainly tendinous.

PROSCIUTTO

Prosciutto in → Italy is the hindlimb or ham of a pork carcass. In which case, smoked ham is prosciutto affumicato, boiled ham is prosciutto cotto, and baked ham is prosciutto dolce. But prosciutto di spallo originates from the shoulder muscles. Prosciutto is now common on the delicatessen counters of many countries where it is usually a dry cured ham.

PROTEN BEEF

A process developed by Swift and Company whereby papain was injected intravascularly a few minutes before stunning. The resulting tenderisation as the meat was warmed in the early stages of cooking changed the potential utilisation pattern of meat beef cuts. Hence, some special names were developed to alter consumer responses and treatment of the meat. The names used in Canada were a curious mixture (Figure 202). The blade roast gave rise to Belmont, Scotch broil, and Niagara steaks. The rump ended up between top and bottom round steaks, whence were derived Manhattan and Denver roasts. Eye of the round produced penthouse steaks, while the sirloin tip gave dixie steaks, and the flank gave Hudson broil. The short rib roast gave Boston roasts and deluxe barbecue ribs, and the brisket gave Miami roll. True creativity was expressed with the appearance of a cut called meaty ribs, while the disappearance of legs and shanks probably went unnoticed.

In Britain, a scattering of American names plus a few new names were used (Figure 203).

PSOAS MAJOR

In mammals, this relatively large muscle (1.7% total muscle weight) occupies a conspicuous position ventral to the transverse processes of the lumbar vertebrae (Figure 204). It is thin anteriorly where it starts under the last few ribs, but thickens to form a major eye of meat posteriorly. It is joined by *iliacus*, and inserts onto the trochanter minor on the femur.

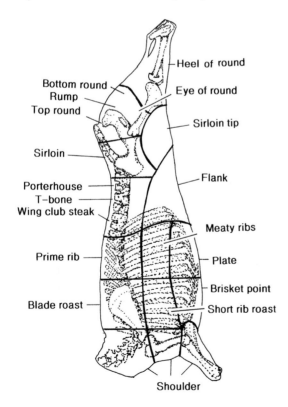

Figure 202. Proten beef cutting in Canada.

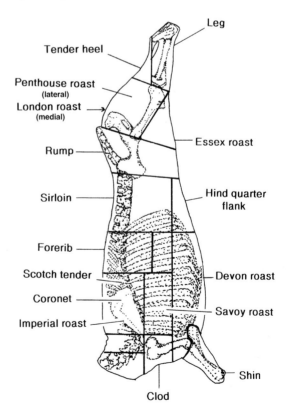

Figure 203. Proten beef cutting in Britain.

It has a low connective tissue content and is stretched by conventional carcass suspension methods so that it is almost always the most tender muscle in a meat animal and commands the highest price.

Figure 204. Lateral view of beef psoas major (solid black) showing its origins on the last rib (1) and transverse processes of the lumbar vertebrae (2), and its insertion on the femur (2).

PSOAS MINOR

In mammals, the *psoas minor* is a long, thin muscle, located medially to the *psoas major*. Thus, it is most readily seen near the midline of the body, running ventrally to the lumbar vertebrae.

PTARMIGAN

Ptarmigans are grouse-like gamebirds (→ Galliformes) found around the world, either up north or up high, wherever it is cold. *Lagopus mutus* is the rock ptarmigan and *L. lagopus* is the willow ptarmigan. The winter plumage is white. Ptarmigans are usually too tough for roasting and are best braised or stewed.

(Source: Vesey-Fitzgerald, 1949)

PUCHERO

Beef for the cooking pot, cut from the anterior rib region in → Argentina.

PUFFIN

→ Lundi

PUHA-HÁTSZÍN

Translated as the soft sirloin, this is a cut containing the dorsal abdominal muscles in a → Hungarian beef carcass.

PUHA-SZEGY

This is the soft-brisket or flank of a → Hungarian beef carcass.

PULPA

In Latin-American Spanish, pulpa refers to a mass of boneless meat. In → Mexico, the pulpa is derived from *supraspinatus.* In → Ecuador, there are several pulpa cuts including the pulpa blanca (*biceps femors*), pulpa negra (*adductor* and *semimembranosus*), pulpa redonda (*quadriceps femoris* group), and pulpa de brazo (*triceps brachii*).

PUNTA DE GANSO

The point of the goose contains the proximal part of *biceps femoris* in → Chilean beef cutting.

PUNTA DE PALETA

The *infraspinatus* in Cuban beef cutting or in → Chile. Literally the point of the shovel.

PUNTA DE PICANO

Tensor fascia lata in → Chilean beef cutting.

PUNTBORST

The anterior part of a beef brisket in the → Netherlands.

PUTH

The lumbar region of a carcass in → Pakistan.

PUUN OK

The postero-ventral thoracic region of a beef or buffalo carcass in → Thailand.

PUUN THONG

The ventral abdominal region of a beef or buffalo carcass in → Thailand.

PYLORIC CAECUM

A major factor affecting the rate of post-mortem autolysis in fish muscle is whether or not the fish are filleted fairly soon after capture. The pyloric caecum, a diverticulum of the intestine, may contain a high concentration of collagenases and trypsin-like enzymes. Leakage from the caecum accelereates autolysis of the musculature.

PYÖRISTETTY KINKKU

In → Finland, this is a centre cut ham on a pork carcass.

PYRIDINOLINE

Pyridinoline is a non-reducible, trifunctional cross-link between three → tropocollagen molecules. Its presence in the urine is a marker of pathological collagen degradation. Pyridinoline contributes to the increased heat stability of connective tissues from older animals. It can be detected on-line with an ultraviolet fibre-optic probe because it is fluorescent.

QUADRATUS FEMORIS

In beef and lamb, this is a small, parallel-fibred muscle originating from the ventral surface of the ischium and inserting on the trochanteric ridge of the femur (Figure 205). It is quite large in pork.

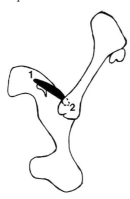

Figure 205. Lateral view of beef quadratus femoris *(solid black and broken line) showing its origin from the ischium (1) and insertion on the femur (2).*

QUADRATUS LUMBORUM

In beef and lamb, this is a thin muscle located laterally and ventrally on the transverse processes of the lumbar vertebrae (Figure 206). With origins as far forward as the eleventh thoracic vertebra, the insertions extend back to a ventral tubercle on the wing of the ilium. The muscle is well developed in pork.

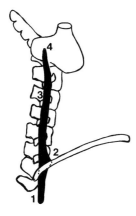

Figure 206. Lateral view of beef quadratus lumborum *(solid black and broken line) showing its origins from thoracic vertebrae (1), the last rib (2), and the ventral sides of the transverse processes of lumbar vertebrae (3), and its insertion on the ilium (4).*

QUADRICEPS FEMORIS

A group of four major muscles that pull on the patella in the mammalian hindlimb (*rectus femoris, vastus lateralis, vastus medialis,* and *vastus intermedius*).

QUAGLIA

Italian for → quail.

QUAIL

The quails are grouped together with the partridges and pheasants in the Family Phasianidae in the Order → Galliformes. The common quail in Europe, *Coturnix coturnix*, is like a miniature partridge with a very short beak. It produces a really small carcass (say 10 cm in length and 150 g in weight from a bird of 6 weeks). There are many different types of quail in North America. Perhaps the most well known is the bob-white, *Colinus virginianus*. Japanese quails are raised commercially and may be browned in butter then braised.

QUARTO

In → Brazil, the forequarter (quarto dianteiro) and hindquarter (quarto traseiro) of a carcass.

QUASI

A cut of beef near the femur in → Luxembourg.

QUERRIPPE

In → German beef cutting, this is a cross-rib cut similar to the Spannrippe.

QUEUE

French for tail. In → Luxembourg it may also include the sacral region of the ham.

QUIJADA

The jowl of an → Argentinean pork carcass, literally the jaw bone.

QUITITEÑA

Beef *serratus ventralis* and *subscapularis* in → Costa Rica.

RAAN

The whole hindleg of beef, veal, camel or lamb in → Pakistan.

RAAS AL-METLEH

Dorsal neck of beef in → Jordan.

RABADA

Oxtail in → Portugal.

RABADILHA

In → Portuguese beef cutting, the rabadilha is a knuckle cut containing the patella.

RABBIT

The rabbit (*Oryctolagus cuniculus*) was used extensively for meat production in Roman times, but did not reach England until the Norman invasion. Rabbit meat seems to go through cycles of popularity, depending on other events. With meat shortages during and after the second World war, and before myxomatosis, a tremendous tonnage of rabbit meat was sold and transported by rail in England. Now customers may have to go searching for rabbit meat at the retail level. Rabbit is common in Spain and southern France (from where the wild rabbit originated) and in some Asian countries.

Rabbit is hard to find at retail outlets in North America, although back-yard rabbit raising was common 50 years ago. Rabbit meat consumption has not recovered from the enormous success of the North American poultry industry in providing attractive, easy-to-cook white meat. Wild rabbits are, of course, extensively hunted in North America, but many are cottontails, various species of *Sylvilagus*. The favourite small game of the western plains of the USA is jackrabbit, which includes various species of the genus *Lepus* (→ hare).

Meat from domestic rabbits is tender and pale in colour, with a mild but unique taste. At carcass weights up to 2.5 kg from rabbits up to about 14 weeks, the rabbit carcass is typically jointed and fried or casseroled. Larger carcasses up to 4 kg are sometimes roasted whole. Much beyond this weight requires stewing. Some breeds, like the Flemish Giant, may reach 8 kg carcass weight, but the future of the industry is to supply small cut-up parts that are easy to cook. A typical 7-part cutting pattern is to detach the whole hindquarter just anterior to the ilium, then to split left from right hindlegs. The lumbar region is then separated from the thoracic region and kept intact (left plus right side as a saddle). Left and right forelimbs are separated from the ribcage, which is then split into left and right sides. As might be suspected of a prolific domestic animal, selective breeding has produced an astonishing number of variants of the basic wild stock. If one had to pick one breed that was most important and widespread commercially for meat production, it would probably be the New Zealand White which, despite its name, was developed in the USA. It has now become a generic white rabbit with numerous commercial strains and hybrids. Yields vary enormously between breeds, but have been reported around 51 to 55% for dressing percentage, and 4.1 to 4.8 for muscle to bone ratio in experimental strains. However, dressing percentages may be unreliable at face value because sometimes the liver, heart and kidneys are left in the carcass, while sometimes they are not. Leaving them in the carcass, a muscular rabbit might reach 60% dressing percentage. Rabbit meat may get more tender as animals grow and increase their level of intramuscular fat.

(Sources: Thompson and Worden, 1956; Portsmouth, 1979; Rue, 1981; Cheeke *et al.*, 1982; Sandford, 1986; Gondret *et al.*, 1998; Pla *et al.*, 1998)

RABO

The tail of a pork carcass in → Portugal or → Brazil.

RACCOURCI DE CÔTES

The plate côtes of a → Belgian beef carcass.

RACK

In many countries, a rack of lamb is a rib cut, including most of the ribs posterior to the scapula. Sometimes left and right sides of an unsplit carcass are used, as in a North American hotel rack, or the rack may be unilateral, as in → Australia and → New Zealand.

RÆKJA

Icelandic for → shrimp.

RAGBA

Camel neck in → Morocco.

RAKABEH

Neck of beef, veal or lamb in → Jordan.

RAQUETE

In → Brazilian beef cutting, the raquete is part of the paleta and includes the *infraspinatus*.

RAM

An entire male sheep.

RATITES

Several orders of flightless birds, including the → emu, cassowary and → ostrich.

RATOŃ DELANTERO

Beef *biceps brachii, brachialis, coracobrachialis*, distal extensors and flexors of the forelimb, and *gastrocnemius* in → Costa Rica.

RATOŃ TRASERO

Beef *gastrocnemius, peroneus, popliteus, tibialis*, and distal extensors and flexors of the hindlimb in → Costa Rica.

RATTAIL

A tapered fish of the cod family, → Gadiformes.

RATTLE

A group of → North American beef primals composed of plate plus chuck, plus brisket, plus shank.

RAUÐMAGI

Icelandic for → lumpfish. Grásleppa is salted lumpfish.

RAY

A flattened, cartilaginous fish, → Chondrichthyes.

REALE

A mid-rib cut medial to the shoulder in an → Italian beef carcass.

REBERNAYA CHAST

A ladder of midrib sections through a → Russian beef or lamb carcass.

REBHUHN

German for → partridge.

REBRA

Rib cuts in varying positions of beef, pork and veal carcasses in → Serbia & Montenegro.

RECTUS ABDOMINIS

A major muscle of the mammalian abdominal wall (≈ 2% total muscle weight) strengthened by bands of connective tissue (Figure 207). It originates from the sternum as far anterior as the fourth costal cartilage, and inserts on tendons and ligaments associated with the pelvis. In pork, the muscle is especially thick.

RECTUS CAPITIS DORSALIS MAJOR

In the mammalian neck, this muscle originates on the spine of the axis and is inserted on the occipital bone of the skull. The belly of the muscle is next to the → ligamentum nuchae.

RECTUS CAPITIS DORSALIS MINOR

Located beneath the *rectus capitis dorsalis major* in the mammalian neck, but originating on the atlas rather than the axis. In pork, this muscle is fused to the *rectus capitis dorsalis major*.

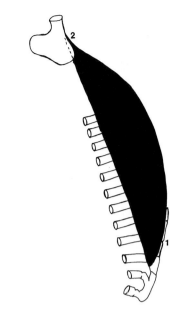

Figure 207. *Lateral view of beef* rectus abdominis *(solid black and broken line) showing its origin from the sternum (1) and insertion on the prepubic tendon and subpelvic ligament (2).*

RECTUS CAPITIS LATERALIS

This is the smallest of the muscles acting around the joint between the atlas and skull in mammals.

RECTUS CAPITIS VENTRALIS

A small neck muscle of mammals, located ventro-laterally to the joint between the atlas and the skull.

RECTUS CAPITIS VENTRALIS MAJOR

Longus capitis.

RECTUS FEMORIS

A major muscle of the mammalian hindlimb (> 2% total muscle weight). Of four large muscles that pull on the patella, this is the most anterior and centrally located (Figure 208). It originates with two tendinous heads on the ilium. The muscle is clearly separated in beef but often fused with the *vastus* muscles in pork.

Figure 208. *Lateral view of beef* rectus femoris *(solid black) showing its origins from the pelvis (1) and insertion on the patella (2).*

RECTUS THORACIS

In mammals, this is a thin muscle originating on the first rib and inserting onto the lateral surfaces of the second to fifth sternebrae (Figure 209). It is flat, broad, has parallel myofibres, and is located immediately below *pectoralis profundus*.

Figure 209. Lateral view of beef rectus thoracis *(solid black) showing its origins from the first rib (1) and its insertion on the sternum (2).*

RED FISH

A spiny bony fish, → Scorpaeniformes.

RED SNAPPER

A perch-like bony fish, also called the pargo, of the family Lutjanidae, → Perciformes.

REINDEER

Reindeer and caribou are very similar animals. Wild caribou provide a local source of meat for Inuit hunters in North America. Semi-domesticated reindeer are equally important to Saami herders in northern Norway, Sweden, Finland and Russia, but there is also a well developed industry providing reindeer meat as a gourmet item to urban supermarkets and restaurants. Caribou and reindeer are often regarded as the same species, *Rangifer tarandus* (but sometimes caribou are treated separately as *R. arcticus*). Both caribou and reindeer are unique in the deer family in that both males and females have antlers (although those of the female are smaller). Caribou tend to be larger and wilder than reindeer of similar age.

Reindeer are typically slaughtered at less than one year of age in central abattoirs to facilitate meat inspection. The meat is tender, relative to other large game animals, and no more than two days aging before cutting and freezing are required. The meat is best in early autumn, after summer feeding. The cull bulls are slaughtered first, then the winter slaughter proceeds, from November through to March. Gathering the animals is the main logistical problem, with modern methods of helicopter and snowmobile gathering comparing favourably with traditional methods by helping to preserve muscle → glycogen up to the point of slaughter (otherwise stress leads to → dark, firm, dry meat).

A wide range of reindeer meat products is produced, with cold-smoked reindeer meat topping the gourmet list. Reindeer steaks and roasts are generally of high quality, although quite small, because carcass weights are typically within a range from about 17 to 28 kg. Tougher cuts are marinated or stewed, or used for novel dishes such as reindeer tongue mousse (no pun intended). Swedish and Saami language names for reindeer cuts are shown in Figure 210. The Swedish names for reindeer cuts share a few similarities with those used for beef in → Sweden : such as lägg (shank), bringa (brisket), högrev (chuck), bog (shoulder) and hals (neck). However, the hindlimb is called a stek, following the general terminology used in Sweden for lamb, as well as in → Norway for both lamb and veal carcasses. Similarly, sadel may be used for either lamb or reindeer. Slaksida is the flank of the reindeer, with connotations of both a slaughter side (slakt = slaughter) and/or a slack side (slak = slack).

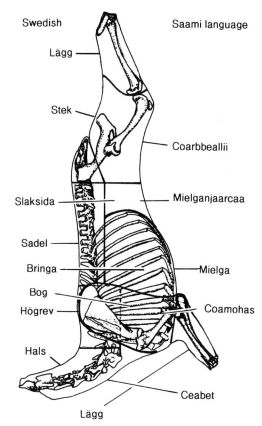

Figure 210. Swedish and Saami language names for reindeer cuts.

(Sources: Rue, 1981; Mäkelä *et al,* 1995; Wiklund, 1996, 1999)

REJE

Danish for → shrimp.

REN

Swedish for → reindeer.

RENO

Spanish for → reindeer.

RETICULAR FIBRES

Small diameter Type III → collagen fibres stained black with silver for light microscopy. They are highly branched and often appear as a network or reticulum. They dominate the → endomysium.

RETRACTOR COSTAE

This muscle retracts the last rib in the mammalian ribcage towards the origins of the muscle on the transverse processes of the first three lumbar vertebrae (Figure 211).

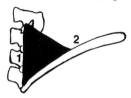

Figure 211. *Lateral view of beef* retractor costae *(black) showing its origins from the transverse processes of lumbar vertebrae (1) and its insertion on the last rib (2).*

REVBEN

Rib-bone cuts of meat in → Sweden.

REVBEN UTAN BRÖSTKAM

The sternal region of a → Swedish pork carcass after removal of the overlying limb bones and muscle of the bog.

REYKTUR LUNDI

→ Lundi

RHEA

Originating from South America, the greater rhea, *Rhea americana*, and the lesser rhea, *Pterocnemia pennata*, are the smallest of the ranched ratites, reaching live weights of about 35 kg and 20 kg, respectively. The lean meat yield of the carcass is about 64%, similar to that of ostriches. However, relative to the ostrich, the rhea has a higher proportion of wings, feet and head. The greater rhea is ranched commercially in the USA for its meat, hide, oil and feathers.

(Source: Sales *et al.*, 1997)

RHOMBOIDEUS

Located deeply between the scapulae, this mammalian muscle originates on the → ligamentum nuchae and the spines of the first five thoracic vertebrae (Figure 212). It inserts onto the costal surfaces of the scapula. In beef, it has two parts: a rounded cervical part, and a thinner thoracic part. The cervical part makes a major contribution to the hump of *Bos indicus* carcasses. Three parts to the muscle appear in lamb and pork.

RIB BLADE MEAT

In → Australian beef cutting, the rib blade meat contains *latissimus dorsi*, *trapezius* and *serratus ventralis* from ribs 6 to 8.

Figure 212. *Lateral view of beef* rhomboideus *(solid black and broken line) showing cervical (1) and thoracic (2) parts.*

RIBCARBONADE

Rib chops of pork in the → Netherlands.

RIBLETS OF LAMB

In North America, the distal parts of the ribs in a breast of lamb may be separated to create small chops called riblets.

RIB STEAKS

In many countries and types of meat, rib steaks from the middle and posterior thoracic region are composed of *longissimus dorsi* plus associated muscles.

RIEDDECKEL

A cut from the dorsal ribs in an → Austrian beef carcass.

RIF

A rib cut of beef in → Iceland.

RIG

An incompletely castrated male, typically a sheep.

RIGOR MORTIS

When ATP is depleted after slaughter (→ adenosine triphosphate), cross bridges develop between the → actin and → myosin molecules of the thin and thick → myofilaments. Filament sliding becomes impossible. Thus, the muscle becomes relatively inextensible. The length of time that rigor mortis takes to develop after a meat animal has been slaughtered or a fish has been caught depends on the amount of glycogen stored in the muscles. Factors such as stress, exercise or shivering that deplete muscle glycogen will accelerate the onset of rigor.

(Sources: Bendall, 1973; Skjervold *et al.*, 1999)

RIGOROMETER

Apparatus for measuring the development of → rigor mortis,

typically by periodically loading a strip of muscle to determine how far it extends. The main problems are that removal of the muscle strip from the carcass damages → myofibre and axonal membranes which then may activate → contraction and ATP depletion, and that the muscle strip must be kept moist, warm and anaerobic to simulate conditions within the carcass. Loading of the strip also damages the first myofibres to develop rigor mortis, giving a complex interaction between rigor development in a population of myofibres and the extent and frequency of loading. Despite these problems, information obtained by rigorometry is the foundation for understanding the conversion of muscle to meat.

(Source: Bate Smith, 1939; Bendall, 1973)

RINDFLEISCH

German for beef.

RIÑONADA

In → Cuban beef cutting, a rack of rib steaks extending back to the kidney.

RINTA

A brisket of beef in → Finland.

RIPPE

Rib of beef in → Switzerland.

RJÚPA

Icelandic for → ptarmigan.

ROAD-KILL EFFECT

Wild animals killed by cars and remaining at the side of the road have a powerful effect on the general public - they do not want to eat anything that might have originated from the species they have just sighted, flattened in the road. In England it may be pheasant or rabbit, in North America it may be deer, or in Australia it may be kangaroo and wallaby. Adverse publicity from road-kill is a major, but generally unrecognised negative factor in attempts to develop specialty markets for game meat.

ROASTBEEF

In beef cutting in → Germany, → Switzerland and → Luxembourg, the Roastbeef is a premium cut in the lumbar and posterior thoracic region.

ROCK CRAB

Western Atlantic shore crab, *Cancer irroratus*, → crabs.

ROCK-DOVE

→ Pigeon

ROCK FISH

A bony fish, often highly coloured with spines, → Scorpaeniformes.

ROCK SALMON

The rock salmon of British fish and chip shops is derived from the lesser spotted dogfish, *Scyliorhinus caniculus*, → Chondrichthyes.

RODAJA

Distal meat in the hindlimb of a beef carcass in → Ecuador.

RÖDSPÄTTA

Swedish for plaice, → Pleuronectiformes.

ROLLARDEN

In North America, a retail cut of beef made by thinly slicing through the inside (top) round (*adductor* and *semimembranosus*).

ROLLED JOINT

After a meat cut has been boned (i.e., its bones have been removed), it may be used to make a rolled joint. Areas of subcutaneous fat are sliced, then fused together by being beaten with a mallet. A sheet of fat matching the size of the lean meat is cut, wrapped around the lean, and tied into place with white butcher's string with a slip knot secured by a half-hitch. The free ends of the string are cut short (so they do not smoulder in the oven). The whole cylindrical rolled joint is cut into variable lengths. The layer of fat bastes the meat as it is roasted, and prevents the lean from drying out. The customer is able to carve the joint very easily, with a transverse cut passing across the grain of the meat to avoid stringiness.

ROMSTECK

The romsteck is a hindlimb cut of → French beef, dominated by the gluteal muscles and overlapped by *biceps femoris*. The posterior portion of the romsteck may be separated as a culotte (including sacral vertebrae three to five, plus the first caudal vertebra).

RONDE

In → French beef cutting , the ronde de tranche grasse is the *rectus femoris*, while the ronde de tranche is the *adductor* muscle. The ronde fricandean is located medially in a → Netherlands veal carcass.

ROSE

In → French beef cutting, sometimes the tranche grasse is called the rose, while the macreuse may be called the rose of the shoulder, rose d'épaule.

ROSE BEEF

Loin of beef in → Jordan.

ROSE FISH

A bony fish of the North Atlantic, with an orange-red body, black eyes, and sharp spines on the head, → Scorpaeniformes.

ROSTBEF

The lumbar region of a beef carcass in → Poland.

ROSTBIFF

In → Australian beef cutting, the rostbiff is part of the sirloin butt and is restricted to the gluteal muscles. In → Sweden, the rostbiff med ben (with bone) is a cut of beef through the ilium.

ROSTBRATEN

Prime rib roast on an → Austrian beef carcass.

ROSTÉLYOS

Mid-thoracic rib roasts from a → Hungarian beef carcass.

RÔTI

French for a roast.

RÔTI À LA TACHE NOIRE

Hindlimb leg roast of pork in → Belgium.

RÔTI D'ÉPAULE

A shoulder roast of pork in → Belgium.

RÔTI DE PORC

Hindlimb leg roast of pork in → Belgium.

ROUELLE DE PORC

Pork shoulder roast or rôti d'épaule in → Belgium.

ROULÉ DE PORC

Pork shoulder roast or rôti d'épaule in → Belgium.

ROULETTE

In France, roulette are small wheel-like rounds of meat, usually taken transversely across the distal part of the hindlimb.

ROUND

In North American beef cutting, the round is part of the hip, and often is sliced perpendicularly to the femur. Although it is seldom sliced whole right across the hindlimb (full-cut

round), this provides a key to identifying muscles when the round is subdivided (Figure 213). The *gracilis* is a thin sheet of muscle spread over the medial face of the hindlimb. The *gracilis*, together with the strap-like *sartorius* muscle anterior to the *gracilis*, may be used for orientation. The key is to decide which is medial (i.e., towards the thin *gracilis*) then, which is anterior (towards the *sartorius*). Once this has been decided, the remaining muscles grouped around the femur may be named fairly easily. The *quadriceps femoris* muscles form a group of four large muscles that pull on the patella when the leg is extended. The *vastus medialis* is medial, the *vastus lateralis* is lateral, the *vastus intermedius* covers the anterior face of the femur, and the *rectus femoris* covers the *vastus intermedius*. The *biceps femoris* is a single large muscle on the lateral face of the hindlimb. In cross section, it often appears divided into two parts because it has a very deep cleft along part of its length. But the *biceps femoris* appears as a single muscle in cuts of meat which miss the cleft, whereas sections through the cleft make the muscle appear double. To add to the confusion, the small segment of *biceps femoris* cut off by the cleft is often more pale than the main part of the muscle. The *semitendinosus* and *semimembranosus* are two large muscles located on the posterior face of the hindlimb. The *semimembranosus* is medial to the *semitendinosus*. The *adductor* and *pectineus* are located in the medial part of the hindlimb, near to the femur. The *pectineus* is anterior to the *adductor*. In lean carcasses, it may be difficult to separate the *adductor* from the *semimembranosus*, and these two muscles may appear as a single muscle. Given the possibility that the *biceps femoris* muscle on the other side of the limb may look like two muscles, caution is needed in the identification of these muscles.

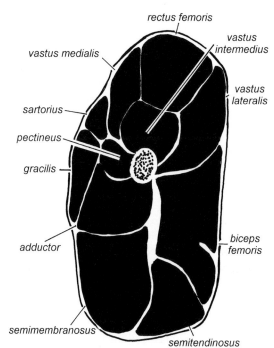

Figure 213. Muscles of a North American whole round steak of beef.

For beef, a round also is cut in → Scotland, Northeast → England, Liverpool, and → Northern Ireland, all dating back to the London 1876 pattern.

ROUND BONE ROAST OR POT-ROAST

In → North American beef cutting, an arm steak or pot roast.

RÖVIDKARAJ

The thoracic region of a pork loin in → Hungary.

ROZBRATEL

In → Poland, a cut of beef from the anterior thoracic region.

RÜCKEN

A cut medial to the shoulder in → Swiss beef cutting. In → Austria it is a loin of lamb.

RÜCKENSPECK

Back-fat of a pork carcass in → Germany or → Switzerland.

RUDD

Scardinius erythrophthalmus, in the → Cypriniformes, has tasty flesh, but with numerous small bones.

RUGSPECK

Back-fat of a pork carcass in the → Netherlands.

RUMP

In → North America, a rump of beef usually is called a standing rump and has a triangular shape. It is cut from the hip of beef and contains the proximal end of the femur (knuckle bone), the ischium (aitch bone), and coccygeal vertebrae (tail bones). It has a triangular shape, with one cut surface matching the sirloin and the other matching the round steak. In Great Britain, however, the beef rump is much larger and extends anteriorly to include the anterior face of the *ilium*. Thus, a modern English rump steak is similar to a North American sirloin steak (Figure 214 and 215).

However, it is possible that the North American pattern of rump and round may have originated in England and reached America via Liverpool. To add to the confusion, an → Australian beef sirloin may be called a rump, while a → New Zealand rump may be called a sirloin.

RUMSTEAK

A major cut of beef including the whole pelvis in → Luxembourg.

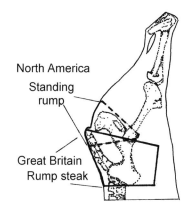

Figure 214. *Positions of a standing rump in North America versus a rump in Great Britain.*

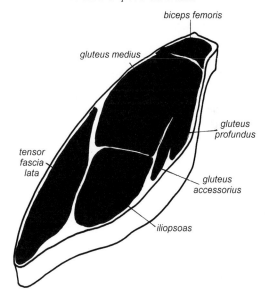

Figure 215. *English rump steak.*

RUNDERACHTERVOET

The hindquarter in → Netherlands meat cutting.

RUNDERVOORVOET

The forequarter in → Netherlands beef cutting.

RUNDSTEK

In → Norwegian beef cutting, the rundstek is anterior to the femur and includes the *quadriceps femoris* group of muscles.

RUSSIA - BEEF CUTS

The separation of the fore- from the hindquarter may leave all the ribs in the forequarter (Figure 216). The main primal cuts of the hindquarter are a full round (tazobedrennaya chast), loin (poyasnichnaya chast) and flank (paschina). The *psoas* fillet is the virezka. The forelimb (lopatochnaya chast) is removed from the ribcage which then is sectioned in an antero-posterior direction to give the spine (spinnaya chast) and neck (scheynaya chast), a ladder of midrib sections (rebernaya chast) and a brisket (groudinka). Both hind- and foreshanks are called golyashka.

(Source: Lissitsyn, 1999)

RUSSIA - LAMB CUTS

Russian lamb cuts similar to those for beef are shown in Figure 217.

(Source: Lissitsyn, 1999)

RUSSIA - PORK CUTS

The trotters (nojka), tail (khvost) and head (golova) are removed with the edible offals in the abattoir and are not found on the commercial carcass (Figure 218). The pattern of carcass cutting is similar to that of beef, except that the paschina is not removed separately. The section of the forelimb containing the ulna and radius is called the roulka rather than the golyashka.

(Source: Lissitsyn, 1999)

RYANODINE RECEPTOR

The ryanodine receptor is the calcium-ion releasing channel of the → sarcoplasmic reticulum, named from its binding of ryanodine, a plant alkaloid poison. A genetic defect causing a single amino acid replacement in the calcium ion release channel protein is responsible for the → porcine stress syndrome and many of the worst cases of → PSE pork.

(Sources: MacLennan and Phillips, 1992; Mickelson *et al.*, 1992; O'Brien *et al.*, 1993)

RYGGBRINGA

On a → Swedish beef carcass, the ryggbringa or back-brisket contains the proximal parts of the ribs removed in trimming an entrecote.

RYGG MED HALS

Back with neck, a cut of lamb in → Sweden.

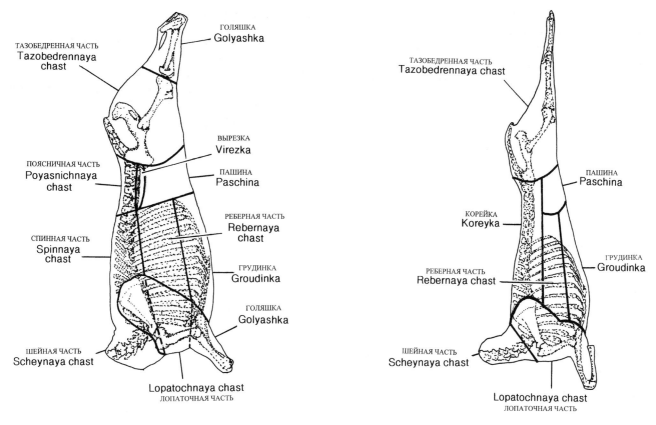

ГОЛЯШКА
Golyashka

ТАЗОБЕДРЕННАЯ ЧАСТЬ
Tazobedrennaya chast

ВЫРЕЗКА
Virezka

ПОЯСНИЧНАЯ ЧАСТЬ
Poyasnichnaya chast

ПАШИНА
Paschina

РЕБЕРНАЯ ЧАСТЬ
Rebernaya chast

СПИННАЯ ЧАСТЬ
Spinnaya chast

ГРУДИНКА
Groudinka

ГОЛЯШКА
Golyashka

ШЕЙНАЯ ЧАСТЬ
Scheynaya chast

Lopatochnaya chast
ЛОПАТОЧНАЯ ЧАСТЬ

Figure 216. Russia – beef cuts.

ТАЗОБЕДРЕННАЯ ЧАСТЬ
Tazobedrennaya chast

ПАШИНА
Paschina

КОРЕЙКА
Koreyka

ГРУДИНКА
Groudinka

РЕБЕРНАЯ ЧАСТЬ
Rebernaya chast

ШЕЙНАЯ ЧАСТЬ
Scheynaya chast

Lopatochnaya chast
ЛОПАТОЧНАЯ ЧАСТЬ

Figure 217. Russia – lamb cuts.

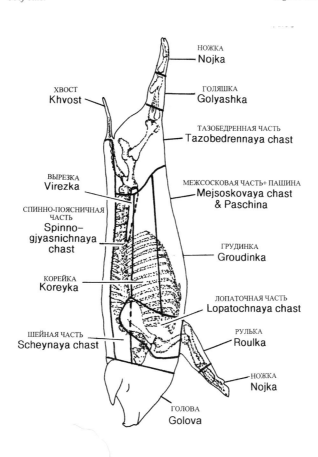

НОЖКА
Nojka

ХВОСТ
Khvost

ГОЛЯШКА
Golyashka

ТАЗОБЕДРЕННАЯ ЧАСТЬ
Tazobedrennaya chast

ВЫРЕЗКА
Virezka

МЕЖСОСКОВАЯ ЧАСТЬ+ ПАШИНА
Mejsoskovaya chast
& Paschina

СПИННО-ПОЯСНИЧНАЯ ЧАСТЬ
Spinno-gjyasnichnaya chast

ГРУДИНКА
Groudinka

КОРЕЙКА
Koreyka

ЛОПАТОЧНАЯ ЧАСТЬ
Lopatochnaya chast

ШЕЙНАЯ ЧАСТЬ
Scheynaya chast

РУЛЬКА
Roulka

НОЖКА
Nojka

ГОЛОВА
Golova

Figure 218. Russia – pork cuts.

SAAM CHAAN

The belly of a pork carcass in → Thailand.

SACROCOCCYGEUS DORSALIS LATERALIS

A long, thin muscle originating from the fifth and sixth lumbar vertebrae and inserting past the sacrum into the tail.

SACROCOCCYGEUS DORSALIS MEDIALIS

The most dorsal muscle along the mammalian tail, pulling the coccygeal vertebral spines towards the sacral spines.

SACROCOCCYGEUS VENTRALIS

A mammalian tail muscle located ventrally to the diminutive transverse processes of the coccygeal vertebrae.

SADDLE OF LAMB

In many countries, lamb carcasses may be cut with transverse cuts across the vertebral column so as to leave right and left sides of the carcass together. Thus, right and left loins of lamb together form a saddle.

SADEL

A saddle of lamb or reindeer in → Sweden.

SADRIA

Sternal region of a beef carcass in → Morocco.

SAEDANIA

Ribs on a camel carcass in → Morocco.

SAIGA

Saiga antelope, *Saiga tatarica*, was once near extinction, but now has been revived to become a major source of meat from commercial hunting in Russia.

(Source: Sokolov and Lebedeva, 1989)

SAITHE

A cod-like bony fish or pollack, *Pollachius virens*, → Gadiformes.

SAK

Distal leg of poultry in → Morocco.

SALAME

→ Salami

SALAMI

A dry → sausage, typically containing large particles of pork

with garlic seasoning and flavoured by the action of lactobacilli.

SALCHICHA

Spanish for → sausage.

SALMON

A salmon is a large bony fish with a pigmented myosystem, → Salmoniformes, but it is also a fanciful name given to the beef *semitendinosus* in Cumbria and some parts of → Scotland and→ Northern Ireland. However, like all fanciful names, this may be unreliable, with other parts of Scotland designating the *supraspinatus* as the salmon muscle. The likeness between the *semitendinosus* and a good-sized salmon can be quite striking, especially after taking a couple of distillery tours. Winking of the salmon's eye is a useful early warning of dipsomania.

SALMONIFORMES

In this order of important bony fish, the family Salmonidae includes various types of salmon, trout, charr and whitefish. Most of them have an adipose fin between the dorsal fin and the tail (Figure 219).Although they cannot tolerate water that is too warm or with a low oxygen concentration, they are able to cope with a wide range in salinity and many species are anadromous (they migrate between sea water and fresh water environments). Typically, salmonids hatch in fresh water, but may migrate to the sea to grow and mature, and then return again to fresh water to spawn. When salmon start on their return journey to spawn in fresh water, their muscles have a high oil content and a bright red colour. After an exhausting journey up-river their muscles become darker and less oily. The dull skin of the sexually mature salmon is called watermarked. The family Salmonidae contains many sought-after sport fish with richly flavored muscle. Smoked salmon is an important luxury food with unusual textural properties that enable it to be thinly sliced. The colour of salmonids is of great importance in marketing. The carotenoid pigment of wild salmon muscle is → astaxanthin. Canthaxanthin (carophyll red) may be added to the feed of farmed fish to enhance colour development, using a reflectance ratio (650/ 510 nm) to estimate the pigment concentration.

Salmo   Trout

Figure 219. Trout.

The Atlantic salmon is *Salmo salar*. The rainbow or steelhead trout is *S. gairdneri*, often used for aquaculture because the larvae will accept artificial food. It may be distinguished by a reddish band along the centre of each side and by regular rows of black dots on its dorsal and tail fins. *S. trutta*, the brown trout may be distinguished by having a mixture of black spots and red, orange and yellow spots with pale borders. *Oncorhyncus* includes various Pacific salmon, with the five main types being chinook salmon (*O. tshawytscha*, the largest, from 6.75 to 25 kg); pink salmon (*O. gorbuscha*, small, ≈ 1.8 kg); sockeye salmon (*O. nerka*, its red myosystem and high oil content make it ideal for canning); chum salmon (*O. keta*, has a pale myosystem with a low oil content and is sold fresh or smoked); and the coho salmon (*O. kisutch*, a saltwater sport fish famed for its agility). The Russian-sounding specific names of these fish suggest, correctly, that these species are found in both western and eastern waters of the North Pacific. The genus *Salvelinus* includes the charr, brook trout, lake trout, and Dolly Varden (*S. malma*). The Arctic charr, *S. alpinus* has a delicious taste with attributes of both trout and salmon, and is farmed with pigment supplementation to produce a pink, carotenoid colour. The brook trout, *S. fontinalis* may be distinguished by red spots with blue surrounding halos along its sides, black edges bordered by white along the front edges of the lower fins, and by light, wavy lines on its back. The lake trout, *S. namaycush*, may be distinguished by its light spots and deeply forked tail. It can grow to a very large size (47 kg). *Coregonus*, the whitefish, is usually found in deep, cold lakes (various species are known as types of cisco).

Fishes of the family Argentinidae look like miniature salmon. They are usually cleaned and served whole after being pan fried or deep fried, such as *Mallotus*, the capelin and *Osmerus* the smelt, with large, easily removed scales. Anadromous smelt are superior in quality to landlocked fish from the Great Lakes. *Thaleichthyes*, the eulachon occurs in the Pacific, and ascends rivers to the north of Oregon. It has a long dorsal fin, weak dentition and is found in very large schools. When pan-fried it has an excellent taste that rivals that of trout. The eulachon contains a large amount of oil with a high melting point, sufficiently high for use in making candles, hence an old name for this fish, the candle fish. The oil has a delicate flavour.

The family Esocidae includes sport fish such as the northern pike, *Esox lucius*, which is circumpolar, occurring in Europe, Asia and North America. The pike is a wolf-like predator with an elongated shape, military-style camouflage, and dorsal and anal fins far back near the tail. It has a large mouth with impressive bands of sharp teeth. The myosystem is firm, white, and finely textured when cooked. Similar relatives include the muskellunge, *E. masquinongy*.

(Sources: Skrede and Storebakken, 1986; Ryan, 1988)

SALON

Semitendinosus cut from a beef carcass in → Ecuador.

SALSICCIA

Fresh → sausage in Italy.

SALT (SODIUM CHLORIDE)

Salt is used extensively in meat processing because it causes the → myofilament lattice to swell, thus increasing → water-holding capacity, and because dilute salt solutions dissolve → myosin which then can be heat-set to bind meat fragments together.

(Source: Wilding *et al.*, 1986)

SALT PORK

Pork fat after → curing and smoking.

SAMKEUBSAL

The abdomen and postero-ventral region of a → Korean pork carcass.

SANDHVERFA

Icelandic for turbot (→ Pleuronectiformes)

SAND LANCE

An eel-like, burrowing fish of the Atlantic, *Ammodytes*, → Perciformes.

SAN KHO MUU

Scapular region of a pork carcass in → Thailand.

SAN LAANG

The posterior thoracic region of a beef or buffalo carcass in → Thailand.

SAN LAY

Forelimb of a pork carcass in → Thailand.

SAN NAY

The lumbar region of a beef or buffalo carcass in → Thailand.

SAN NOOK

In → Thailand, the sacral region of a beef or buffalo carcass or a pork loin.

SAPO

Spanish for a toad, but more likely the beef *gastrocnemius* muscle, as in → Cuban beef cutting.

SA POOK

In → Thailand, the sa pook is the hindlimb of a beef or pork (san sa pook) carcass. Beef is cut into a medial sa pook nay and a lateral sa pook nook.

SARATOGA CHOPS

In North America, muscles medial to the scapula in a shoulder of lamb may be removed as a sheet, rolled, then cut into chops (Figure 220). A wooden skewer prevents the chop from unrolling. The major muscle is usually *serratus ventralis*.

Figure 220. Canadian Saratoga lamb chop.

SARCOLEMMA

The → myofibre, being a giant cell, is bounded by a plasma membrane, on which is a viscous layer called the → basement membrane. Embedded in the basement membrane are the → reticular fibres of the → endomysium. The term sarcolemma is usually used for all three layers (plasma membrane, basement membrane plus endomysium).

SARCOMERE

Muscles contract by thick → myofilaments pulling themselves past thin myofilaments. The thin myofilaments are anchored in the → Z-line. Thus, the smallest contractile unit is from one Z-line to the next, which has the following sequence of parts: Z-line, thin myofilament, thick myofilament, thin myofilament, next Z-line, as shown in Figure 221. Sarcos means muscle. Meros means part.

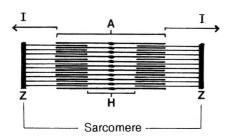

Figure 221. A sarcomere composed of an A band, plus half an I band on each side, and extending from one Z-line to the next.

The orderly arrangement of sarcomeres along the length of a myofibril creates transverse striations. This, plus the → cytoskeleton keeping sarcomeres of laterally adjacent myofibrils in register, creates transverse striations along the length of myofibres. Sarcomere length decreases when a muscle contracts and increases when a muscle is stretched.

In general, meat with short sarcomeres tends to be tougher than equivalent meat with long sarcomeres (except sometimes at the extremes of the range where sarcomere structure becomes disrupted).

SARCOMERE LENGTH

New → sarcomeres are added at the ends of → myofibrils as → myofibres grow in length. But if muscle usage becomes restricted (as in a patient with a broken arm kept too long in a sling) then sarcomeres are removed to shorten the whole muscle (myostatic contracture). Because the relationship between sarcomere length and muscle length is fairly dynamic and regulated biologically, it is difficult to be precise about the absolute length for a sarcomere at rest length. For working with beef, pork and lamb, the following approximation is fairly robust (but is not applicable to all food myosystems, some of which have a radically different sarcomere structure).

If the length of a thin myofilament is about 1 μm, and the length of a thick myofilament is about 1.5 μm, then a sarcomere length of 1.5 μm means the ends of the thick myofilament are rammed into the Z-line. Thus, sarcomeres are fully contracted at near minimum length. If, on the other hand, a muscle is severely stretched when it sets in rigor mortis, then thick and thin myofilaments are pulled right past each other to give a maximum sarcomere length of approximately 1 + 1.5 + 1 = 3.5 μm (for thin + thick + thin myofilaments). A key point to bear in mind when working with meat is that, if some myofibres are sinuously folded with long sarcomeres while adjacent myofibres are contracted with short sarcomeres, then mean sarcomere length rapidly becomes meaningless. Two methods are in common use for measuring sarcomere length - light microscopy and laser diffraction (→ transverse striations). The presence of sinuously folded myofibres is obvious under the microscope, and causes fuzzy, poorly defined diffraction patterns with the laser method. Another problem which also interferes with the correlation of sarcomere length with meat toughness is that part of the length of a single myofibre may form a contraction node with very short sarcomeres, which then stretches the myofibre somewhere else along its length to produce very long sarcomeres. Other common errors include improperly calibrated micrometer devices or image analysis systems, and shrinkage of prepared sections by fixation or embedding procedures. Beware - not all transverse striations have a periodicity equal to the length of one sarcomere, especially at the extremes of the range in sarcomere length or with optical apparatus used improperly.

The relationship between sarcomere length and meat tenderness involves the refrigeration history of the meat. At the onset of rigor mortis, muscles may be stretched or contracted, depending largely on their position in the hanging carcass. Stretched muscles tend to produce meat that is more tender than that from contracted muscles. Rapid cooling before the start of rigor mortis may enhance the differences between muscles that are stretched and those that are unrestrained because cold shortening may occur in the latter. Freezing of meat before the completion of rigor mortis leads to extreme shortening when meat is thawed (thaw shortening). As well as its effect on sarcomere length, muscle shortening also causes a decrease in the number of myofibres transected per unit of muscle cross sectional area and changes the angles of collagen fibres in the endomysium and perimysium.

(Source: Marsh and Carse, 1974; Tornberg, 1996; Koh and Herzog, 1998)

SARCOPLASM

The cytoplasm within a → myofibre is given a special name - sarcoplasm. It has all the properties of ordinary cytoplasm, plus some special proteins characteristic of the myofibre. In typical meat, the proteins dissolved in the sarcoplasm amount to almost half the weight of the myofibrillar proteins. The sarcoplasmic proteins are dominated by enzymes for → glycolysis, by creatine kinase (→ creatine phosphate) and by → myoglobin. The sarcoplasm provides the immediate energy for muscle contraction. The major enzymes of the sarcoplasm are too large to fit between the thick and thin → myofilaments of → myofibrils so that the aqueous space between myofilaments has few dissolved proteins (→ myofilament lattice). Glycolytic enzymes in the sarcoplasm are concentrated at the level of the → I-bands of myofibrils, and the three proteins forming the bulk of the thin myofilament (F-actin, tropomyosin and troponin) all may bind to a number of the glycolytic enzymes. The glycogenolytic enzymes are bound to parts of the sarcoplasmic reticulum. The localized distribution of these various enzymes and substrates is important and, when they are mixed in dilute solutions for biochemical analysis, the results may be quite different from those that actually occur in meat. In intact myofibres, biochemical reactions may occur much more rapidly since all the appropriate components are tightly fitted together at a high concentration.

SARCOPLASMIC RETICULUM

The sarcoplasmic reticulum is a membranous system surrounding the → myofibrils within a → myofibre (Figure 222). It occupies about 4% of the volume of the myofibre, but is better developed in fast-contracting than in slow-contracting muscles (→ histochemical types of myofibres). It is called the sarcoplasmic reticulum because sarcoplasm is the special name for cytoplasm inside a myofibre and because the sarcoplasmic reticulum was first discovered by the way it precipitates silver salts. Thus, the silver was precipitated in the spaces between the myofibrils to form a reticulum or net-like structure.

The sarcoplasmic reticulum uses energy from → ATP to collect or sequester calcium ions. Thus, any calcium ions close to the sarcoplasmic reticulum membrane are grabbed and moved to the interior of the sarcoplasmic reticulum. Because ions are constantly in motion, bumping into each other and spreading themselves uniformly through the space they occupy, after one calcium ion is grabbed, its place is soon taken by another which, in turn, ends up inside the

sarcoplasmic reticulum. But, when prompted to do so by events initiated from the central nervous system, the sarcoplasmic reticulum releases calcium ions. Calcium ions are the signal, detected by troponin and relayed to tropomyosin, for the start of muscle → contraction caused by myosin molecule heads.

(Source: Briggs *et al.*, 1977; Schmalbruch, 1979)

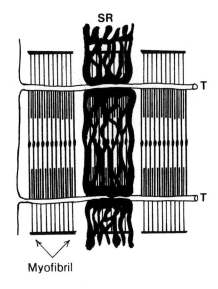

Figure 222. Diagram of the sarcoplasmic reticulum (SR) covering one myofibril inside the myofibre.

SARCOSPAN

A protein embedded in the plasma membrane of → myofibres, allowing tension generated within the contracting myofibre to be transmitted laterally outside the cell. Figure 223 shows the lateral linkages that may be involved.

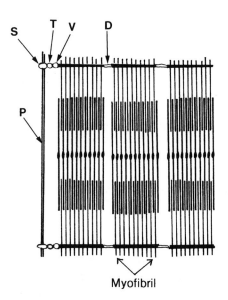

Figure 223. Lateral transmission of the force of myofibrillar contraction to the plasma membrane (P) of the myofibre, via desmin (D) between adjacent Z lines, via vinculin (V) attached to the α-actinin of Z lines, and via talin (T) to sarcospan (S).

(Source: Crosbie *et al.*, 1997)

SARDINE

A small herring-like fish such as *Sardinia*, in the → Clupeiformes , but also a French term in beef cutting for the muscular part of the diaphragm.

SARTORIUS

A long strap of muscle on the medial face of the proximal part of the mammalian hindlimb (Figure 224). It is not particularly large (0.4% of total muscle weight) but is important for orienting the other muscles of the ham or beef round. Proximally, it has two heads, one originating on the tendon of the *psoas minor*, and the other from the ilium. Distally, it thins down to an insertion on the tibia and medial patellar ligament. The *sartorius* is an ideal muscle for the study of muscle growth in meat animals because of its relatively simple structure with parallel myofibres.

Figure 224. *Medial view of beef* sartorius *(solid black) showing its origin with two heads (1) and insertion to the tibia (2).*

SATAE

The foreshank of a → Korean beef carcass.

SAUCISSON

French for sausage.

SAURY

An Atlantic bony fish with a long beak, *Scomberesox saurus* → Beloniformes.

SAUSAGE

Chopped meat and seasoning packed into a → casing which becomes cylindrical or bulbous in shape. There is tremendous variation in: (1), the size of the product; (2), the source of the meat, (3), the size of the meat particles; the (4), fat, (5), salt, (6), water, (7), binder and (8), filler content; (9), whether the sausage is cooked during manufacturing or by the consumer (sometimes both); (10), the type and strength of spicing; (11), the type of casing, primarily whether synthetic or from some

part of the alimentary tract; and whether or not the product has been (12), cured, (13), fermented, or (14), smoked. The order in which ingredients are added, the methods of chopping and mixing, and time-temperature interactions all may affect the final nature of the product.

Fresh sausage These contain particles of fresh meat and are not fermented, smoked or cooked during manufacturing. They are usually identified by the species from which the meat was derived.

Semidry sausage A fermented product containing coarsely chopped beef and/or pork with a final water content of 40 to 45%. Smoking and cooking during manufacture, combined with a high salt content and low pH (< 5.3), produce a strong taste and good keeping properties. Examples are cervelat, göteborg, holsteiner, summer sausage and thuringer.

Dry sausage A fermented product which owes its good keeping properties to a high salt content and very low water content. Some are air-dried without cooking (e.g., genoa salami), some are cooked (e.g., hard salami), and some are either air-dried or cooked (e.g., pepperoni).

Emulsion sausage These are based on the creation of a meat → emulsion which is set by cooking. Examples are frankfurters and bologna in small and large casings, respectively. Liver sausage and braunschweiger have a high content of liver, but also usually some skeletal muscle, such as pork jowl with pork liver. Cotto salami is an emulsion sausage which is smoked.

(Source: Pearson and Tauber, 1984; Kutas, 1984).

SAVOY ROAST

Name used for the leg of mutton cut in British → Proten beef.

SCALE FISH

A flat fish, *Arnoglossus*, → Pleuronectiformes.

SCALENUS DORSALIS

A thin sheet of muscle in mammals, originating from the transverse processes of the fifth to sixth cervical vertebrae and inserting on the fourth rib (Figure 225).

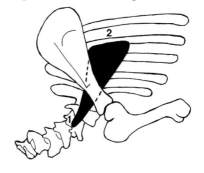

Figure 225. *Lateral view of beef* scalenus dorsalis *(solid black and broken lines) showing its origins from the transverse processes of cervical vertebrae (1) and insertion on the fourth rib (2).*

SCALENUS PRIMAE COSTAE

Scalenus ventralis.

SCALENUS VENTRALIS

A fan-shaped mammalian muscle, originating from the transverse processes of the third to seventh cervical vertebrae, and inserting on the manubrium and first rib (Figure 226).

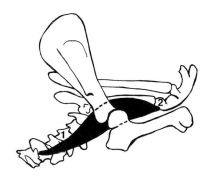

Figure 226. *Lateral view of beef* scalenus ventralis *(solid black and broken lines) showing its origins from the transverse processes of cervical vertebrae (1) and its insertion down the first rib to the manubrium (2).*

SCALLOP

Many types of scallops are exploited commercially, such as the giant, *Placopecten magellanicus*, and Iceland scallops, *Chlamys islandicus*, taken in Atlantic Canada, as well as the bay, *Aequipecten irradians*, and calico, *A. gibbus*, scallops taken along the American east coast. In Europe, the main species are the scallops *Pecten maximus* and *P. jacobaeus*. Scallops may be reared in a hatchery. All are filter-feeding bivalve molluscs, with a flat lower valve, which is smooth and pale, hinged against an upper valve which is arched and reddish brown. Scallops can move quite energetically by sharply closing their valves and jetting out water. The single large white adductor muscle between the two valves produces a myosystem with an excellent taste and texture, best appreciated when cooked in white wine. Scallops grow slowly, normally reaching a commercial size after five to seven years. Older scallops, up to 20 years, may occur, but have poor meat quality. Scallops are shucked as soon as they are caught by dredging.

Scallop → myofibres are thin and ribbon-like (about 1 x 10 μm in cross section and 0.6 mm in length). Sarcomere lengths are highly variable, and thick myofilaments have a core of paramyosin. When hardness develops in cooked scallop muscle, initially it is caused by dehydration, followed by denaturation of myofibrillar proteins.

(Sources: Findlay and Stanley, 1984; Robert, 1988; Laing and Psimopoulous, 1998)

SCAMONE

An → Italian cut of beef containing the *gluteus* muscles.

SCAMPI

Abdominal muscle from a clawed → lobster, *Nephrops norvegicus*.

SCHAB

A truncated loin containing posterior thoracic and anterior lumbar parts of the vertebral column in a → Polish pork carcass.

SCHALE

A cut from the medial part of the hindlimb in an → Austrian beef, lamb or veal carcass.

SCHAUFELSTÜCK

The shovel-piece or Schaufelstück is a → German beef cut from the middle of the Schulter, including part of the scapula

SCHENKEL

The shank of a carcass in the → Netherlands or → Switzerland.

SCHENKELVLEESCH

→ Schenkel

SCHEYNAYA CHAST

The neck of a beef, pork or lamb carcass in → Russia.

SCHIJF

A lateral cut from the ham of a → Netherlands pork carcass.

SCHINKEN

A pork ham in → Germany or → Switzerland. Schinkenspeck is a pork ham wrapped in a pork belly and cured and smoked together. Schinkenwurst is an emulsion → sausage containing chunks of ham.

SCHLEPP

The tail of a carcass in → Austria.

SCHLÖGEL

Leg of lamb in → Austria.

SCHLUSSBRATEN

A pelvic cut from an → Austrian pork, lamb or veal carcass.

SCHMER

Belly of pork in → Switzerland.

SCHNEPFE

German for → snipe.

SCHOPFBRATEN

Neck and anterior thoracic region of an → Austrian pork carcass.

SCHOUDER

The shoulder of a beef carcass in the → Netherlands.

SCHOUDERLAP

A beef cut from the *subscapularis* of a → Netherlands beef carcass.

SCHOUDERMUIS

A beef cut from the *supraspinatus* of a → Netherlands beef carcass.

SCHULTER

In → German beef cutting, the shoulder cut or Schulter is composed of all three heads of *triceps brachi*. In German and → Swiss pork cutting, the Schulter (or Schultern) includes most of the scapula, humerus and associated muscles. In Swiss beef cutting, the whole shoulder is the Schulter. The same term in used in → Austria for a shoulder of lamb.

SCHULTERBLATTKNORPEL

In → German beef cutting, this is a cut that includes the scapular cartilage.

SCHULTERDECKEL

The dorsal part of the shoulder in → Swiss beef cutting.

SCHULTERSCHERZEL

Infraspinatus on an → Austrian beef carcass.

SCHWANZSTÜCK

In → German beef cutting, the hindlimb or Keule may be split four ways, with the tail-piece or Schwanzstück containing the *biceps femoris* and *semitendinosus*.

SCHWEINEFLEISCH

German for pork.

SCORPAENIFORMES

These marine fishes have heavy body armour and often a variety of spines like perches. Many species live in rocky habitats, hence their common name, rockfish. The body armour and spines may be camouflaged. Some species have defensive poison spines and are brightly coloured, hence another common name, scorpionfish. Despite this, some species are extremely highly valued for the taste of their myosystem and are the object of extensive commercial fisheries. *Trigla* is the gurnard. *Sebastes* includes the ocean perch or rosefish of the North Atlantic, with an orange-red body, black eyes, and sharp spines on the head (Figure 227). It is a not usually a very large fish (0.5 to 1 kg) and may be baked or poached whole. The myosystem is firm with a distinctive flavor, and fillets are sold with or without skin. Numerous species of *Sebastes* and a couple of *Sebastodes* account for a variety of rockfish in the North Pacific, including the yelloweye rockfish (*Sebastes ruberrimus*, with a low oil content), canary rockfish (*Sebastes pinniger*, with a high yield of firm myosystem), china rockfish (*Sebastes nebulosus*, with small fillets), copper rockfish (*Sebastes caurinus*, with flaky fillets ideal for battering and deep frying) and tiger rockfish (*Sebastes nigrocinctus*).

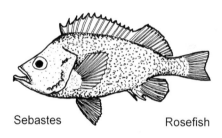

Sebastes   Rosefish

Figure 227. Rosefish.

Also included in the order Scorpaeniformes are fishes of the family Cyclopteridae, such as the lumpfishes or lumpsuckers, and the family Hexagrammidae, such as the lingcod, *Ophiodon elongatus*. It is not a cod. Lingcod are found in shallow waters from California to Alaska, and reach a large size (to 36 kg). The musculature may be blue-green but becomes white on cooking, and the taste is rated as delicious.

(Sources: Nagtegaal, 1985; Lamb and Edgell, 1986; Beamish and Cass, 1990; McKone and LeGrow, 1990)

SCORPION FISH

A bony fish with spiny fins, → Scorpaeniformes.

SCOTCH-CUT FORERIB OF BEEF

In → England, a Scotch-cut forerib (Figure 228) is the most posterior rib cut of a forequarter of beef and has five ribs (ribs 6 to 10) and part of the blade of the scapula, unlike Figure 57 which shows a four-rib forerib.

SCOTCH ROLL OR TENDER

In → North America, a cut of beef prepared from the *supraspinatus* muscle , but also used for → Proten beef in England.

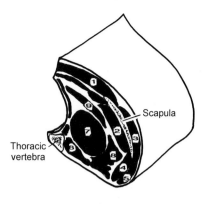

Figure 228. English Scotch-cut forerib of beef with landmark muscles ;
latissimus dorsi *(1)*, subscapularis *(2)*, infraspinatus *(3)*, rhomboideus *(4)*,
trapezius *(5)*, complexus *(6)*, longissimus dorsi *(7)*, spinales dorsi *(8)*,
and serratus ventralis *(9)*.

SCOTLAND - BEEF CUTS

Beef cutting in Scotland is similar to that in England, but has some unique names (Figure 229). The hind and foreknap may originate from the use of the word knap to indicate a sharp break achieved with a snapping sound, as in knapping a flint. Thus, in the abattoir when the cannon bone and foot are removed, the appropriate joint capsules are cut, and the cannon bone is used as a lever to open the joint before completing the operation. Thus, the cannon bone is removed by knapping. Hough is an alternate spelling of hock, emphasising the distinctive Scottish pronunciation of the word. Similar in sound is the nearby cut, the heuk bone. A heuch is a mountain glen with steep overhanging sides - rather like the shape of the pubis above the pelvic cavity on a side of beef.

The curved shape of a lyre, as a stringed musical instrument like a harp, may account for the naming of a curved cut of beef removed from the anterior sternal region, medial to the hough. Similarly, based on visual analogy, thin and thick runners remind one of a stair runner, or thin carpet in the centre of a stairway. In Scotland, the same image of a runner as a series of parallel ribs along a cut of meat is used for the shoulder runner of a lamb carcass (Figure 230). Nine holes is an ancient board game in which balls are rolled into holes or under arches to score points, depending on the difficulty of the play. Figure 229 is not sufficiently accurate to determine exactly which of the cross-cut ribs or costal cartilages of beef may have reminded old Scottish butchers of this game. A final point to note is that the Edinburgh cutting pattern given by Gerrard (1949) has a round in the hindlimb, not a topside and silverside.

SCOTLAND - LAMB CUTS

The Scottish pattern of lamb cutting (Figure 230) given by Gerrard and Mallion (1977) is fairly similar to the → North American cutting pattern in that the lateral part of the shoulder is not detached from the middle neck. Interestingly, the French name for the leg, gigot, may be used, plus a shoulder runner of ribs.

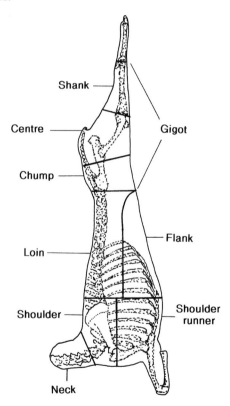

Figure 230. Scotland – lamb cuts.

(Sources: Gerrard, 1947; Gerrard and Mallion, 1977; Moore *et al.*, 1983).

SCRAG

Neck of lamb or mutton or veal in → England.

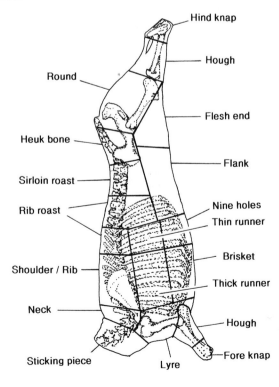

Figure 229. Scotland – beef cuts in Edinburgh.

SDAR

Breast of poultry in → Morocco.

SEA BASS

Perchlike fish of the family Serranidae, → Perciformes.

SECONDARY MYOFIBRE

In the early stages of muscle development, myoblasts fuse to form myotubes which later mature into myofibres. In the later stages of muscle development, however, the myotube stage is missed and new secondary myofibres with peripheral nuclei are formed directly by myoblast fusion.

SEENA

The sternal region of a carcass in → Pakistan.

SELKÄ

In → Finland, selkä is the back of a carcass, as in the loin of a pork carcass.

SELLE

In → French lamb cutting, the selle is cut across the carcass level with the ilium.

SEMIMEMBRANOSUS

A major muscle (> 5% total muscle weight) of the mammalian hindlimb. It originates on the ischium and has two heads, one inserting on the medial epicondyle of the femur, and the other on the tibia (Figure 231). The fasciculi are large and loosely-packed.

Figure 231. Lateral view of beef semimembranosus *(solid black and dotted) showing its origin from the ischium (1) and insertions on the tibia (2) and femur (3).*

SEMISPINALIS CAPITIS

A large, flat muscle in the dorsal neck region of mammals (> 1.5% of total muscle weight). It originates on vertebral processes from the fourth cervical to tenth thoracic vertebrae and inserts onto the occipital bone of the skull (Figure 232). In pork, it is divided into dorsal (*biventer cervicis*) and ventral parts (*complexus*).

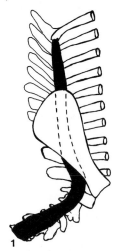

Figure 232. Lateral view of beef semispinalis capitis *(solid black and dotted) showing its origins along the vertebral column and severed insertion to the skull (1).*

SEMITENDINOSUS

A major fusiform muscle of the mammalian hindlimb (≈ 2.5% total muscle weight). It originates from the ischium and has one insertion on the tibial crest and a second to a conspicuous secondary tendon which, combined with *gastrocnemius* and superficial flexor muscles, inserts onto the tuber calcis along with the Achilles tendon (Figure 233).

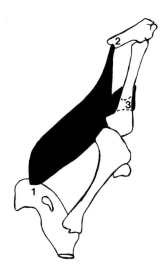

Figure 233. Lateral view of beef semitendinosus *(solid black and broken lines) showing its origin from the ischium (1) and insertions to the tuber calcis (2) and tibia (3).*

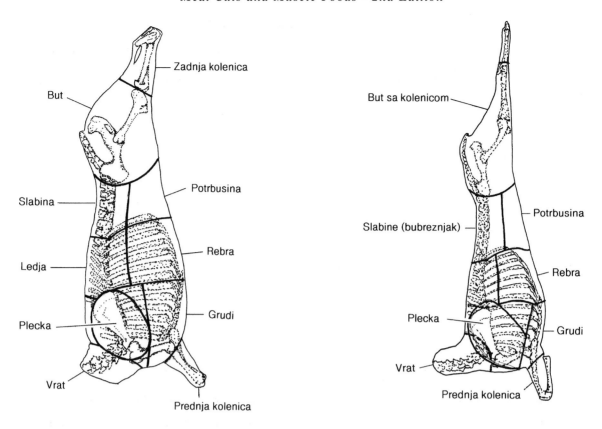

Figure 234. *Beef cuts in Serbia & Montenegro.*

Figure 235. *Lamb cuts in Serbia & Montenegro.*

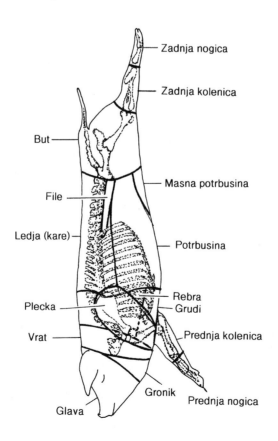

Figure 236. *Pork cuts in Serbia & Montenegro.*

SERBIA & MONTENEGRO - BEEF CUTS

Beef is govedina (Figure 234). From the but or thigh is removed a flank cut (slabina) which includes the lumbar vertebrae. The back or ledja contains the posterior thoracic vertebrae. A round shoulder or plecka is centred on the scapula (possibly the remaining ribcage is the greben). Ventrally, from anterior to posterior are the grudi (chest), rebra (ribs) and potrbusina (stomach). Fore and hindshanks are the prednja kolenica and zadnja kolenica, respectively. The neck is the vrat.

(Source: Alexic, 1999)

SERBIA & MONTENEGRO - LAMB CUTS

The terminology and cutting pattern for lamb (janje) is very similar to that for beef (Figure 235).

(Source: Alexic, 1999)

SERBIA & MONTENEGRO - PORK CUTS

The terminology for pork (svinjetina) is very similar to that for beef, except for the division between the grudi and rebra, and the outline shape of the plecka (Figure 236). The masna potrbusina means the "fat stomach".

(Source: Alexic, 1999)

SERBIA & MONTENEGRO - VEAL CUTS

The terminology and cutting pattern for veal (teletina) is identical to that for beef shown in Figure 234.

(Source: Alexic, 1999)

SERRATUS DORSALIS CAUDALIS

In mammals, a series of small muscles originating on the lumbodorsal fascia and inserting onto ribs 9 to 13, approximately.

SERRATUS DORSALIS CRANIALIS

In mammals, a thin muscle originating on the lumbodorsal fascia and usually inserting onto the dorsal parts of the sixth to ninth ribs (Figure 237).

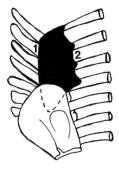

Figure 237. *Lateral view of beef* serratus dorsalis cranialis *(solid black and broken line) showing its origin from the lumbodorsal fascia (1) and insertions to the ribs (2).*

SERRATUS VENTRALIS CERVICIS

A large triangular muscle located laterally in the neck and thorax of mammals (Figure 238). It originates on the scapula and inserts onto the transverse processes of the last four cervical vertebrae, and onto the first three ribs. In *Bos indicus* carcasses it contributes to the hump. The muscle is large and well developed in pork.

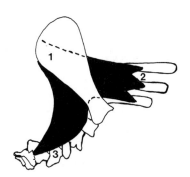

Figure 238. *Lateral view of beef* serratus ventralis cervicis *(solid black and broken lines) showing its origin from the scapula (1) and insertions to ribs (2) and cervical vertebrae (3).*

SERRATUS VENTRALIS THORACIS

An important muscle of the mammalian forequarter, where it functions as the main component of the muscular sling by which the scapula is attached to the rib cage (Figure 239). The muscle originates from the costal surface of the scapula and inserts onto the lateral surfaces of the fourth to ninth ribs. The wedges to each of the costal insertions give the muscle a serrated edge ventrally.

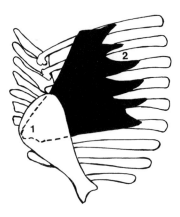

Figure 239. *Lateral view of beef* serratus ventralis thoracis *(solid black and broken line) showing its origin from the medial surface of the scapula (1) and insertions to the ribs (2).*

SERTÉSCOMB

Leg of pork in → Hungary.

SERTÉSFAROK

The tail of a → Hungarian pork carcass.

SERTÉSFEJ

The head of a → Hungarian pork carcass.

SERTÉSLAPOCKA

The whole shoulder of a → Hungarian pork carcass.

SERTÉSOLDALAS

Ribs of a → Hungarian pork carcass.

SERTÉS-SZŰZPECSENYE

Filett of pork in → Hungary.

SERTÉSTARJA

A collar of pork, including cervical and anterior thoracic vertebrae in a → Hungarian pork carcass.

SEZIONE

Italian for a section or very thick slice through the carcass, as in sezione di coscia (thigh slice through the gluteal region) or sezione e muscolo di spalla (shoulder slice) on an → Italian beef carcass.

SHAD

A North Atlantic herring-like fish with a deep body, *Alosa*, → Clupeiformes.

SHANK

Distal extremities of limbs in beef and lamb (level with the tibia or with the ulna and radius) are often called shank, sometimes with a designation such as hindshank and foreshank. In → North America, shank may be used as an alternative name for a pork hock.

SHARK

A cartilaginous fish, → Chondrichthyes.

SHEBOYGAN SAUSAGE

A summer → sausage in the USA, typically containing beef and pork, flavoured with pork heart muscle.

SHIELD

→ Collagen accumulation in shoulder backfat of pork from heavy boars. Slicing is difficult.

(Source: Wood *et al.*, 1985)

SHIN BONES

Ulna and radius.

SHINTAMA

A → Japanese beef cut taken anteriorly to the femur.

SHORT HIP OF BEEF

In → North American beef cutting, an alternative name for the primal sirloin.

SHORT LOIN OF BEEF

In → North American beef cutting, an alternative name for the primal loin.

SHORT PLATE OF BEEF

In North American beef cutting, the plate shown as a primal cut in Figure 167 is also a short plate, in distinction from a long plate which is the brisket plus the plate shown in Figure 167.

SHORT RIBS OF BEEF

In → North America, short ribs of beef are sections of rib and adjacent muscles from the dorsal part of the plate. In → Australian beef cutting, the short rib is more dorsal in position.

SHOTT

A male pig castrated at a relatively early age.

SHOULDER CLOD

In Canada, a shoulder clod may be a rolled joint of veal (pointe d'épaule) prepared from the large shoulder muscles after removal of the scapula. This is similar to the shoulder clod of beef cut in → New Zealand and → Denmark.

SHOULDER OF BEEF

A primal cut of beef in → Northern Ireland.

SHOULDER OF LAMB

In → North America, a shoulder of lamb is cut as a solid block of meat including the vertebrae, scapula and much of the humerus. It usually includes parts of the first five ribs.

The whole shoulder may be sold as a square cut roast, boned and rolled, or subdivided into blade roasts or blade chops with a muscle grouping very similar to that seen in beef chuck steaks. Arm chops can be made by cutting perpendicularly to the humerus, with a muscle grouping following that seen in Figure 240. In English lamb cutting, however, the scapula and associated muscles are released from the underlying ribs (the middle neck), as shown in Figure 241. Thus, the North American shoulder of lamb is equivalent to the English middle neck plus shoulder.

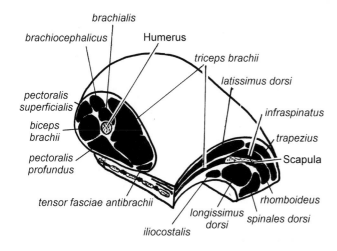

Figure 240. *North American shoulder of lamb.*

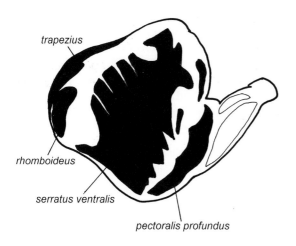

Figure 241. *Medial view of a shoulder of lamb from the left side of the carcass in England.*

SHOULDER OF PORK

In → North America, the shoulder is a primal cut of pork which is subdivided into a Boston butt and a picnic (Figure 242).

SHOULDER PIECE OF BEEF

In English beef cutting, a shoulder piece in the scapular region was noted by Smith (1876), probably in London, and survived to the Liverpool cutting pattern known to Gerrard (1949), as seen in Figures 58 and 60.

SHOULDER RUNNER

In → Scotland, the antero-ventral part of the lamb carcass is removed as a shoulder runner.

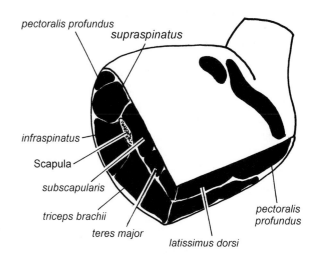

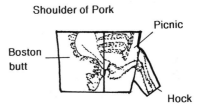

Figure 242. *North American pork picnic and its location in a shoulder of pork.*

SHRIMP

Shrimps are decapod crustaceans of the suborder Macrura-Natantia (→ Crustacea). Many different species of shrimps are caught in commercial fisheries around the World. There may be a clear-cut distinction between shrimps and prawns in many British seaside markets, but certainly not in the rest of the World. The common brown shrimp around the British Isles is *Crangon crangon* and the common prawn is *Palaemon serratus*. The latter species is usually larger than the former and is also distinguished by its laterally compressed body and its well developed rostrum, not to mention its higher price. In Commonwealth countries where this English etymology still prevails, but where the species of Crustacea are different, a prawn might simply be any large species of shrimp. In the USA and in many other countries, however, the term prawn is rarely used and, if it is used, may be used haphazardly to indicate some local species of crustacean.

Crangon crangon (Figure 243) is caught in large numbers by nearly all the European countries with an Atlantic coastline. It burrows in sandy bottoms where the water is shallow (0 to 20 m) and it tolerates brackish water in estuaries. It is a translucent grey or brown in colour and reaches a length of 9 cm, although the commercial catch is usually composed of smaller individuals. *Palaemon serratus*, the common prawn (Figure 243), also inhabits inshore waters (0 to 40 m) of a similar East Atlantic range, with both *C. crangon* and *P. serratus* extending into the Mediterranean. *P. serratus* is typically about 7 or 8 cm in length but may reach 11 cm.

Other species of *Crangon* and *Palaemon* are commercially important in other areas. In the Northwest Atlantic, for example, the grey shrimp (*Crangon septemspinosa*) is fished on both marine and estuarine sandy bottoms (0 to 90 m) from Baffin Bay down to Florida. In shallow waters (1 to 10 m) of the East Atlantic, and with a range that also includes the Mediterranean, the Baltic and the Black Sea, is found the Baltic shrimp, *Palaemon adspersus*. With a similar geographical range but with the shoreline and rockpools as its habitat is found the rockpool prawn, *Palaemon elegans*.

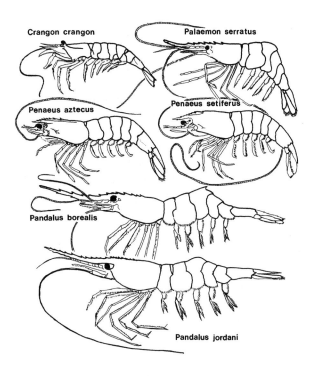

Figure 243. *Some commercially important shrimps and prawns (not to the same scale).*

Penaeus aztecus (Figure 243) is caught in vast numbers in the West Atlantic from Massachusetts down to Texas, and off Mexico. Although this species is generally called the brown shrimp, its colour may also be red, orange or green. It is caught on muddy or sandy bottoms, mostly between 27 and 54 m, and is often fished at night. The female grows larger than the male (maximum length 236 versus 195 mm), and a typical market length is about 5 to 16 cm.

Penaeus setiferus has a later season than *P. aztecus* and is caught in shallow water (2 to 9 m). It is also a West Atlantic species frequenting muddy or sandy bottoms. Fishing for this species is centred off Louisiana and Mexico. The common name for this species, the white shrimp, describes its usually translucent white appearance, but its colour may also range from grey to greenish-blue. The female grows larger than the male (maximum length 200 versus 175 mm) and the typical

market length is from 5 to 15 cm. *P. setiferus* has very long antennae (Figure 243).

Pandalus borealis is extensively fished with otter trawls over muddy or sandy bottoms at a wide range of depths in both the Atlantic and Pacific Oceans (Figure 243). There is a very high demand for the cooked and peeled tails of this species and most of the catch is cooked soon after capture. Its pink-red colour when alive is due to countless fine red dots on a translucent background, hence one of the common names for this species - the pink shrimp. Alternate names are the deep-water or great northern prawn. The average commercial length is from 75 to 100 mm, with about 120 to 130 shrimps/ kg. The female grows larger than the male (maximum length 165 versus 120 mm).

Pandalus jordani, the smooth pink shrimp of the East Pacific from Alaska to California, is a relative new-comer to the shrimp menu (Figure 243). In British Columbia, fishing for this species was only commenced in the 1950s, but it is presently one of the most important commercial species of shrimp in that province. It is generally caught at depths between 70 and 230 m over sandy or muddy bottoms, and is generally larger than *P. borealis*. Females grow larger than males (maximum length 290 versus 210 mm). The distinction between these two species of *Pandalus* is best made at the point where the third abdominal segment forms a humped back to the animal. *P. borealis* has an upward projecting spine while *P. jordani* does not (Figure 243).

Countless other species of shrimps are fished around the World. A geographical sample is given in Table 1.

Some trade terms used for shrimp include fantail (abdominal muscles retaining telson or tailfan), vein (intestine beneath the abdominal muscles) and green-headless (raw, head removed, shell still in place). Apart from odour problems associated with spoilage, a major problem with preservation of shrimp meat is softening. This occurs as a result of proteases from the hepatopancreas diffusing into the musculature. Protease activity depends on the diet of the shrimp before capture.

(Sources: Holthuis, 1980; Butler *et al.*, 1989; Ezquerra *et al.*, 1999)

SIDA

The belly of a pork carcass in → Sweden.

SIÐA

Abdominal region of a beef, pork or horse carcass in → Iceland.

SIDE

One half of a carcass, either right or left. In → Norwegian meat cutting, side has the same meaning as in English, but refers to the flank of a lamb or veal carcass.

Table 1. Some commercially important species of shrimp.

Location	Species	Common Name
Africa (E)	Penaeus indicus	White prawn
Argentina	Pleoticus muelleri	Argentine red shrimp
Australia	Metapenaeus bennettae	Greentail shrimp
Australia	Penaeus esculentus	Brown tiger prawn
Brazil	Penaeus notialis	Southern pink shrimp
Brazil	Penaeus schmitti	Southern white shrimp
British Guiana	Nematopelaemon schmitti	Whitebelly prawn
Canada	Pandalus platyceros	Spot shrimp
China (N)	Acetes chinensis	Northern mauxia shrimp
India	Atypopenaeus stenodactylus	Periscope shrimp
India (NW)	Metapenaeus brevicornis	Yellow shrimp
India	Palaemon styliferus	Roshna prawn
India (W)	Parapenaeopsis stylifera	Kiddi shrimp
Japan	Penaeus chinensis	Fleshy prawn
Japan	Penaeus japonicus	Kuruma prawn
Mediterranean	Aristeus antennatus	Crevette rouge
Mediterranean	Parapenaeus longirostris	Deepwater rose shrimp
Mexico	Macrobrachium carcinus	Painted river prawn
Mexico	Penaeus californiensis	Yellowleg shrimp
Mexico	Penaeus vannamei	Whiteleg shrimp
Pacific (W)	Metapenaeus ensis	Greasyback shrimp
Pakistan	Metapenaeus affinis	Jinga shrimp
Pakistan	Penaeus penicillatus	Redtail prawn
Persian Gulf	Penaeus merguiensis	Banana prawn
USA (Atlantic)	Penaeus duorarum	Northern pink shrimp
USA (Pacific)	Xiphopenaeus kroyeri	Seabob

SIDEFLESK

Ventral ribs and belly of a → Norwegian pork carcass, including flatribbe and buklist.

SIDE OF PORK

In many parts of the World, a side of pork could be one half of a pork carcass, either the right or left side, but the term also has more specific usages. In → North America, the belly of pork is separated from the loin following the curvature of the vertebral column. After removal of the ribs and adjacent muscles as spareribs, the remaining muscles of the abdomen, such as *rectus abdominis* and *obliquus internus abdominis* may be called side of pork. Side of pork may be cured and smoked to make slab bacon.

SII KHROONG MUU

Ribs of a pork carcass in → Thailand.

SILD

Danish and Icelandic for a herring, → Clupeiformes.

SILL

Swedish for a herring, → Clupeiformes.

SILUNGUR

Icelandic for → trout.

SILVERSIDE

A bony fish, such as *Atherina*, the Atlantic silverside, → Mugiliformes.

SILVERSIDE OF BEEF

The beef *semitendinosus* muscle, which is located laterally in the hip, has a natural, silvery seam of epimysium along which to define the start of the English silverside of beef (Figure 244). The silverside is located laterally in the hip and contains both parts of *biceps femoris* plus *semitendinosus*. Silverside may be sliced as braising steak or used to make a large, or two small rolls of beef.

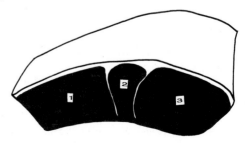

Figure 244. *English silverside of beef with some landmark muscles;* biceps femoris *(1), ischiatic head of* biceps femoris *(2), and* semitendinosus *(3).*

SILVERSIDE OF PORK

In English pork cutting, a silverside of pork may be a boned and rolled cut prepared from the lateral muscles of the ham.

SIRLOIN

Our old English word for sirloin is derived from the French surlonge (on or above the loin). But the English sirloin contains lumbar and posterior thoracic vertebrae and straddles the loin and rib cuts of North America, so on or above means anterior in England and in France (where the surlonge is set far forward medial and dorsal to the scapula). However, in North America the sirloin is posterior to the loin, so on or above means posterior in → North America. → Australia is similar to North America in placing the sirloin posteriorly level with the ilium (but the sirloin may also be called a rump in Australia).

In North American beef cutting the delineation of a sirloin usually is made by two almost horizontal planes in the hanging carcass. The upper or posterior plane touches the anterior, proximal tip of the femur while the lower plane touches the anterior face of the ilium. But this primal cut also may be called a short hip. In England, a sirloin extends from the ilium to the last few ribs, and sometimes this sirloin is called a loin. Figure 245 contrasts the different positions of the sirloin in Great Britain versus North America.

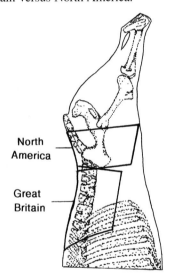

Figure 245. North American beef sirloin versus the sirloin in Great Britain.

SIRLOIN BUTT

An → Australian primal cut of beef with a curved shape containing the ilium.

SIRLOIN STEAK

In North American beef cutting, sirloin steaks are cut from the primal sirloin, but this may be missing some of the less tender muscles contained in the sirloin tip. The most anterior sirloin steak is the pinbone sirloin steak. It resembles a porterhouse steak and has an attractive, tender section of

iliopsoas, but may contain a hidden disadvantage from the massive bone content of the anterior edge of the ilium. Next in posterior sequence is the double-bone sirloin with a high bone content associated with the ilium and sacrum. This is followed by the round-bone sirloin as the shaft of the ilium narrows. Then, finally, as the shaft of the ilium becomes triangular in cross section there is the wedge-bone sirloin steak (Figure 246).

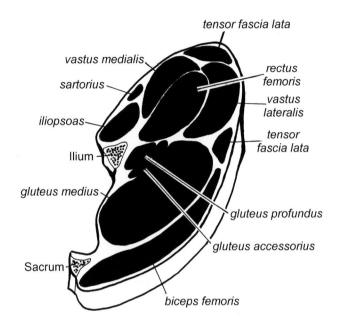

Figure 246. North American wedge-bone sirloin steak of beef.

The *longissimus dorsi* is the dominant muscle of an English sirloin steak (Figure 247), which is essentially the same as a North American strip-loin steak.

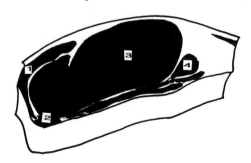

Figure 247. English sirloin steak of beef with landmark muscles; spinales dorsi (1), multifidus dorsi (2), longissimus dorsi (3), and longissimus costarum (4).

SIRLOIN TIP

In North American beef cutting, the sirloin tip (Figure 248), sometimes called a knuckle roast, is cut from the meat anterior to the femur. If tender it may be roasted, but braising is its usual fate. It may be included in the → diamond round.

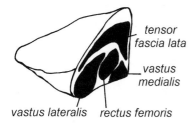

Figure 248. *North American sirloin tip of beef.*

SISÄFILEE

The *psoas* muscles of a pork carcass in → Finland.

SISÄPAISTI

In → Finnish beef cutting, the sisäpaisti (mostly *semimembranosus, adductor* and *pectineus*) is the medial part of the long round roast or pitkä paisti.

SKANKASNEIÐAR

Shank slices in → Iceland.

SKARKOLI

Icelandic for plaice (→ Pleuronectiformes)

SKATE

Cartilaginous flat fishes such as *Raja*, → Chondrichthyes.

SKELETON

The major bones are identified in the beef carcass skeleton shown in Figure 249. The vertebral column (also called the spine, backbone or chine bone) is the main axis of the skeleton. It protects the spinal cord located in a neural canal formed by a long series of neural arches, each contributed by a different vertebra. The neural arch of each vertebra is supported on the body or centrum of the vertebra. In some types of vertebrae, the neural arch extends dorsally as a prominent dorsal spine (also called the neural spine or a spinous process). Where not fused, the centra are separated by cartilaginous intervertebral discs. In mammals, the anterior and posterior faces of the centra are almost flat. The numbers of vertebrae are quite variable (Table 2), because of the selective breeding that has been applied to modify carcass length. Sometimes there are additions to, or subtractions from a typical number of vertebrae, while sometimes one type of vertebra is replaced by another.

Table 2. Numbers of vertebrae in meat animals.

Name	Region	Beef	Pork	Lamb
Cervical	Neck	7	5-7	6-7
Thoracic	Ribcage	13	13 - 17	13 - 14
Lumbar	Loin	6	5 - 7	6 - 7
Sacral	Sirloin	5	4	4
Caudal	Tail	18 - 20	20 - 23	16 - 18

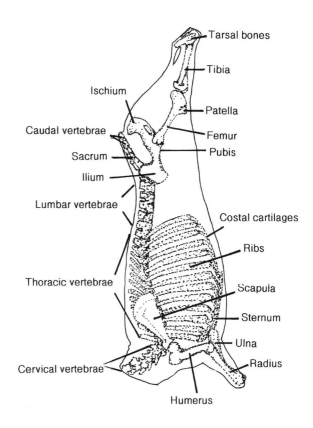

Figure 249. *Major bones of a beef skeleton.*

Thoracic vertebrae are distinguished by their tall dorsal spines, many of which point posteriorly and are often called feather bones. The ribs are joined to the vertebral column dorsally so that the head of each rib articulates with the bodies of two adjacent vertebrae. It is not uncommon to find a skeleton with more ribs on one side than the other. Each rib has a tubercle that articulates with the transverse process of the more posterior of its two vertebrae. Ventrally, the anterior ribs articulate with the sternum and are termed sternal ribs (Table 3). The more posterior ribs are called asternal ribs and they only connect to the sternum indirectly via costal cartilages. Some of the costal cartilages are very hard and may appear more like bones than typical cartilage. The sternum is formed by a number of closely joined bones, the sternebrae. When split through the midline, the interior structure of the sternebrae resembles that found in the centra of the vertebrae.

Table 3. Ribcage structure in meat animals.

	Beef	Pork	Lamb
Total pairs of ribs	13	13 -17	13 - 14
Pairs of sternal ribs	8	7	8
Pairs of asternal ribs	5	7 - 8	5 - 6
Number of sternebrae	7	6	6 - 7

The lumbar vertebrae have flat, wing-like transverse processes which broaden the abdominal cavity dorsally, provide a strong attachment for the muscles of the abdominal wall that carry the weight of the viscera, and appear in T-bone steaks or chops through the lumbar region. The propulsive thrust generated by the hindlimb during locomotion is transmitted to the sacral vertebrae by the pelvis. To strengthen the sacral vertebrae, they are fused together to form the sacrum. Fusion is incomplete in young animals and provides an important clue to animal age in the dressed carcass.

The pelvis is formed by three bones on each side. The most anterior bone on each side is the ilium. The shaft of the ilium expands anteriorly to form a flat wing attached to the sacrum. This joint is called the slip joint. When seen in a sirloin steak, the ilium may appear round, triangular or flat. The anterior edges of the ilia form the hooks of the live animal. The most posterior bone of the pelvis on each side is the ischium. The pelvis and the sacrum form a ring of bone completed ventrally by the pubes. The left pubis is separated from the right pubis by fibrocartilage which, at parturition, may soften to allow movement between the bones of the pelvis. The pubes are separated when carcasses are split into left and right sides in the abattoir. The pubic bone exposed on a carcass is called the → aitch bone. The aitch bone is curved in steer and bull carcasses, is moderately curved in heifers, but is straight in cow carcasses. Only a couple of caudal or coccygeal tail vertebrae usually remain on a commercial beef carcass.

The most proximal bone of the forelimb is the blade bone or scapula. It is not fused to the vertebral column (like the pelvis in the hindlimb), and this allows muscles that hold the scapula to the ribcage to function as shock absorbers during locomotion. The scapula has a shallow socket (glenoid cavity) for the humerus. Along the dorsal edge of the scapula there is a flexible scapular cartilage. On the lateral face of the scapula is a prominent ridge of bone called the spine of the scapula. In beef carcasses, the scapular spine is extended distally as a prominent acromion process. The humerus has a relatively flat knob or head to fit into the glenoid cavity of the scapula. Two well defined condyles on the distal end of the humerus contribute to the hinge joint at the elbow. The elbow is formed by the olecranon process, an extension of the ulna. The radius is joined to the ulna and is the shorter and more anterior bone of the pair.

Beef and lamb carcasses may have a set of six compact carpal bones remaining on the carcass after slaughter. Before slaughter, the forefeet of cattle and sheep have a large cannon bone located distally to the carpal bones. Beef cannon bones are removed with the feet at slaughter since there is virtually no meat on them. Cannon bones sometimes are left on lamb carcasses in the abattoir to prevent the meat contracting proximally up the limb, as shown on most of the lamb cutting charts. This is rather variable, even within individual countries, so no great significance should be attached to this feature in the lamb diagrams. Each forelimb cannon bone in ruminants is derived by the enlargement and fusion of the third and fourth metacarpal bones.

In pigs, digits 3 and 4 on each foot bear most of the body weight and are larger than the lightly loaded digits 2 and 5. The first digit is absent. The evolutionary trend towards lifting of the foot and reduction of digits is even more extensive in cattle and sheep. In the ruminants, digits 2 and 5 are reduced to dew claws behind the fetlock. Weight-bearing digits 3 and 4 are enlarged, and their metacarpals are fused to form a long cannon bone. The small bones in the toes of both fore and hind feet are called phalanges.

The proximal bone of the hindlimb is the femur or round bone. Its articular head is deeply rounded and it bears a round ligament that holds it into the acetabulum. Another distinctive feature of the femur is the broad groove between the two trochlear ridges located distally. The patella or knee cap slides in this groove. The tension generated by muscles above the knee is transmitted over the knee or stifle joint by the patella to avoid having an important tendon in a vulnerable position over the anterior edge of a joint. In beef and lamb carcasses there is a single major bone, the tibia or shank bone, located distally to the femur. In the corresponding position in a pork carcass there are two parallel bones, a large tibia and a more slender fibula. In cattle and sheep, the fibula has lost its shaft and only a remnant of the head remains. In pigs, the fibula retains its shaft and the bone is mobile at birth. After a few years, however, the fibula becomes fused to the tibia.

Distal to the tibia are the tarsal bones of the hock. The structure of the tarsals, metatarsals and phalanges of the hindlimb is similar to that of the carpals, metacarpals and phalanges in the forelimb. Pork carcasses normally are suspended by a gambrel or hooked bar placed under the tendons of the hind feet. Beef carcasses normally are suspended by a hook under the tuber calcis, for the insertion of the Achilles tendon at the hock.

In poultry, the long double curved neck contains 14 cervical vertebrae, and the ring-like atlas articulates to the skull with only a single occipital condyle. The axis has a large odontoid process that projects anteriorly. There are 7 thoracic vertebrae, but numbers 2 to 5 are fused. Thoracic vertebrae → can move freely, but the last thoracic vertebra is fused to the synsacrum. The synsacrum is a fused length of the vertebral column that contains thoracic vertebra 7, 14 lumbo-sacral vertebrae, and the first coccygeal or caudal vertebra, but skeletal fusion in the vertebral column does not occur until some weeks after hatching. There are six caudal vertebrae that, apart from the first, are free and mobile. However, only numbers 2 to 5 are normal vertebrae, since the last one is formed into a three sided pyramidal bone called the pygostyle.

In poultry, there are seven ribs: the first two are free while the last five are attached to the sternum. There are no costal cartilages. Ribs 2 to 6 each have an uncinate process which overlaps the next posterior rib. The sternum is extremely large. It has a conspicuous ventral ridge in the midline, the carina, which increases the area available for attachment of the flight muscles (→ *pectoralis*). The dorsal surface of the expanded sternum is concave and forms the floor of a continuous thoracic and abdominal cavity.

In the wing of poultry, distal to the humerus are the widely spaced radius and ulna. The carpals, metacarpals and digits are reduced to form a stiff skeletal unit for the anchorage of the primary flight feathers. The three digits of the wing are equivalent to digits 2, 3, and 4 in other animals. The wing articulates with the body at the glenoid cavity which is strengthened by the convergence of three bones, the scapula, the coracoid and the clavicle. In birds the coracoid is a separate bone, whereas in mammals it has been reduced to a small integral part of the scapula. The clavicles of right and left sides are fused ventrally to form the furcula or wishbone. Although many mammals have a pair of clavicles, they are absent in cattle, sheep and pigs. In poultry, the distal end of the coracoid is braced against the sternum.

In the legs of poultry, distal to the femur, the fibula is reduced to leave the tibia as the major bone. The proximal tarsal bones are fused to the distal end of the tibia to increase its length, and the whole skeletal unit is called the tibiotarsus. The distal tarsal bones are incorporated into the proximal end of a single bone, the tarsometatarsus, which also includes the fused metatarsals 2, 3 and 4. Of the four digits which form the bird's claw, digit 1 is directed posteriorly while digits 2, 3 and 4 are anterior, which enables the bird to perch. The ilium is fused to the synsacrum. Instead of being fused in the midline, the pubic bones are separate, and they project backwards as thin rods. The open structure of the pelvis in the ventral region facilitates the laying of eggs.

SKINÉE

The petites côtes, a cut from the anterior part of the sternum in a → Belgian pork carcass.

SKINKA

→ Swedish for ham, either as cured pork or the hindlimb of a pork carcass. In → Iceland, skinka is more likely to mean just cured ham.

SKINKE

→ Danish and → Norwegian for the ham of a pork carcass.

SKIRT STEAK

The muscular part of the diaphragm. Inside skirt is *transversus abdominis*.

SKÖTUSELUR

Icelandic for → monk fish.

SLABINA

Lumbar region of a beef or lamb carcass in → Serbia & Montenegro.

SLAKSIDA

The flank of a → reindeer in Sweden.

SLAG

Posterior flank of a → Danish beef carcass.

SLAGSIDE

Posterior flank of a → Norwegian beef carcass.

SKELEMIN

A cytoskeletal protein with a high molecular weight (200 to 220 kD) found at the level of the → M-line. Skelemins may form threads wrapped around the → myofibrils at the M-line so as to link the M-lines of laterally adjacent myofibrils.

(Source: Price, 1987)

SLIDING FILAMENT THEORY

Striated muscles contract by the → myosin molecules of thick → myofilaments pulling themselves past the → actin molecules of thin myofilaments, but exactly how they do it is a continuous debate. The myosin molecule → cross-bridges that swing so nicely in textbook diagrams of muscle contraction are difficult to detect actually rowing in real muscles.

(Source: Huxley, 1965; Rayment *et al.*, 1993; Oplatka, 1997)

SLIP VAN DE LENDE

In the → Netherlands this may be the tail of the loin, essentially the *tensor fascia lata* on a beef carcass.

SLOAT

A cut of beef anterior to the humerus in Northeast → England, possibly named from an old usage of the word slot, to pierce through (i.e., near where the animal was exsanguinated in the abattoir).

SLÖG

Breast of lamb in → Iceland.

SŁONINA

Polish for pork lard. Slonina paprykowana is cured, smoked and flavoured with paprika.

SLOVAKIA - BEEF CUTS

The stehno or round is cut to include the whole of the pelvis, femur and associated muscles, with the zadná nožina of the hindshank including the tibia and associated muscles (Figure 250). The lumbar and posterior rib steaks are in a single primal cut called the nízka roštenka. The plece or shoulder is lifted

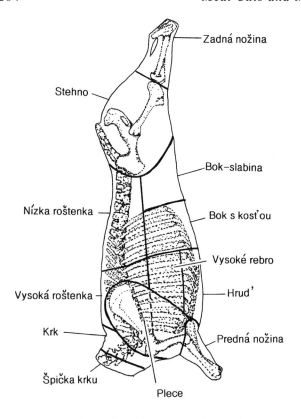

Figure 250. Slovakia – beef cuts.

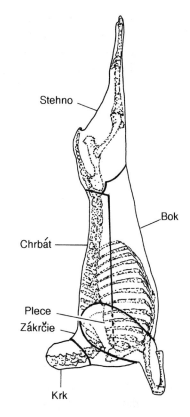

Figure 251. Slovakia – lamb cuts.

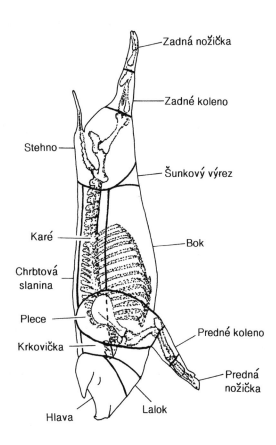

Figure 252. Slovakia – pork cuts.

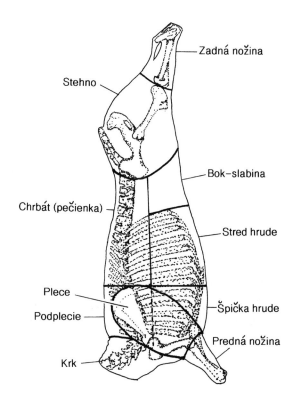

Figure 253. Slovakia – veal cuts.

from the ribcage to leave the anterior rib steaks (vysoká roštenka) and neck meat (krk and špička krku). There is a ladder of midlength rib sections (vysoké rebro) dorsal to the brisket (hrud'). The abdominal muscles (bok-slabina) are separated from the postero-ventral ribs (bok s kost'ou).

(Source: Lahucky, 1999).

SLOVAKIA - LAMB CUTS

The leg or stehno cuts sharply around the anterior face of the ilium to give a square cut end to the loin or chrbát (Figure 251). The breast and flank is taken as a single large cut called the bok. The shoulder or plece is lifted from the ribcage to leave an anterior rib roast (zákrčie).

(Source: Lahucky, 1999).

SLOVAKIA - PORK CUTS

The hindlimb is broken into a large leg (stehno), a hock (zadné koleno) and trotter (zadná nožička) as shown in Figure 252. Removal of a triangular šunkový výrez in the region of the *tensor fascia lata* straightens the trunk for a long loin (karé) and square-cut belly (bok). The backfat is removed as chrbtová slanina. The plece or shoulder is lifted from anterior ribs and neck (krkovička). The jowl (lalok) is removed from the head (hlava).

(Source: Lahucky, 1999).

SLOVAKIA - VEAL CUTS

For beef, the separation of the forequarter from hindquarter is already located quite far anteriorly, but in veal (Figure 253) it is right against the scapula to give a cut containing a long length of lumbar and posterior thoracic vertebrae (chrbát or pečienka). The bok-slabina contains the abdominal muscles of the flank, while the corresponding rib cut is the stred hrude. The shoulder or plece is lifted from the ribcage to leave a podplecie or under-shoulder roast.

(Source: Lahucky, 1999).

SMELT

A small salmonid fish, → Salmoniformes. *Osmerus eperlanus* from European coastal waters. It tastes a little like cucumber.

(Source: Pivnička and Černý, 1987)

SMITHFIELD HAM

In the USA, complete hams (skin on) from pigs fed exclusively on peanuts are dry cured and cold smoked for a long period with hickory and sassafras.

SMOKED CALLIE

A fanciful name, sometimes used in North America for smoked picnic of pork.

SNAPPER

A perch-like bony fish of the family Lutjanidae, → Perciformes.

SNIPE

The common snipe in Europe and North America, *Capella gallinago*, is a tiny bird with a long, thin beak. It is grouped with the sandpipers and curlews in the Family Scolopacidae. It is fairly tender and can be roasted.

SNOW CRAB

Long-legged deep-sea crabs from the West Atlantic, *Chionoecetes*, → crabs.

SOBRECOSTILLA

Literally over the ribs, a beef cut in → Chile.

SOC

A Boston butt of pork in Quebec.

SOLE

A small-mouthed flatfish, → Pleuronectiformes.

SOLEUS

In the mammalian hindlimb, a small slip of muscle originating on the femur and contributing to the lateral head of *gastrocnemius*.

SOLOMILLO

Solomillo is Spanish for the tenderloin. On an → Argentinean pork carcass it is *psoas* plus *iliacus* muscles.

SOOK

Female → blue crab.

SOPRESSATA

Cured, smoked and dried pork → sausage distinguished by large chunks of fat.

SOTTOFESA

An → Italian cut of beef containing the *biceps femoris*.

SOTTOSPALLA

A mid-rib cut medial to the shoulder in an → Italian beef carcass.

SOUSE

To souse means to immerse or pickle, and the latter includes anything from gherkins to fish. In the USA, the term has become more restricted in reference to low-grade meat, such as pork snouts and skins. These are first cured in brine then, with a generous helping of gelatin, are made into sausages which are cooked then cooled.

SOUTH AFRICA - ANTELOPE CUTS

Carcass cuts for small antelope such as the blesbok (*Damaliscus dorcas*), impala (*Aepyceros melampus*) and springbok (*Antidorcas marsupialis*) are given in Figure 254. The shoulder may be removed, by first lifting the shin away from the ribcage, then cutting medially through the muscles holding the scapula to the ribcage. Alternatively, the flank, brisket and shin may be removed together with a straight cut, then separated later. Antelope meat may be called venison in South Africa. See Bothma (1989) for a full list of species suitable for game ranching.

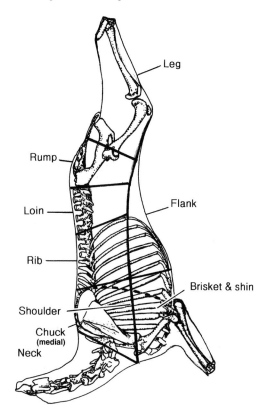

Figure 254. *South Africa – antelope cuts.*

(Source: van Rooyen, 1989; Bothma, 1989)

SOUTH AFRICA - BEEF CUTS

Excellent documentation is provided by ADSRI (1981) for beef cutting in South Africa, as well as numerous other countries. For meat cuts named in Afrikaans (Figure 255), the hindshin or agterskenkel contains all the distal extensors and flexors, plus part of *gastrocnemius* and *popliteus*. The medial muscles of the round are contained in the binneboud, including the major muscles *semimembranosus*, *adductor* and *pectineus*, plus all the lesser medial muscles such as *gracilis*

and *sartorius*, as well as overlapping parts of adjacent muscles such as parts of *biceps femoris* and *gluteus profundus*. Most of the lateral muscle mass, essentially most of *biceps femoris* and all of *semitendinosus*, plus part of *gastrocnemius* and *flexor digitorum superficialis*, are contained like an English silverside in a cut called the dy. The *quadriceps femoris* group of muscles is contained in a cut anterior to the femur called the diklies. The rump or kruis containing the whole of the ilium is dominated by the three *gluteus* muscles and *tensor fascia lata*, plus overlapping parts of other muscles.

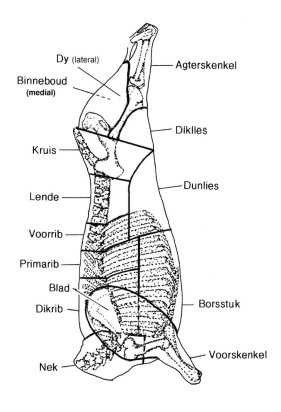

Figure 255. *South Africa – beef cuts.*

Anterior to the kruis are three primal cuts containing the best steaks: the lende in the lumbar region, the voorrib or wing rib containing the last three ribs, and the primarib or prime rib containing three more ribs. The dunlies or flank contains abdominal muscles, the diaphragm, and muscles attached to the ribs. The blad contains all the major muscles relating to the scapula and humerus, the bulk of which are the *triceps brachii, infraspinatus, supraspinatus* and *subscapularis*. *Serratus ventralis, trapezius* and *rhomboideus* remain on the underlying dikrib or chuck. This leaves the nek, the brisket or borsstuk, and the foreshin or voorskenkel.

For English rather than Afrikaans names, the cutting pattern is very similar, but with some variations (Figure 256). The ischium dominates an aitchbone cut, as it does in London and Midlands cutting of England. This leaves a medial topside, a lateral silverside and an anterior thick flank following the London pattern. However, there are some interesting variations in the forequarter, most likely to accommodate carcasses from heat-tolerant Zebu-type cattle with their characteristic hump.

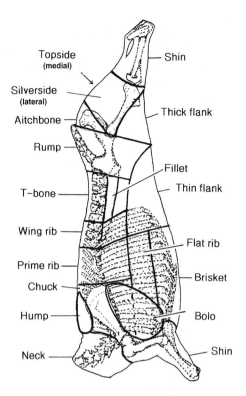

Figure 256. South Africa – beef cuts.

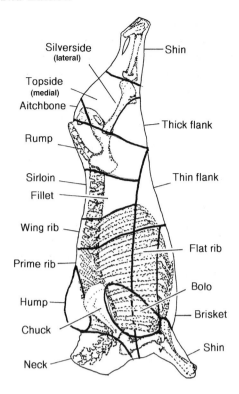

Figure 257. South Africa – big game cuts.

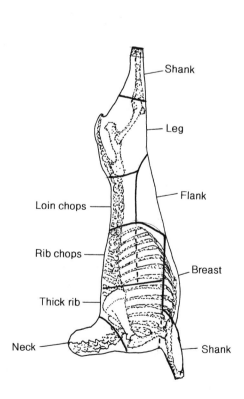

Figure 258. South Africa – lamb cuts.

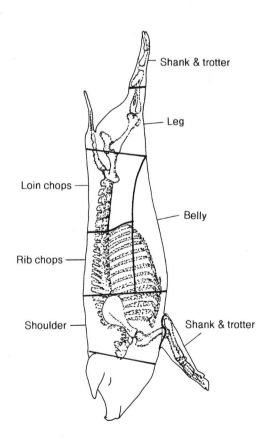

Figure 259. South Africa – pork cuts.

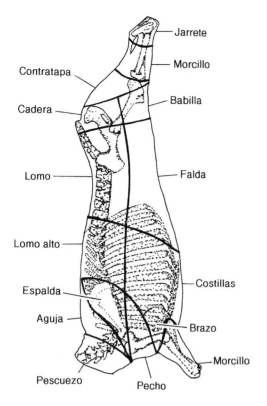

Figure 260. Spain – beef cuts.

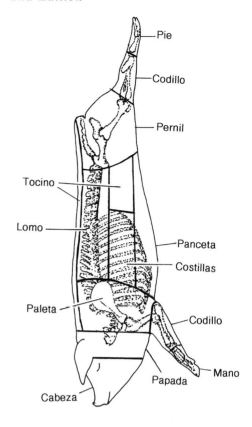

Figure 261. Spain – pork cuts.

Thus, the *trapezius* dominates a cut called the hump. *Triceps brachii* is removed as the bolo, rather than sectioning in a typical English manner.

(Sources: ADSRI, 1981; Hoffman, 1999)

SOUTH AFRICA - BIG GAME

South African cuts for big game (Figure 257), such as eland (*Taurotragus oryx*), gemsbock (*Oryx gazella*) and kudu (*Tragelaphus strepsiceros*) are similar to those used for beef. The hump is removed first, down to the vertebral column, before the side is quartered. The bolo is dominated by *triceps brachii*. The dorso-medial border of the shin goes through the ball and socket joint of the scapula. The prime rib may contain three or four ribs. If the *longissimus dorsi* is removed from the hindquarter it is called a scotch fillet. The thin flank is smaller than in beef. Numerous other species (see Bothma, 1989) are suitable for meat production by game ranching in South Africa.

(Source: van Rooyen, 1989)

SOUTH AFRICA - LAMB CUTS

The cutting pattern for lamb in South Africa is shown in Figure 258. The scapular region of the shoulder is included in the thick rib.

(Source: Hoffman, 1999)

SOUTH AFRICA - PORK CUTS

The cutting pattern for pork in South Africa is shown in Figure 259. The shoulder is cut square, rather than lifting the scapula from the ribcage.

(Source: Hoffman, 1999)

SOW

A female pig that has farrowed (produced a litter of piglets).

SPAIN - BEEF CUTS

The major division across the round of a Spanish beef carcass produces the babilla (perhaps named from the apron-like pattern of the *quadriceps femoris* muscles), and the cadera or hip (Figure 260). Located "against" these "top cuts" is the contratapa, and the more distal morcillo and jarrete (hock). Blood sausage also may be called morcillo. The loin is the lomo, and the prime rib region is the high loin or lomo alto. The flap or skirt of flank muscles is the falda. The mid and ventral rib region is kept intact as a primal cut called the costillas. The scapular region follows an ancient pattern of naming (→ spaud) and is called the espalda, and removal of the scapular cut leaves a needle-shaped cut, the aguja. The cut described as the breast is taken anteriorly through the humerus, leaving the neck (pescuezo), foreleg (brazo), and yet more morcillo.

(Sources: OEEC, 1961)

SPAIN - PORK CUTS

As shown in Figure 261, the ham may be divided into a foot (pie), knee (codillo) and upper leg (pernil). Tocino, salted fresh lard may be taken from the back-fat and abdominal region of heavy carcasses. The primal loin or lomo extends from the ilium anteriorly to the scapula, and separates off the ribs (costillas) and belly (panceta). Distally in the forelimb are located the elbow (codillo) and hand (mano). The jowl or double-chin (papada) is separated from the head (cabeza).

(Sources: OEEC, 1961)

SPALLA

The shoulder of an → Italian lamb carcass.

SPALLAROLO

Italian for the shoulder of an animal.

SPANNRIPPE

In → German beef cutting, the Spannrippe (spanning the ribs) is a long cut that includes the midlengths of many of the ribs.

SPARERIBS OF LAMB

In North America, the subcutaneous fat may be trimmed from breast of lamb to create lamb spareribs.

SPARERIBS OF PORK

In → North America, spareribs of pork are taken from the belly (Figure 262). Initially, the whole spareribs includes the sternum and ventral parts of the ribs, but the sternum often is removed. The muscles are diaphragm, *transversus thoracis* and intercostals.

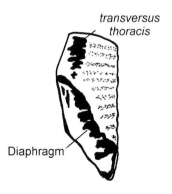

Figure 262. North American pork spareribs.

SPAUD OF BEEF

An archaic cut of beef in → England, relating to the *triceps brachii* in the shoulder region.

SPECTRIN

One of a trio of similar proteins (the α-actinin, spectrin, dystrophin family) involved in binding F-actin (→ actin).

(Source: Thomas *et al.*, 1998)

SPEGEPOELSE

Danish summer sausage.

SPENCER ROLL

In → Australian beef cutting, a spencer roll is equivalent to the chuck roll, plus overlying muscles (*trapezius, infraspinatus* and *latissimus dorsi*).

ŠPIČKA HRUDE

Brisket of a veal carcass in → Slovakia.

ŠPIČKA KRKU

Anterior region of a beef neck in → Slovakia.

SPIDER CRAB

Atlantic long-legged crab, *Maia squinado*, → Crabs.

SPIDSBRYST

In → Danish beef cutting, the spidsbryst or point of the breast is the anterior brisket.

SPIERSTUCK

For beef cutting in the → Netherlands, the spierstuck or muscle-piece is the *quadriceps femoris* group of muscles anterior to the femur.

SPINALES CERVICIS ET THORACIS

Spinales dorsi.

SPINALES DORSI

A compound muscle, with a long strap-like shape overall. A part of the muscle may be called *spinalis dorsi*. It extends along the mammalian vertebral column from the lumbar region to the neck, where often it is immediately ventral to the → ligamentum nuchae. Subunits of the muscle originate from the dorsal spines of all the lumbar vertebrae, from the eleventh to thirteenth thoracic vertebrae, and from the tenth to fifteenth ribs. Subunits insert on the lateral surfaces of vertebrae, from the third cervical to the tenth thoracic. In many rib steaks, the *spinales dorsi* forms a crescent-shaped cap to the *longissimus dorsi*.

SPINNAYA CHAST

A cut containing the thoracic vertebrae in a → Russian beef carcass.

SPIRLING

Côte découverte of a → Belgian beef carcass or the anterior thoracic ribs from a pork carcass in Belgium.

SPJALDHRYGGUR

Lumbar region of a carcass in → Iceland.

SPLENIUS

A thin, flat muscle on the lateral surface of the mammalian neck (Figure 263). It originates from a fascia connected to the dorsal spines of the anterior thoracic vertebrae, and inserts on an extensive fascia to the first four cervical vertebrae and temporal crest of the skull (which is severed). In pork, the muscle is quite thick and has a three-part insertion (occipital, temporal and to the wing of the atlas).

Figure 263. Lateral view of beef splenius (solid black and broken line) showing its origin from dorsal spines of thoracic vertebrae (1) and extensive insertion on cervical vertebrae (2).

SPLIT PIECE OF BEEF

In West of → England beef cutting, the split piece is the *quadriceps femoris* group of muscles, similar to toprump or thick flank in other parts of England or the posterior part of the sirloin tip in → North America.

SPRAT

A herring-like fish of the North Atlantic, such as *Sprattus*, → Clupeiformes. The myosystem is not particularly pleasant when fresh, but is tasty after marinading and smoking.

SPRING

Part of a pork shoulder in → England, as in a hand and spring of pork.

SPRINGBOK

Meat production from springbok, *Antidorcas marsupialis*, is well established in South Africa where springboks coexist with domestic cattle. The springbok grows to about 80 cm at the shoulder and to a live weight of about 35 kg. Both sexes have horns, but those of the doe are smaller and straighter

than in the buck. The → South African cutting pattern is shown.

SPUNTATURA DI COSTA

Also known as the lista, this is a long cut level with the midlength of the ribs on an → Italian beef carcass.

SQUAB

Fledgling → pigeon.

SQUARE CUT CHUCK OF BEEF

In North American beef cutting, the chuck shown in Figure 167 is a square cut chuck.

SQUID

Occurrence

Squids are cephalopod molluscs of the Subclass Coleoidae (Figure 264). The tentacles surrounding the head are derived from the muscular foot. The foot also gives rise to the muscular mantle and siphon which play a major role in escape locomotion by jet propulsion. Some of the commercially important squid of the Atlantic are *Loligo patagonica*, in the Southwest; *Loligo pealei*, the long-finned squid of the Northwest; *Loligo vulgaris*, the long-finned squid of the Northeast; and *Illex illecebrosus*, the short-finned squid of the Northwest. Occurring in the Pacific are: *Doryteuthis singhalensis*, Gulf of Tonkin; *Dosidicus gigas*, the jumbo squid of California; *Loligo opalescens*, also caught off California; *Loligo edulis*, ichibansu-rume, Japan; *Omastrephes sloani*, suruma-ika, Japan; various species of *Nototodarus* or arrow squids, New Zealand; *Todarodes pacificus*, Japan and Russia; and *Watasenia scintillans*, Japan. *Sepioteuthis lessoniana*, *Loligo edulis*, and *Loligo duvauceli* are caught in the Indian Ocean. Hundreds of other species are taken in many other seas as well.

Illex illecebrosus

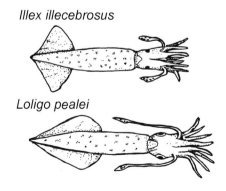

Loligo pealei

Figure 264. Squid.

Myosystem and processing

The mantle contains both circular and radial → myofibres. In *Illex illecebrosus* and *Loligo pealei*, the myofibres are

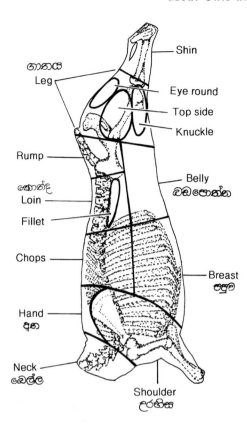

Figure 265. Sri Lanka – beef cuts.

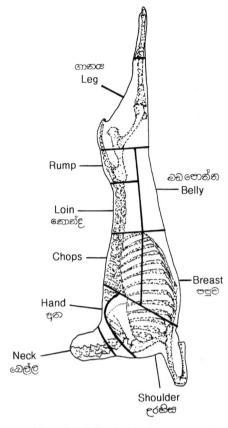

Figure 266. Sri Lanka – lamb cuts.

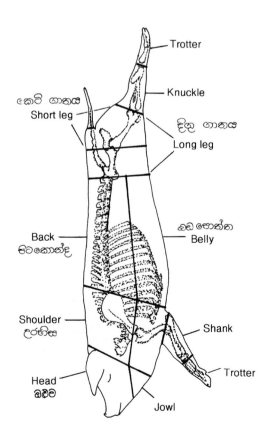

Figure 267. Sri Lanka – pork cuts used in large processing plants.

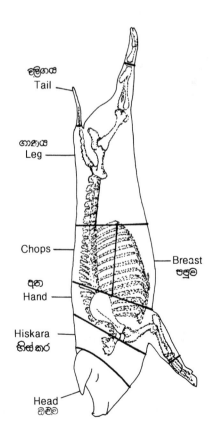

Figure 268. Sri Lanka – pork cuts used in retail shops.

roughly triangular in cross section, with a diameter ≈ 3.6 µm. Myofibrils are → obliquely striated and rectangular in cross section, and are twisted in a left-handed helix around an axial core of sarcoplasm at a contraction-dependent angle of 16 to 17° to the longitudinal axis of the myofibre, with a repeat distance of 38 µm. The major food myosystem prepared from squid (≈ 16% protein, 1 to 5% triglyceride) usually consists of strips or rings of mantle muscle, or the whole mantle can be eaten on a stick in local markets. Freshness is assessed by → TVB or an ammonia test. The texture of mantle, which is often perceived as being too tough or resilient, is mainly determined by the degree of cooking-induced gelatinization of the inner and outer tunics of woven → collagen fibres in which the radial muscles of the tunic are attached. Two types of cooking have been proposed to minimize toughness. Either cooking should be very quick, such as frying or sautéing for 2 to 3 minutes, or else prolonged, such as stewing for longer than 16 minutes. Gelatinization of the outer tunic starts at around 70°C and is continued at higher temperatures.

Automated processing may exploit friction to drag the tentacles behind the mantle, thus orientating the squid for the automated excision of the eye-bearing region of the head, and separating the tentacles from the mantle. The hollow mantle then may be forced onto a peg where water jets eviscerate it. In the Atlantic, *Loligo pealei* is preferred to *Illex illecebrosus*, which has greater fluid losses in refrigerated storage and a more rubbery texture when cooked. The pH of cooked mantle increases with the duration of refrigerated storage.

Colour

The colour of squid is an important commercial attribute and is determined by chromatophores and their associated structures (radial myofibres, axonal innervation, glial cells and sheath cells). Pigment granules are contained within a filamentous compartment within the chromatophore cell and are attached to its plasma membrane. Thus, when the chromatophore is flattened by neurally regulated contraction of the radial muscles, colour intensity is increased. Loss of colour, which is commercially undesirable, is accelerated by prolonged exposure to seawater or meltwater from ice. This causes pigment loss from the filamentous compartment.

(Sources: Cloney and Florey, 1968; Lipiński, 1973; Ampola, 1980; Singh and Brown, 1980; Otwell and Giddings, 1980; Gutworth *et al.*, 1981; Amaria, 1982; Stanley and Hultin, 1982; Stanley and Smith, 1984; Hincks and Stanley, 1985; MacGillivray et al., 1999)

SQUINÉE

The same as the petites côtes, a cut from the anterior part of the sternum in a → Belgian pork carcass.

SRI LANKA - BEEF CUTS

The cutting pattern used in large processing plants is shown in Figure 265. Three main parts of the leg are the eye round,

top side and knuckle. The nomenclature for the primal cuts of beef is unusual because it borrows terms from other species, such as hand and belly (from pork), and breast (from lamb). Thus, the sections through the posterior thoracic vertebrae are called chops rather than steaks. In retail shops, the cutting pattern is similar, but the loin is shorter.

(Source: Cyril, 1999)

SRI LANKA - LAMB CUTS

The pattern for lamb cutting in Sri Lanka (Figure 266) resembles that used in England, but with a couple of differences. The cut around the anterior thoracic vertebrae, the middle neck in England, is called the hand in Sri Lanka. The cutting line along the posterior edge of the shoulder is straight in Sri Lanka instead of being curved as it is in England.

(Source: Cyril, 1999)

SRI LANKA - PORK CUTS

The differences in cutting between large processing plants (Figure 267) and retail shops (Figure 268) are greater for pork than for beef.

(Source: Cyril, 1999)

STAARSTUCK

A beef cut prepared from the *biceps femoris* in the → Netherlands.

STAG

A male beef animal (sometimes a pig) that has been castrated, typically quite late so that secondary sexual characteristics have started to develop.

STANDING RIB ROAST OF BEEF

In North America, the rib may be separated into standing rib roasts (Figure 269) or rib steaks by cutting perpendicularly to the vertebral column. Rib-eye or delmonico steaks contain sections of *spinalis dorsi* as well as *longissimus dorsi* muscles.

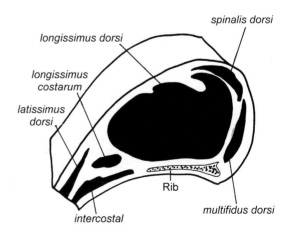

Figure 269. *North American standing rib roast of beef.*

STANDING RUMP OF BEEF

In North American beef cutting, the standing rump (Figure 270) is cut from a hip of beef and contains the proximal end of the femur (knuckle bone), the ischium (aitch bone), and coccygeal vertebrae (tail bones). With a triangular shape, one face of the standing rump resembles a sirloin, while the other face resembles part of a round steak. Because of its high bone content, it is usually boned and rolled, although it can be cooked as a standing roast if the meat is tender.

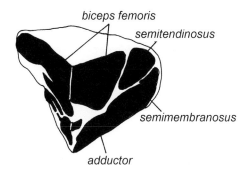

Figure 270. *North American standing rump of beef.*

STEAKMEAT

This is a common term used in → English beef cutting, but the cut sounds better than it is. Usually it is from the forequarter.

STEAKPIECE OF BEEF

In → English beef cutting, the steakpiece is in the sacral region in the West of England and Liverpool.

STEER

A castrated male beef animal.

STEHNO

The hindlimb in → Slovakia.

STEINBÍTUR

Icelandic for a catfish, a freshwater bony fish, → Cypriniformes.

STEK

In → Norway, the stek is the leg roast of a lamb or veal carcass. In Sweden, the whole leg of lamb or stekar med lägg becomes stek utan lägg after removal of the leg, and stek utan lägg och bäckenben after boning out. → Reindeer hindlimb in Sweden is also called stek.

STELZE

On → Austrian pork, lamb and veal carcasses the Hintere Stelze and Vordere Stelze are the shanks of the hind and forelimbs, respectively.

STERNOCEPHALICUS

A major, parallel-fibred muscle of the mammalian throat composed of *sternomandibularis* plus *sternomastoideus* (Figure 271). Both parts originate on the manubrium of the sternum and the first costal cartilage, but separate in their insertions on the head. Once the head is removed from the carcass the only way to separate the parts of the *sternocephalicus* is that the *sternomandibularis* is superficial to the *sternomastoideus*. In pork the *sterno-cephalicus* is undivided.

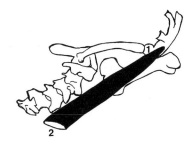

Figure 271. *Lateral view of beef* sternocephalicus *(solid black) showing its origin on the sternum and first costal cartilage (1), with insertions to the head severed (2).*

STERNOHYOIDEUS

A thin ribbon of muscle ventral to the trachea in mammals, usually obliterated when the carcass is dressed.

STERNOMANDIBULARIS

The most superficial part of *sternocephalicus*, the muscle originates on the manubrium of the sternum and first costal cartilage, and inserts on the lower jaw. In pork, the muscle has a long tendon to the mastoid process.

STERNOMASTOIDEUS

A broad strap of muscle forming the deep part of *sternocephalicus*. It shares it origins on the manubrium and first costal cartilage with *sternomandibularis* but inserts onto the occipital bone and temporal crest of the skull.

STICH

In German beef cutting, this is a cut from the neck region, near to where the animal was exsanguinated, as in the English sticking cut.

STICKING

British cuts of beef from the neck region, ranging from the cervical region down to the humerus.

STITHOPLEVRES

A ladder of ribs from a → Greek beef carcass.

STITHOS

In → Greece, a cut from the pectoral region.

STONE CRAB

An Atlantic shore crab with massive claws, *Menippe mercenaria*, taken in the southern states of the USA, → crabs.

STOTZEN

A whole hip of beef in → Switzerland.

STRED HRUDE

Postero-ventral rib region of a veal carcass in → Slovakia.

STRIP-LOIN

In →Australia, → England, → Ireland and → North America, strips of beef *longissimus dorsi* and associated muscles may be removed as a strip-loin which then is sliced transversely into steaks. Thus, in North America, strip-loin originates from the primal loin while, in England, it is from the primal sirloin.

STROGILO

In → Greece, a cut of beef taken anteriorly to the femur.

STURGEON

Elongated fish with rows of shield-like scutes, *Acipenser*, → Osteichthyes.

SUBSCAPULARIS

A large, flat muscle originating on the costal surface of the mammalian scapula and inserting onto the medial tuberosity of the humerus (Figure 272).

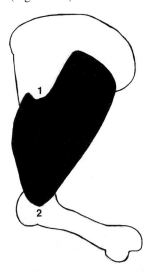

Figure 272. Medial view of beef subscapularis *showing its origin (1) from the medial face of the scapula and the surfaces of adjacent muscles* (supraspinatus and teres major), *and its insertion to the medial tuberosity of the humerus (2).*

SUCADESTUK

A cut based on the *infraspinatus* of a Netherlands beef or pork carcass.

SUCUK

A fermented dry sausage produced in Turkey and traditionally using air drying of the meat.

(Source: Bozkurt and Erkmen, 2002).

SUET

Perirenal beef fat with a relatively high melting point so that it is hard at room temperature.

SULDO

The outside round and rump of a → Korean beef carcass.

SUMMER SAUSAGE

A semidry → sausage.

ŠUNKOVÝ VÝREZ

Around the *tensor fascia lata* on a → Slovakian pork carcass.

SUORA KINKKU

The ham of a pork carcass in → Finland.

SUPRACORACOIDEUS

In birds, a muscle that elevates the wing (→ chicken). It is located between the sternum and the → *pectoralis* muscle.

SUPRASPINATUS

A large muscle filling the fossa above the spine of the mammalian scapula and inserting onto the medial tuberosity of the humerus (Figure 273).

Figure 273. Lateral view of beef supraspinatus *showing its origin from the supraspinous fossa and adjacent scapular spine (1) and insertion to the tuberosities of the humerus (2).*

SURIMI

Food products using gelled myofibrillar proteins, usually from low-grade seafood myosystems.

(Source: Park *et al.*, 1997)

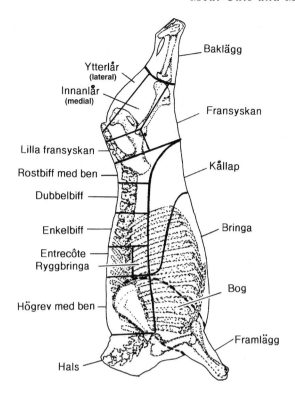

Figure 274. Sweden – beef cuts.

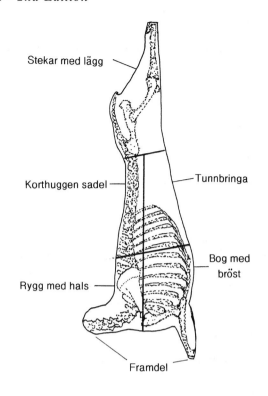

Figure 275. Sweden – lamb cuts.

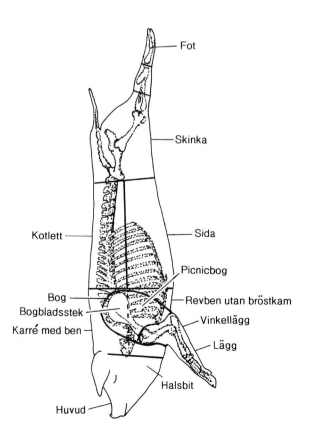

Figure 276. Sweden – pork cuts.

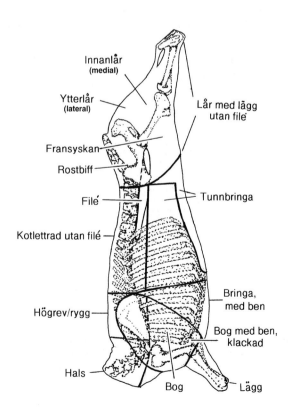

Figure 277. Sweden – veal cuts.

SURPRISE

French term for the subscapularis muscle.

SUWARI

A set → surimi gel with a globular microstructure.

(Source: Alvarez et al., 1999)

SVARTFUGL

In the northern hemisphere, the auks, guillemots and puffins of the family Alcidae have a similar role to that of the penguins in the southern hemisphere. The guillemot (*Uria aalge*) breeds in enormous numbers in Iceland, where it is known as the svartfugl, and is commonly taken as a game bird (as is the puffin or → lundi).

(Sources: Bertin, 1967; IAIS)

SVÍNAKJÖT

Pork in Iceland.

SVINJETINA

Pork in Serbia & Montenegro.

SWEDEN - BEEF CUTS

The whole hindlimb of beef (nötkött) from the pubis posteriorly (lår med lägg) is split into a hindshank (baklägg); the lateral ytterlår dominated by *biceps femoris*; the medial innanlår dominated by *pectineus, adductor* and *semimembranosus*; and the fransyskan containing the *quadriceps femoris* group of muscles (Figure 274). In this case, the sirloin cut through the shaft of the ilium becomes the lilla fransyskan. But the lilla fransyskan also may be included in a larger primal cut called the rostbiff med ben (roast beef with bone).

The cut containing the lumbar steaks is the dubbelbiff, and anterior to this are the prime rib steaks of the enkelbiff. Thus, double (dubbel) indicates the two eyes of meat (*longissimus dorsi* plus *psoas major*) while simple (enkel) indicates a single rib-eye (*longissimus dorsi*). The former is used for T-benstek (T-bone steak) and the latter for clubstek (club steak). Ribs 7 to 9 may appear in entrecôte, but not if the ryggbringa (back-brisket) is removed separately. After removal of the scapula and distal limb bones as a shoulder or bog, the thoracic vertebrae may be known as högrev med ben (high rib with bone) or tjoka reven (thick rib). The bog may be cut in thick slices across the bones (as in the → Norwegian pattern), in which case they are called märgipan (marrow slices), or the bog may be deboned and sliced (bog, benfri). The neck or hals is continuous with the brisket or bringa remaining after the removal of the shoulder or bog. The flank is called the kållap.

(Sources: SKI; OEEC, 1961)

SWEDEN - LAMB CUTS

The whole hindlimb of lamb (lammkött) is removed as a roast with leg (stekar med lägg), becoming stek utan lägg after removal of the leg, and stek utan lägg och bäckenben after boning out. Leg slices may keep the latter name, or may be identified as fransyska (*quadriceps femoris* muscle group). The sadel med tunnbringa is a bilateral cut through the carcass, including both left and right sides of the korthuggen sadel (short saddle) and tunnbringa (thin breast) shown in Figure 275.

The forequarter or framdel is split through the scapula into a bilateral back with neck (rygg med hals) and two shoulders with breast (bog med bröst). Single rib chops are called kotlett, enkel, and bilateral double chops are kotlett, dubbel. Scapular muscles may be sold as bog.

(Sources: SKI)

SWEDEN - PORK CUTS

As in many other countries, the Swedish pattern for cutting pork (griskött) has changed from a curvilinear hand-cut pattern to a rectilinear pattern suitable for the bandsaw (Figure 276). Similarly, with modern lean pork, the thick cuts of lard have largely disappeared, such as the späck or back-fat and ister or side lard.

The ham or skinka may be removed as a large straight cut (skinka rak) after removal of the foot (fot). The whole round ham with bone as a primal cut is called skinka, rund med ben, becoming a shop-ham or butiksskinka when boned out for retail slicing. The muscle layers of the abdomen which attach laterally to the posterior ribs, form the sida. The sida may be separated into a major primal for manufacturing side bacon (sidfläsk), plus a long ventral strip for small bacon (småfläsk). After lifting off the shoulder or bog with the scapula and limb bones (bog med ben), the remaining dorsal part of the ribcage is the karré med ben, and the ventral part is the revben utan bröstkam (rib without brisket). The bog is separated into the scapular portion (bogbladsstek), *triceps brachii* (picnicbog) and the corner of the leg with distal limb bones (vinkellägg), which gives flintastek when sliced transversely. The neck-bit or halsbit is the fleshy part of the neck removed from the head (huvud).

The bone-in loin of pork (kotlett med ben) becomes the crown cutlets or kronkotlett after boning out, with *longissimus dorsi* being presented in lengths as kotlett, benfri, or slices, as kronkotlett. The *psoas* muscles may be sold as a long filet of kotlettfilé.

(Sources: SKI; OEEC, 1961)

SWEDEN - VEAL CUTS

The whole hindlimb with leg but not filet (lårr med lägg utan filé) is a primal cut which is subdivided like a Swedish beef carcass into a medial innanlår, a lateral ytterlår, and anterior fransyskan and a rostbiff around the ilium. With the *psoas*

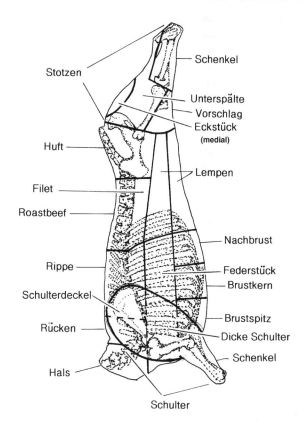

Figure 278. Switzerland – beef cuts.

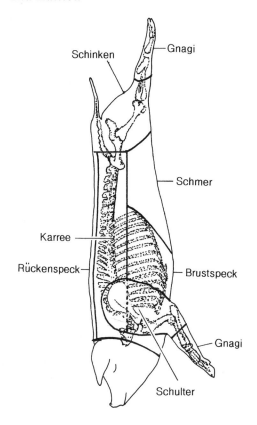

Figure 279. Switzerland – pork cuts.

major and *iliacus* removed as a whole fillet (filé), the loin is the kotlettrad utan filé. The abdominal muscles or tunnbringa are divided as shown in Figure 277.

The shoulder or bog is lifted from the ribcage to leave the high-rib back or högrev/rygg and underlying brisket with bone (bringa med ben). The foreshank (lägg) is removed from the bog, as the bone-in heel of the shoulder (bog med ben, klackad). The scapula is removed so that the bog can be sliced, but marrow bone slices are made across the humerus. Retail preparation results in a typical range of veal products such as schnitzels, filet slices and butterfly cutletts (fjärilskotlett). The brisket may be sold in slices or rolled (bringa, rullad).

(Source: SKI)

SWEET RIB

A cut of beef in the postero-ventral region of the ribcage from the Midlands of → England.

SWISS STEAKS

In North American beef cutting, Swiss steaks usually are prepared from the round, and are presented as solid flat pieces of muscle with all connective tissues removed. Reforming from small parts is not usually allowed.

SWITZERLAND - BEEF CUTS

As shown in Figure 278, the hindlimb (Stotzen) is composed of an anterior cut containing the *quadriceps femoris* group of muscles (Vorschlag, literally, the forestrike), a lateral cut dominated by *biceps femoris* (Unterspälte, literally, under the fissure), an angle-piece containing *semitendinosus, semimembranosus, adductor* and *pectineus* (Eckstück), and the shank (Schenkel).

The loin and prime rib region is the Roastbeef, anterior to which is a rib cut (Rippe), back (Rücken), and neck (Hals). The flank (Lempen) may be divided longitudinally into a dorsal thick flank (dicker Lempen) and a ventral thin flank (dünner Lempen). The feather-piece or Federstück contains a long ladder of rib sections. The brisket is divided, from anterior to posterior, into a breast-point (Brustspitz), centre-breast (Brustkern), and after-breast (Nachbrust). The shoulder (Schulter) is divided into a dorsal region (Schulterdeckel), from which is taken a *supraspinatus* shoulder-filet (Schulterfilet). The *triceps brachii* is separated as thick shoulder or dicke Schulter.

(Sources: Schön, 1958; Pomeroy and Williams, 1976)

SWITZERLAND - PORK CUTS

The whole hindlimb or ham is the Schinken, and may be used for roast with crackling (Braten und Platzli). The whole long loin is called the Karree, after removal of the back fat or bacon (Rückenspeck). The ventral thoracic region is the breast bacon or Brustspeck. The Schulter is the shoulder, and the trotters are Gnagi (Figure 279).

(Source: OEEC, 1961)

SWORDFISH

A bony fish without scales and with an upper jaw forming a sword-like projection, *Xiphius*, → Perciformes.

SYZNKA

The ham of a → Polish pork carcass.

SZEGY

Szegy is the brisket or ventral region of a → Hungarian beef carcass. A cut of this name includes most of the costal cartilages. The szegyfő contains the anterior part of the sternum, while the puha-szegy is the most posterior part of the szegy.

SZEGYFŐ

The most anterior part of the brisket or szegy in a → Hungarian beef carcass.

SZPONDER

Flank or plate, this is a cut with a ladder of rib sections from a → Polish beef carcass.

SZYJA

In → Poland, a cut of beef or veal from the anterior cervical region.

TACHE NOIRE

The pièce levée of a → Belgian beef carcass.

TAFELSPITZ

An → Austrian beef cut containing the sacrum.

TAFELSTÜCK

Biceps femoris on an → Austrian beef carcass.

TAGLIO DI SOTTOSPALLA

A cut under the shoulder in an → Italian beef carcass.

TAIL

The posterior end of a fish fillet is called a tail. Whole tails of pork with skin-on, or subdivided oxtails are both edible byproducts with an appreciable muscle content (*coccygeus*).

TAILEND

An → Irish cut of beef containing the ischium and sacrum.

TAIWAN - PORK CUTS

The primal cuts are shown in Figure 280. Sides of pork without the head may be marketed skin-on plus fore and hind trotters, or with the trotters removed, or with trotters plus skin removed.

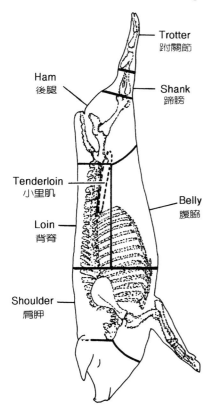

Figure 280. Taiwan - pork cuts.

(Sources: TMDF; Kuo, 1999)

TAKAPOTKA

In → Finland, the hindshank of a beef or pork carcass.

TAKASORKKA

The hindfoot of a pork carcass in → Finland.

TALIN

A protein involved in the attachment of → myofibrils to the plasma membrane of the → myofibre.

TÅNGRÄKA

Swedish for → shrimp.

TAPABARRIGA

The top of the paunch is a → Chilean beef cut.

TAPA DE ESPALDILLA

A → Mexican cut of beef (top of the blade) taken from the espaldilla y costillar.

TAPA DE PALETA

In → Cuban beef cutting, a boneless cut dominated by the subscapularis.

TAPA DE PECHO

Brisket of beef in → Mexico.

TAPAPECHO

The "lid" of the beef brisket in → Chile.

TARF JNAH

Distal region of the poultry wing in → Morocco.

TARJA

Beef from the anterior thoracic region in → Hungary.

TAUBE

German for → pigeon.

TAZOBEDRENNAYA CHAST

The main hindlimb cut from a → Russian beef, pork or lamb carcass.

TARGET

In English mutton or lamb cutting, a target is the forequarter with the shoulder removed. The exposed muscles over the ribcage appear as a round, red target.

T-BONE STEAK

In North America and England, T-bone steaks are beef steaks taken through the lumbar region. The muscle groups are similar to the porterhouse steak shown in Figure 197, but with less *psoas major*. The porterhouse steak is a posterior T-bone steak with especially large areas of muscle.

TEAL

Several species of teal (→ duck) are hunted for their tasty meat in North America: blue-winged teal (*Anas discors*), cinnamon teal (*Anas cyanoptera*) and green-winged teal (*Anas carolinensis*).

TEAWURST

Soft, cooked → sausage for spreading in sandwiches.

TEEWURST

→ Teawurst

TEG

A castrated male sheep, sometimes any young sheep or deer.

TELEOST

Bony fish (→ Osteichthyes).

TELESCOPE

Saving space by cutting an unsplit carcass, typically lamb, with a transverse cut in the lumbar region so that the anterior part of the carcass can be telescoped inside the posterior part.

TELETINA

Veal in → Serbia & Montenegro.

TENCH

The tench, *Tinca tinca*, → Cypriniformes, has desirable moist, white muscle and is extensively farmed in Europe.

TENDE DE TRANCHE

The *semimembranosus* in the tranche of → French beef cutting.

TENDERETTS OF PORK

In North America, cubed scraps of meat, typically from a fresh leg of pork.

TENDER HEEL

A cut from the *gastrocnemius* region in British → Proten beef.

TENDERLOIN

In North America, slices of *psoas major* and *minor* muscles from beef and pork may be sold separately as tenderloin. Beef tenderloin is often wrapped with bacon.

TENDERSIDE OF BEEF

In West of → England beef cutting, the tenderside is the equivalent of the topside on the medial part of the round.

TENSOR FASCIAE ANTIBRACHII

A thin slip of muscle on the posterior edge of the long head of *triceps brachii*.

TENSOR FASCIA LATA

A large triangular muscle in the angle between the belly and hindlimb of a mammal. It originates from the pelvis and inserts onto the fascia lata. The shape of this muscle is changed in cuts of meat when the carcass is suspended by the pelvis rather than the hock. The muscle appears in Figures 74, 215, 246 and 248.

TERES MAJOR

In mammals, a thin muscle originating on the scapula and inserting onto the humerus.

TERES MINOR

In mammals, a small muscle originating on the scapula and inserting onto the humerus. It is parallel with the *infraspinatus*, but below it. It is a large, rounded muscle in pork.

TERMINAL CISTERNA

An expanded part of the → sarcoplasmic reticulum where it meets a → transverse tubule to form a → triad. The plural is cisternae.

TERNERA

Spanish for veal.

TESTA SENZA GUANCIALE

Jowl or cheek muscle from the head of an → Italian pork carcass.

TÊTE DE NOIX

The petite tête d'aloyau of a → Belgian beef carcass.

TÊTE DE PHALAN

The petite tête d'aloyau of a → Belgian beef carcass.

T-FILAMENT

A name proposed for → titin filaments.

(Source: Locker, 1987)

THAILAND - BEEF AND BUFFALO CUTS

In Thailand, the traditional cutting patterns for buffalo are similar to those used for beef, but North American meat cuts may appear in modern supermarkets (Figure 281). The hindlimb (sa pook) is cut into a medial sa pook nay and a lateral sa pook nook, with the ischial region being separated as nua look ma praaw. From posterior to anterior, the primal cuts are san nook (sacral), san nay (lumbar) and san laang (posterior thoracic). A square shoulder (lay) is cut to include both scapula and anterior thoracic vertebrae. From anterior to posterior are the ok (sternum), poon ok (costal cartilages) and puun thong (ventral abdominal muscles). The foreshank is the khaa naa.

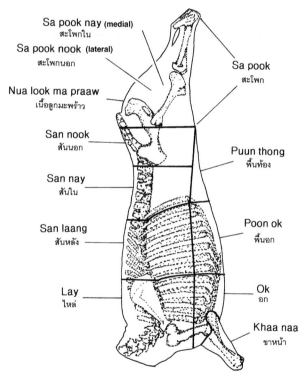

Figure 281. Thailand – beef cuts.

(Source: Hongsprabhas, 1999; Suwattana, 1999)

THAILAND - PORK CUTS

The feet or trotters are removed from the hind and forelimb as the kho khaa laang and kho khaa naa, respectively (Figure 282). The ham or hindlimb is the san sa pook. The back-fat (man khang) is stripped from both the loin (san nook) and the scapular shoulder cut (san kho muu). The san lay contains humerus, ulna, radius and associated muscles. The spare ribs (sii khroong muu) are separated from the remainder of the belly (saam chaan).

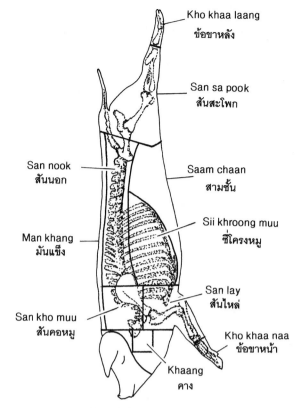

Figure 282. Thailand – pork cuts.

(Source: Hongsprabhas, 1999; Suwattana, 1999)

THAW SHORTENING

Meat that is frozen before the onset of → rigor mortis still contains ATP. The → sarcoplasmic reticulum is disrupted by ice crystal growth. When the meat is thawed, it undergoes a violent → contraction because of the release of calcium ions from the sarcoplasmic reticulum. The meat becomes extremely tough, looses a massive amount of fluid and is virtually useless.

THEAVE

A female sheep a couple of years old that has not yet produced a lamb.

THICK CHINE AND NECK

A cut of beef from the scapular and cervical region in Northeast → England.

THICK FLANK OF BEEF

In English beef cutting, the thick flank is an alternative name for the → toprump.

THICK RUNNER

A → Scottish cut of beef in the anterior-dorsal region of the ribcage.

THIN RIB

An → English cut of beef from the posterior thoracic region, known in Liverpool and Northeast England.

THIN RUNNER

A → Scottish cut of beef in the postero-dorsal region of the ribcage.

THRESHER

A shark (→ Chondrichthyes), *Alopias vulpinus*, the myosystem of which is like white tuna.

(Source: Pivnička and Černý, 1987)

THURINGER

Traditionally, a fermented, semidry → sausage.

TIBIALIS ANTERIOR

Tibialis cranialis.

TIBIALIS CAUDALIS

Part of *flexor digitorum profundus* in the hindlimb.

TIBIALIS CRANIALIS

A thin, two-headed muscle on the dorso-lateral face of the mammalian tibia, although it is quite large in pork.

TIBONE

T-bone steaks in → Brazil.

TIP OF BEEF

The knuckle tip of a North American hip or → diamond round of beef (*vastus lateralis, vastus medialis, vastus intermedius, rectus femoris*, plus some of *tensor fascia lata*). Pulling the tip from a diamond round starts with a medial cut along the anterior edge of the femur, posterior to the *vastus medialis*, followed by a lateral separation cutting into the anterior edge of the *biceps femoris*. With medial and lateral separations already cut, the distal face of the patella is freed, allowing the patella and tip to be pulled away from the anterior face of the femur, stripping off the periosteum from the anterior face of the femur. The tip of beef may be split in half longitudinally (from proximal to distal) and used for a rolled joint (tip French rolls) or simply sliced as tip London broil. A variety of steaks and kabobs also may be cut from the tip.

TITIN

For practical purposes, titin can be regarded as a synonym for connectin. It is an elastic protein extending from the → Z-line to the → M-line at the midlength of the thick → myofilament. Titin has an exceptionally high molecular weight (around 3000 kD) and is a major source of series → elasticity in the → sarcomere since it may be stretched to four times its rest length. In parallel with the thick and thin myofilaments, and in series with thick myofilaments, titin holds Z-lines together and is involved with → nebulin in anchoring the terminal Z-lines at → myotendon junctions. Titin filaments run alongside the thick myofilaments, starting near the center of the A-band and extending to the Z-line, near which they may be attached to thin myofilaments. In series with the thin myofilaments, titin binds to the myofibre plasma membrane at muscle-tendon junctions as well as to the rod-like part of the myosin molecule. Doubtless, titin is important in understanding sarcomere structure, but titin content does not appear to be directly related to variation in meat toughness.

(Source: Horowits *et al.*, 1989; Maruyama, 1997; Fritz *et al.*, 1993; Labeit and Kolmerer, 1995b)

TJOKA REVEN

On a → Swedish beef carcass, the thick rib remains in the anterior thoracic region after removal of the bog or shoulder (högrev med ben).

TOCINO

Fresh lard from the back-fat or abdomen of a Spanish pork carcass, but also Spanish for bacon.

TOMOBARA

A long flank or plate of beef from a → Japanese beef carcass.

TOPKNOT

A flatfish, *Zeugopterus*, → Pleuronectiformes.

TOP RIBS OF BEEF

In → English beef cutting, anterior to a 4-rib forerib is a 4-rib middle ribs cut. The top ribs cut is the ventral part of the middle ribs. A similar cut was noted by Gerrard (1949) in Liverpool while, in London, it tends to be more dorsal in position.

TOP ROUND

In North American beef cutting, the top or inside round is from the medial face of the hindlimb and is dominated by *adductor* and *semimembranosus*.

TOPRUMP OF BEEF

In English beef cutting, the *quadriceps femoris* group of muscles inserting onto the patella is removed as toprump or thick flank with a cut parallel to the femur (Figure 283). When thin sliced and flattened, sections of toprump are known as beef olives.

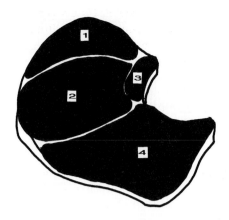

Figure 283. English toprump or thick flank of beef with landmark muscles; vastus medialis (1), rectus femoris (2), vastus intermedius (3), and vastus lateralis (4).

TOPSIDE OF BEEF

In English beef cutting, the topside is a major cut located medially in the hip. It contains the weathered exposure of the *adductor* muscle ventral to the pubis on a side of beef (Figure 284).

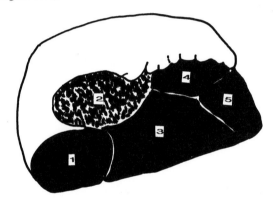

Figure 284. English topside of beef with some landmark muscles; semimembranosus (1), exposed face of adductor originally ventral to the pubis on a side of beef (2), cut surface of adductor (3), gracilis (4) and pectineus (5).

TORSK

Danish and Swedish for cod, → Gadiformes.

TORTUGUITA

Gastrocnemius of an → Argentinean beef carcass.

TOUGHNESS

Meat toughness is a complex subjective term that is difficult to translate into exact, measurable, mechanical terms. When typical samples of meat are cooked, the myofibres decrease in diameter by up to 15% and → collagen is gelatinized to varying degrees. The degraded collagen around muscle fasciculi traps molten fat droplets released from any marbling fat within the meat. Toughness may decrease as the product of time and temperature increases, but the time-temperature optimum differs between muscles, mainly in relation to the amount and strength of connective tissue. Beef semitendinosus, for example, becomes increasingly tender up to 67°C and then becomes tougher at higher temperatures. Some of the strongest factors influencing the toughness of meat are: (1) → sarcomere length, (2) → collagen content, and (3) the degree of → conditioning post-mortem.

TOURNEDOS

Small steaks taken across beef *longissimus dorsi*, level with or just posterior to the scapula.

TRAIN-DE-CÔTES

In → France, a primal cut of beef through the rib region to include the surlonge, and train-de-côtes couvert and découvert.

TRANCHE

Beef cuts around the femur in → France and → Belgium.

TRANCHE Á LA CÔTE

Rib of beef, entrecôte avant, in → Belgium.

TRANS

A cut taken medially to the femur in → Greek beef cutting.

TRANSVERSALIS COSTARUM

Iliocostalis.

TRANSVERSE STRIATIONS

Under a microscope, both skeletal and cardiac → myofibres are transversely striated (whereas the smooth muscle cells of the viscera are not). As shown in Figure 285, the transverse striations of a myofibre are perpendicular to the long axis of the myofibre (although there are some exceptions in food myosystems, → obliquely striated myofibres). The transverse striations visible in a whole myofibre are caused by the underlying transverse striations of the → myofibrils within the myofibre. With a light microscope at its highest magnification, it is often possible to see the transverse striations of myofibres and isolated fragments of myofibrils by: (1), reducing the size of the substage diaphragm (gaining contrast but losing resolution); (2), staining the specimen with a histological dye; (3), using a phase contrast or differential interference contrast microscope; or (4), using a polarising microscope. The naming of transverse striations originates from their appearance in polarised light.

Polarising filters (like those of Polaroid sunglasses) are arranged with one in front of the light source and the other within the microscope eyepiece. By rotating one filter relative to the other, the transmittance of light through the microscope may be blocked. The first filter only transmits light waves vibrating in a certain plane (polarized light), but if these are unable to get through the second filter whose transmitting plane is at 90° to the first filter, then the field of view is dark, except for any transverse striations. Alternate striations are able to rotate the plane of the polarized light strongly enough for the light to get through the second filter. Thus, these striations appear bright in an otherwise dark viewing field and are said to be birefringent (having two refractive indices). Rotation of the specimen is required to allow this property to be observed in any particular myofibre or myofibril, because the effect is best seen when the long axis of the specimen is at 45° to the polarising filters. Striations that appear bright in polarized light are termed anisotropic and, hence, are called A-bands. Between the A-bands, are striations that appear relatively dim because they are isotropic and, hence, are termed I-bands.

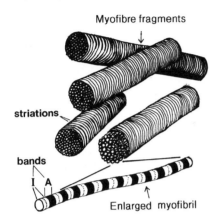

Figure 285. *Transverse striations on lengths of myofibres (top) are caused by the banding of myofibrils within the myofibres (bottom),*

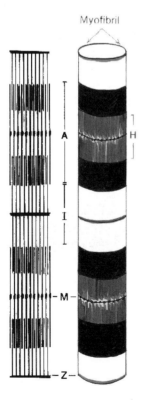

Figure 286. *Arrangement of thick and thin myofilaments creating transverse striations on a stained myofibril.*

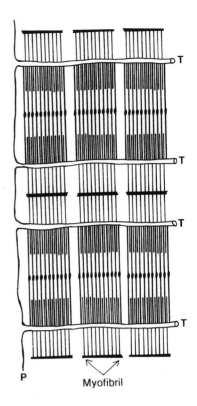

Figure 287. *Transverse tubules (T) pushing into the interior of a myofibre from the plasma membrane (P).*

In summary, if a transparent structure is composed of randomly arranged molecules, it has no directional interactions with polarised light and is said to be isotropic. But, if a structure contains neat alignments of elongated molecules, it may interact with polarised light and is said to be anisotropic. Ordinary window glass is generally isotropic, whereas many mineral crystals are anisotropic. Both A and I-bands contain neatly aligned protein filaments - the A-band is more birefringent than the I-band because its filaments are thicker. Thus, the difference between A and I-bands is relative (the A-band appears bright while the I-band is dim, but not totally dark). With a stained sample viewed under an ordinary light microscope, however, it is the A-band (containing the thicker myofilaments) that generally becomes more darkly stained than the I-band, as shown in Figure 286. Thus, relative to I-bands, A-band are brighter in polarised light but darker when stained.

A thin → Z-line occurs at the middle of the → I-band. The repeating unit of a regular series of transverse striations is termed the sarcomere, and is usually considered to be the structural unit from Z-line to Z-line. The Z-line resembles a woven disk, like the bottom of a wicker basket. Some authors call it a Z-disk.

TRANSVERSE TUBULES

Each transverse tubule or T-tubule is a finger-like inpushing of the plasma membrane of the → myofibre (Figure 287). Experimentally, an extracellular marker such as horseradish peroxidase can move from the extracellular space around the myofibre into its T-tubules, thus proving that the T-tubules open onto the myofibre surface (even though this is a rare sighting in typical electron micrographs).

The T-tubules come into close contact with the → sarcoplasmic reticulum around each → myofibril. Thus, T-tubules can conduct action potentials initiated at the neuromuscular junction deep into the interior of the myofibre. Communication between a T-tubule carrying an action potential and the sarcoplasmic reticulum is mediated by protein-containing bridges or feet. The bridges are composed of four subunits arranged in an offset pattern and they extend through the membrane of the sarcoplasmic reticulum to form channels for calcium ions. However, at the ends of myofibres where T-tubules are scarce, the sarcoplasmic reticulum may be directly connected to the plasma membrane as well. In meat from farm mammals, T-tubules occur at the edges of the A-band, so that there are two per sarcomere. In other food myosystems there may be only one T-tubule per sarcomere, often at the level of the Z-line. In poultry meat, some myofibres may have T-tubules at the A-I junction while others have T-tubules level with the Z-line.

(Sources: Eisenberg and Eisenberg, 1968; Andreev and Wassilev, 1986; Kawamoto *et al.*, 1988; Takekura *et al.*, 1993)

TRANSVERSUS ABDOMINIS

A large flat muscle, deep in the mammalian abdominal wall. Its widely spread origins include ribs 8 to 13 and, indirectly, the transverse processes of the first five lumbar vertebrae. It inserts to the linea alba in the midline of the belly.

TRANSVERSUS COSTARUM

Rectus thoracis.

TRANSVERSUS THORACIS

A flat muscle, inside the mammalian rib cage, dorsal to the sternum.

TRAPEZIUS

A relatively thin, but extensive and superficial muscle, from the neck to the back, between the scapulae. In beef, but not in lamb or pork, it is divided into cranial and thoracic parts.. The *trapezius* muscle appears in Figures 37 and 240.

TRAVERS

In French pork cutting, the travers is a very long strip of flank and ribs, sometimes taken as far forward in the neck as in Figure 78, but not if the distal shoulder region is removed as a square-cut hachage.

TRIAD

In electron micrographs of muscle, a triad is where → transverse tubules running through the → myofibre make contact with two terminal cisternae of → sarcoplasmic reticulum (Figure 288).

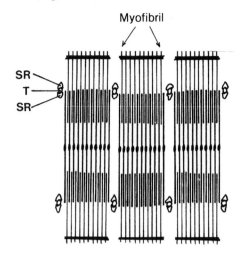

Figure 288. *Triads composed of sarcoplasmic reticulum (SR) each side of a transverse tubule (T).*

TRIANGLE OF BEEF

In North American beef cutting, a triangle is the shank plus brisket plus chuck plus plate shown in Figure 167. However, a triangle roast is the *infraspinatus*.

TRICEPS BRACHII

A large, three-headed muscle filling the triangular space ventral to the scapula and posterior to the humerus in mammals (Figure 289). The lateral head (*caput laterale*) originates on the humerus and inserts on the lateral surface of the olecranon process of the ulna. The long head (*caput longum*) originates on the scapula and inserts on the tip of the olecranon process. The middle head (*caput mediale*) originates on the humerus and inserts on the medial surface of the olecranon process. The long head is the largest (> 3% total carcass muscle weight) while the other two heads are both small.

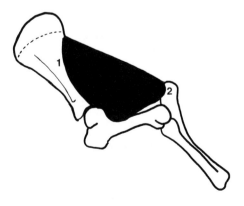

Figure 289. *Lateral view of beef* triceps brachii *showing its origins (1) along the ventral edge of the scapula and* infraspinatus, *and its insertion (2) onto the olecranon and adjacent fascia.*

TRICEPS SURAE

A muscle group composed of the heads of *gastrocnemius* plus the *soleus* muscle.

TRIMETHYLAMINE OXIDE

Trimethylamine oxide (TMAO) is a post-mortem metabolite that affects fish texture by cross-linking myofibrillar proteins, primarily in the cod family ($\rightarrow$ Gadiformes). Formaldehyde is formed from TMAO by a demethylase. At a high pH, formaldehyde may also be formed non-enzymatically.

TROIS CÔTES

Côte découverte of a $\rightarrow$ Belgian beef carcass.

TROPOCOLLAGEN

Tropocollagen is the protein from which collagen and reticular fibres are composed (Figure 290). It has a high molecular weight (300 kD) and is formed from three polypeptide strands twisted into a triple helix. Each strand is a left-handed helix twisted on itself, but the three strands are twisted into a larger right-handed triple helix. The triple helix confers stability to the molecule and allows self-assembly into fibrous components. Telopeptides projecting beyond the triple helix are responsible for cross-linking between adjacent molecules, with cross-links to the shaft of the adjacent molecule.

Within the polypeptide strands of tropocollagen, glycine occurs at every third position, and proline and hydroxyproline account for 23% of the total residues. Hydroxyproline is scarce in other proteins of the body, so that an assay for this imino acid provides a measure of the collagen or connective tissue content of meat. Tropocollagen also contains a relatively high proportion of glutamic acid and alanine, as well as some hydroxylysine.

Although tropocollagen molecules are composed of three alpha chains, many (19) unique alpha chains have been identified, giving rise to 11 different types of collagen. These are categorized into three classes; molecules with a long ($\approx$ 300 nm) uninterrupted helical domain, molecules with a long ($\geq$ 300 nm) interrupted helical domain, and short molecules with either a continuous or an interrupted helical domain.

The main types of collagen fibres in food myosystems are as follows. Type I collagen fibres form striated fibres between 80 and 160 nm in diameter in blood vessel walls, tendon, skin, and muscle. Type II collagen fibres are less than 80 nm in diameter and occur in hyaline cartilage. Type III (reticular) fibres occur in the endomysium and around adipose cells. Type IV collagen occurs in basement membranes where, instead of being arranged in a staggered array, the tropocollagen molecules are linked at their ends to form a loose diagonal lattice. Type V collagen also occurs in myofibre basement membranes. Type VI collagen, a tetramer of Type V, forms a filamentous network in muscle and skin.

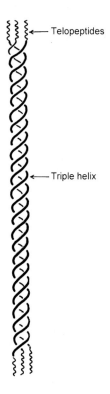

Figure 290. A tropocollagen molecule.

Because tendons often extend some distance into a muscle, both Types I and III collagen may be extracted from meat. Even within tendons, there may be some Type III collagen forming the endotendineum or fine sheath around bundles of collagen fibrils. Small diameter Type III collagen fibres are called reticular fibres because, when stained with silver for light microscopy, they often appear as a network or reticulum of fine fibres. Larger diameter Type I collagen fibres are not blackened by silver, but are stained yellow.

During $\rightarrow$ conditioning of meat after slaughter, collagen fibres undergo degradation, although detection of this change had to await development of refined analytical methods. Early investigators were unable to find any breakdown of collagen during conditioning. In other muscle foods, such as fish, the post-mortem degradation of collagen may be all too rapid under acidic conditions and may cause softness or mushiness.

(Source: Hall and Hunt, 1982; Light *et al.*, 1985; Burson and Hunt, 1986; Stanton and Light, 1990; Aidos et al., 1999)

TROPOMODULIN

A protein involved in $\rightarrow$ actin polymerisation and the assembly of thin $\rightarrow$ myofilaments.

(Source: Almenar-Queralt *et al.*, 1999)

TROPOMYOSIN

Tropomyosin is an elongated protein located near the grooves of thin → myofilaments, which are formed from two strands of F-actin twisted into a double helix, as shown in Figure 291. When triggered by troponin, tropomyosin moves deep into the thin myofilament groove, thus allowing myosin molecule heads to interact with the active sites of → actin molecules as the muscle → contracts.

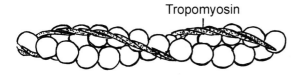

Figure 291. A short part of the length of a thin myofilament formed from a double helix of actin molecules showing tropomyosin molecules near the grooves of the double helix.

(Source: Cohen, 1975; Ohtsuki, 1999)

TROPONIN

Troponin is a protein that detects the presence of calcium ions when muscle → contraction is activated. Troponin is located on the thin → myofilament (which is composed mostly of actin molecules). Troponin has three subunits. TnT or troponin T binds the whole troponin complex to tropomyosin. TnI or troponin I inhibits actomyosin ATPase activity. TnC or troponin C changes shape when it binds to four calcium ions. Troponin C has a dumbbell shape and is about 7 nm long, but this is not shown in Figure 292, which is highly diagrammatic.

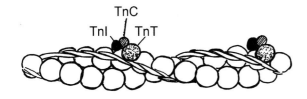

Figure 292. A short part of the length of a thin myofilament formed from actin and tropomyosin molecules, showing the locations of subunits of troponin.

(Source: Parmacek and Leiden, 1991; Hinkle *et al.*, 1999; Perry, 1999)

TROUT

A salmonid fish, → Salmoniformes.

TULIPA

The tranche grasse of a → Belgian beef carcass.

TUNA

A large mackerel-like fish of the family Thunnidae, → Perciformes.

TUNNBRINGA

Thin breast (abdominal muscle) of lamb or veal in → Sweden.

TUP

An entire male sheep.

TURBOT

A large flatfish, → Pleuronectiformes.

TURKEY

Originally, an English turkey-cock was a → guinea fowl. Guinea fowl had been introduced from Africa and were a paradigm of showiness, as in Shakespeare's Twelfth night, "Contemplation makes a rare turkey-cock of him." With the exploration of North America, the name was applied to the ancestors of the birds we now call turkeys, *Meleagris gallopavo*. The geographical range of wild *M. gallopavo* is from southern Canada, through the eastern and southern USA, down to Mexico. After the Spanish invasion of Mexico, Mexican strains of *M. gallopavo* were introduced into Europe where they were gradually improved and assumed the English name of turkey. Thus, when the English and French settled farther north in North America, turkeys (*M. gallopavo*) were re-introduced as domestic birds which were smaller and more compact than their wild ancestors. Agricultural development then restricted the range of wild turkeys, although they are now successfully conserved in various national parks and protected areas. In the old days, turkeys were eaten only on special occasions such as Christmas. But turkeys are now eaten every day, partly because of the invention of numerous processed turkey products, ranging from cooked breast meat slices to turkey sausage, and partly because customers have been introduced to a range of relatively small turkey cuts.

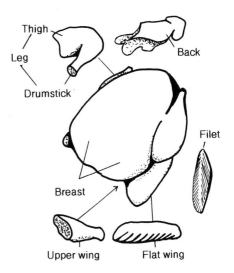

Figure 293. Retail cuts of turkey meat.

The nomenclature for turkey cuts is very simple (Figure 293). The only term that may not be immediately obvious is the filet or fillet, which is the *supracoracoideus* muscle. This muscle elevates the wing and is located between the sternum and the main muscle which depresses the wing, the *pectoralis* (→ chicken).

TURNEDÓ

Slice of *psoas* muscles in → Mexico.

TVÆRREB

Cross-cut beef roast of a → Danish beef carcass.

TVB

A determination of total volatile base nitrogen, including ammonia and other volatile amine compounds. TVB is used to assess the freshness of marine myosystems. In squid, similar results may be obtained with a rapid test for ammonia using tetrazolium.

(Source: LeBlanc and Gill, 1984)

TYKBRYST

In → Danish beef cutting, the tykbryst or thick breast is in the ventral region of the ribcage.

TYKKAM

In → Danish beef cutting, the tykkam or thick crest contains the anterior thoracic vertebrae.

TYKSTEG

In → Danish beef cutting, the tyksteg or thick roast contains the last lumbar and first four sacral vertebrae.

TYNDBOV

In → Danish beef cutting, the tyndbov or thin shoulder cuts across the anterior ribs just ventral to the vertebral column.

TYNDBRYST

In → Danish beef cutting, the tyndbryst or thin breast is in the posteroventral region of the ribcage.

TYNDSTEG

In → Danish beef cutting, the tyndsteg or thin roast contains the lumbar vertebrae.

UDUN

A medial cut of the beef hindlimb in → Korea.

UDZIEC

Polish for haunch or leg, this may be the round of a beef or veal carcass in → Poland or a leg of lamb.

ULKOFILEE

Longissimus dorsi muscle from a → Finnish pork carcass.

ULKOFILEESELKÄ

In → Finnish beef cutting the ulkofileeselkä is a prime rib roast in the posterior thoracic region.

ULKOPAISTI

In → Finnish beef cutting, the ulkopaisti (mostly *semitendinosus* and *biceps femoris*) is the lateral part of the long round roast or pitkä paisti.

ULNARIS LATERALIS

Extensor carpi ulnaris.

UNITED ARAB EMIRATES - CAMEL CUTS

Young camels produce a meat rather like beef, the better cuts of which can be cooked quickly with dry heat. The meat from older animals is tougher and requires moist cooking. The distal extremities of the fore and hindlimbs are removed as Al Gaimah Alamamia and Al Gaimah Alkhalfia, respectively (Figure 294). The neck is Al Regaba. The Al Luah wa Alkatif contains the scapula and antero-dorsal rib region, the Al Dahar wa Adhlla contains the remainder of the posterior thoracic vertebrae, the Al Gatan contains the lumbar region, and the proximal hindlimb forms the Al Fakhed. The Al Sadr contains most of the sternum and ventral ribs, while the Al Shakela contains the abdominal muscles.

(Source: Alhadrami, 1999)

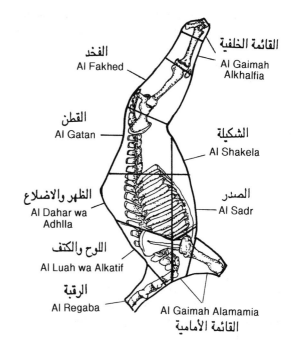

Figure 294. *United Arab Emirates - camel cuts.*

UNTERSCHALE

In → German beef cutting, the Unterschale is on the lateral surface of the round, comparable to an English → silverside. A similar cut may be taken from the medial part of the ham in German pork cutting.

UNTERSPÄLTE

In → Swiss beef cutting, a lateral cut dominated by *biceps femoris*.

USA

→ North American meat cuts.

UTROPHIN

A homologue of → dystrophin involved in the structure of the neuromuscular junction.

(Source: Crosbie *et al.*, 1999)

VACA

Spanish for a cow but used in a culinary context to refer to beef.

VACIO

A void or empty space in Spanish, and used to denote an → Argentinean beef cut containing the abdominal muscles.

VÁGOTT BÁRÁNY

A lamb carcass in → Hungary.

VÄLIKYLKI

In → Finnish beef cutting, the välikylki or gap-flank contains the middle part of a six-rib forerib of beef.

VANG

The flank or abdominal muscles of a beef or veal carcass in the → Netherlands.

VANGLAP

→ Vang

VAST DEEL

A cut containing the *semimembranosus* plus *adductor* of a beef carcass in the → Netherlands.

VASTE

French for the *vastus* muscles - interne (*internus*), intermédiaire (*intermedius*) and externe (*externus*). However, the long vaste is *biceps femoris*.

VASTUS INTERMEDIUS

A medium sized muscle of the mammalian hindlimb (Figure 213). It originates around the anterior face of the humerus and inserts onto the patella.

VASTUS LATERALIS

A large muscle of the mammalian hindlimb (Figure 213). It originates on the femur and inserts onto the lateral patellar ligament.

VASTUS MEDIALIS

A medium sized muscle of the mammalian hindlimb (Figure 213). It originates from the femur and inserts laterally on the patella.

VAZIA

The lumbar region of a → Portuguese beef carcass.

VAZIO

The flank of a → Brazilian beef carcass.

VEIN OF BEEF

An → English beef cut from the neck, noted by Smith (1876) and known by Gerrard (1949) in the Liverpool area.

VEINY PIECE

An ancient beef cut from → England containing the proximal *quadriceps femoris* muscles.

VENADO

Spanish for →venison.

VENAISON

French for → venison.

VENISON

Meat from deer is called venison and there are several species of deer (family Cervidae) of commercial importance. Foremost in commercial imagery is the red deer, *Cervus elaphus*, depicted by Edwin Landseer as the "The Monarch of the Glen". This is virtually the same animal as the North American elk or wapiti (sometimes called *Cervus canadensis*). It is a large animal. Stags sometimes reach 450 kg live weight. The geographical range of *Cervus elaphus* (if numerous subspecies are included) is vast, extending from Asia, through Europe and Africa, to North America. From a butcher's perspective the main point to know is that the animals which produce a magnificent trophy rack of antlers, also produce the toughest, leanest old meat. Thus, meat from young ranched animals is preferable, provided it is properly aged. However, because of pre-slaughter stress, the ultimate pH of venison from ranched deer may be above pH 6 - which means that the meat may verge on being DFD (→ dark, firm and dry) and may be vulnerable to microbial spoilage. For venison obtained from deer shot in the wild, some animals may start their post-mortem metabolism with their muscles fully loaded with glycogen, giving a wide range in pH (for example, pH 5.3 to 5.9). For wild deer, there is a lag period of a couple of months between the seasonal availability of food and the muscularity of the carcass. In Canada, ranched wapiti are finished on good quality alfalfa or a part-grain diet for at least 60 days before slaughter. This gives dressing percentages from 59 to 66% with more than 77% saleable meat cuts from the carcass.

In Europe, the most common deer seen in parklands is the fallow deer, *Dama dama*. It is much smaller than *C. elaphus*, with the male reaching 100 kg live weight and, hence, being demoted in the terminological hierarchy of maleness from stag to buck. To reach the rank of buck, an aspiring male red deer must pass through five yearly ranks, from fawn, through pricket, sorel, sore, and bare back. The lucky ones reaching six years and beyond become great bucks. Fallow deer have been hunted or semi-domesticated for thousands of years, and it was fallow deer that gave rise to the Roman cult of Diana the huntress. The Latin name, dama, is very descriptive, being derived from Indo-Germanic and Persian roots meaning tame or subdued. Venison from farmed fallow deer is typically very tender with → Warner-Bratzler shear forces close to those of lamb. The roe deer, *Capreolus capraea*, is even smaller than the fallow deer, with bucks reaching only 25 kg live weight.

In North America, the most plentiful deer is the whitetail, *Odocoileus virginianus*, which is intermediate in size, reaching over 200 kg live weight. The mule deer, *Odocoileus hemionus*, is heavier and more stocky than the whitetail.

Venison cutting is highly variable, ranging from expedient cutting by hunters in the bush, to centralised cutting with portion control for commercial specifications. Some degree of standardisation is certainly very useful, as reported by Field *et al.* (1973a, 1973b) in Figures 295 and 296. The use of beef names for the elk reflects the more blocky shape of this carcass. In Scotland and England, the hindleg is usually called the haunch, and the left and right loins typically remain together as a saddle (Figure 297). Sometimes the scapula is lifted off the ribcage with the shoulder meat, thus enabling braising escalopes to be taken from the anterior end of *longissimus dorsi*. The shoulder meat may be diced or boned and rolled. Price differentials are a good guide to connective tissue content and the need for long, moist cooking or comminution. In order of decreasing value, names such as the following are in use in England for ranched deer: fillet, loin, haunch steaks, whole saddle, rolled haunch, fondue/kebab cubes, stroganoff strips, casserole, burgers, and mince.

Venison from different sources sometimes receives a special name. Thus, venison from roe deer is called chevreuil in France, reh in Germany and capriolo in Italy, whereas venison from red deer is called cerf, Hirsch and cervo, respectively.

A venison industry is well established in New Zealand, following the introduction of typical meat-producing species. The major export markets are Germany, USA, Japan and Switzerland. Premium carcasses are from 50 to 70 kg and are discounted if they have excess fat. Red deer from 12 to 27 months may give carcasses with from 8 to 12% fat, having 0.23 kg fat for each 1 kg carcass weight. For carcasses from animals over 27 months, this increases by 0.34 kg fat per 1 kg carcass weight.

The Javan rusa (*Cervus timorensis russa*) is now farmed in tropical and subtropical areas of southeast Asia. At 13 to 25 months carcasses from entire males may weigh from 37 to 57 kg. Carcass growth from 19 to 25 months is minimal. Castration reduces carcass size.

(Sources: Vesey-Fitzgerald, 1946; Burton, 1962; Rue, 1981; Field *et al.*, 1973a, 1973b; Fletcher, 1983; Muir, 1989; Renecker, 1989; Reinken, 1997; MDF, 1998; Weber and Thompson, 1998; Grigor *et al.*, 1999; Brodowski and Beutling, 1999; Sookhareea et al., 2001; Volpelli *et al.*, 2003)

VERDERES

Cuts from the anterior rib and flank region on a beef carcass in → Austria.

VESEPECSENYE

Psoas muscles of a → Hungarian beef carcass.

VIENNA SAUSAGE

→ Wiener

VILLIGÆS

Icelandic for a wild → goose.

VILLIN

A protein involved in → actin polymerisation.

VILLIÖND

Icelandic for a wild → duck.

VINCULIN

Vinculin is a lipid binding protein involved in the attachment of → myofibrils to the inner surface of the plasma membrane. In skeletal muscle, vinculin forms clustered patches on the inner surface of the plasma membrane in a pattern that matches the I-bands of the underlying sarcomeres. Thus, vinculin is arranged in costameres or rib-like bands around the myofibre, capable of transmitting force to the extracellular environment via glycoproteins such as sarcospan that extend through the cell membrane. Each costamere has the appearance of a double line astride the position of the → Z-line.

(Source: Wang and Ramirez-Mitchell, 1983; Danowski *et al.*, 1992)

VINKELLÄGG

A → Swedish cut of pork including the humerus, ulna and radius.

VIREZKA

The *psoas* filet of a beef or pork carcass in → Russia.

VOLAILLE

French for → chicken.

VOORPOOT

The front trotter of a → Netherlands pork carcass.

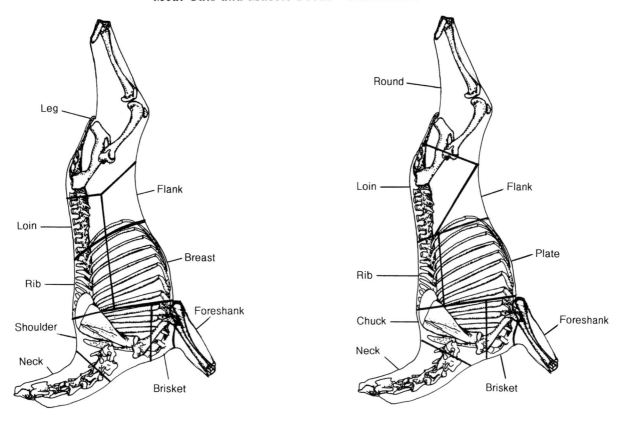

Figure 295. *USA - mule deer cuts.*

Figure 296. *USA - elk cuts.*

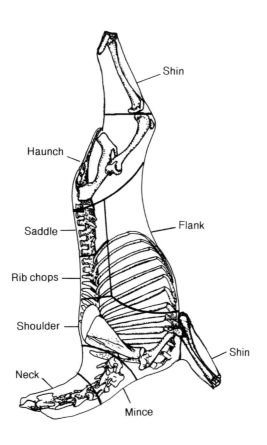

Figure 297. *Scotland - venison cuts.*

VOORRIB

A forerib, but from the posterior thoracic region of a $\rightarrow$ South African beef carcass.

VOORSCHENKEL

The foreshank of a beef carcass in the $\rightarrow$ Netherlands.

VOORSKENKEL

The foreshank of a $\rightarrow$ South African beef carcass.

VORDERER WADSCHINKEN UND BUGSCHERZEL

The foreshank of an $\rightarrow$ Austrian beef carcass.

VORDERHESSE

In $\rightarrow$ German beef cutting, the foreshank may be called the Vorderhesse.

VORSCHLAG

In $\rightarrow$ Switzerland, the Vorschlag is a beef cut containing the *quadriceps femoris* groups of muscles.

VRAT

The neck of a beef, pork or lamb carcass in $\rightarrow$ Serbia & Montenegro.

VYSOKÁ ROŠTENKA

The anterior rib region of a beef carcass in $\rightarrow$ Slovakia.

VYSOKÉ REBRO

A cut of beef in $\rightarrow$ Slovakia through the midlength of the ribs.

WACHTEL

German for → quail.

WALDSCHNEPFE

German for → woodcock.

WALLEYE

Stizostedion vitreum, a major north American sport fish with a conspicuous eye (large and highly reflecting), → Perciformes.

WAMME

In → German pork cutting, the ventral part of the belly or Bauch may be removed as the paunch or Wamme.

WAMMERL

In → Austria, this refers to the belly of a veal carcass.

WAPITI

The Shawnee name for elk, → venison.

WARMED-OVER FLAVOUR

The characteristic taste of meat that has been cooked, allowed to cool, then re-heated some time later originates with the oxidation of polyunsaturated fatty acids (PUFA), typically from plasma membranes. This gives rise to small molecules such as pentanal, hexanal and 2,4 decadienal which are all catalysts for further oxidation. Denatured myoglobin may release Fe^2 which is oxidised to Fe^3 which then enhances PUFA oxidation. Minimisation strategies include the use of nitrite, vitamin E, ascorbate, rosemary, and Maillard browning products, combined with the exclusion of light and oxygen.

WARNER-BRATZLER SHEAR

Invented around 1930 by K.F. Warner and L.J. Bratzler in the USA, the Warner-Bratzler shear test is still the most widely used method to measure the toughness of cooked cores of meat (Figure 298). In strict engineering terms it is not really a shear test, but actually measures the tensile strength of various microstructural components of meat as they are pulled apart between closing blades (tensile test). Preparation of the sample with respect to size, myofibre arrangement and temperature history is very important, as is the condition of the apparatus with respect to the thickness of the moving plate, the degree of wear around the sample aperture, and the separation of the fixed plates. Interpretation of the results requires a knowledge of the sample microstructure. Shear forces should be reported in Newtons not pounds or kilograms.

WATER-BINDING CAPACITY

→ Water-holding capacity

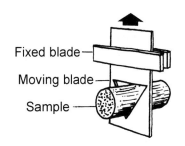

Figure 298. *Warner-Bratzler shear.*

(Source: Voisey, 1976)

WATER-HOLDING CAPACITY (WHC)

Water-holding capacity is a widely used term in the meat industry. In meat processing, it is important to know how much added water can be absorbed by the meat. Similarly, it is useful to be able to measure the amount of fluid likely to be released from a muscle after it has been packaged. Thus, some measurements of WHC deal with water binding capacity (added water) while others deal with fluid release (drip loss or purge). Unfortunately, there is little or no standardisation of methodologies.

Water binding capacity is the ability of meat to bind its own water or, under the influence of external forces such as pressure and heat, to bind added water. Water absorption or gelling capacity is the ability of meat to absorb water spontaneously from an aqueous environment. Water binding capacity and water absorption are closely related. Water absorption may be found from the increase in weight and volume of meat samples placed in an aqueous solution, and water binding capacity may be determined by centrifugation. Another method is to press small meat samples (about 0.3 g) on a filter paper between two plates in a press (typically 35 kg/cm² for 5 minutes). The areas covered by the flattened meat sample and the stain from the meat juice are used to derive a measure of WHC.

WHC is modified by pH and drops from a high around pH 10 to a low at the isoelectric point of meat proteins between pH 5.0 and 5.1. Below pH 5, a value only attained if the pH of a processed meat product is deliberately lowered, water binding capacity starts to increase again. Water absorption follows water binding capacity in this regard. Thus, as the pH of pork declines post-mortem, its water binding capacity decreases, and much of the water associated with muscle proteins is free to leave the → myofilament lattice.

WAZEH

Anterior hindlimb muscles of a beef carcass in → Jordan.

WEASAND

A → casing originating from the oesophagus.

WEDDER

A castrated male sheep.

WEISSES SCHERZEL

Semitendinosus on an → Austrian beef carcass.

WEISSWURST

A cooked, white, emulsion → sausage made from veal and pork. Not smoked.

WESTERN BUTT

A → Boston butt of pork.

WETHER

A castrated male sheep.

WHEEL

A whole-body transverse section of a shark or swordfish, subsequently subdivided into steaks.

WIDGEON

→ Duck.

WIENER

An emulsion → sausage. The name derives from Wien (German for Vienna). In North America, there is now little, if any, difference between frankfurters, wieners and hot dogs.

WILDEBEEST

The blue wildebeest, *Connochaetes taurinus*, reaches 270 kg live weight, but with a massive head, large forequarters, and slender hindquarters.

WING OF BEEF

In → North American beef cutting, a wing is the rib plus the plate. In → English beef cutting for London, the wing end is the anterior of the loin.

WING RIB OF BEEF

In → English beef cutting and patterns derived from it , a wing rib is from the anterior part of the sirloin of the hindquarter. The trend to a reduction in the number of ribs left on the hindquarter is moving the wing rib from the hindquarter to the forequarter. A wing rib needs at least two to three ribs to roast well.

WOLFFISH

A large North Atlantic bony fish with conspicuous teeth for crushing molluscs, family Anarhichadidae, → Perciformes.

WOODCOCK

The European woodcock, *Scolopax rusticola*, is a light-weight bird without long tail feathers and with a straight, sturdy beak. The snipe, looks similar, but is even lighter in weight and has a very thin, long beak. The woodcock is tender enough to roast.

WOOD-PIGEON

→ Pigeon

WUYA

Hausa for the neck in → Nigeria.

YAK

The domesticated yak of China and Mongolia is *Poephagus* or *Bos grunniens*. The wild yak may be called *Bos mutus*. The domesticated yak of Asia is smaller (maximum around 550 kg live weight for bulls and 350 kg for cows) than the wild yak. Yak hair is long and shaggy and is underlain by fine wool. The tail is a long brush, which is unusual for bovines. Yak meat is of secondary importance. Milk and hide products are more important.

(Source: NRC, 1983).

YAKOW

A hybrid of cattle and → yak.

YANGJEE

The ventral region of a → Korean beef carcass including the sternum to the abdominal muscles.

YANKEE POT-ROAST

In → North American beef cutting, yankee pot-roasts are taken perpendicularly through the vertebrae, level with the blade bone in the chuck.

YDERLÅR

Outside round in → Danish beef cutting.

YELLOWTAIL FLOUNDER

Limanda ferruginea of the Northwest Atlantic, → Pleuronectiformes.

YELT

A gilt or female pig that has not yet farrowed (produced a litter of piglets).

YEMA DE PALETA

The *biceps* and *triceps brachii* in → Cuban beef cutting.

YILT

A female pig that has not yet farrowed (produced a litter of piglets).

ÝSA

Icelandic for haddock (→ Gadiformes).

YTRALÆRI

Biceps femoris in → Iceland.

YTREFILET

The outer filet or *longissimus dorsi* of a → Norwegian beef carcass.

YTTERLÅR

A lateral cut from the round of a → Swedish beef carcass, dominated by *biceps femoris*.

ZADNÁ NOŽIČKA

The trotter of the hindlimb on a → Slovakian pork carcass.

ZADNÁ NOŽINA

The hindshank of a beef or veal carcass in → Slovakia.

ZADNÉ KOLENO

A cut containing the tibia and fibula of a pork carcass in → Slovakia.

ZÁKRČIE

An anterior rib roast medial to the scapula in a → Slovakian lamb carcass.

ZAMPI

Distal parts of the fore- or hindlimb on an → Italian pork or lamb carcass.

ŻEBERKA

Ribs and intercostals from a → Polish pork carcass, equivalent to → North American spareribs of pork.

ZEBRA

There are several species of African zebra in the genus *Equus* which produce meat much like horse meat, with a dark colour and high myoglobin content. Dressed carcass weights range from around 114 to 187 kg, with dressing percentages from 53 to 60%. Cooking losses are low.

(Sources: Eltringham, 1984; Onyango et al., 1998)

ZEUGMATIN

Part of the → titin molecule found at the → Z-line.

(Source: Turnacioglu *et al.*, 1997)

ZIMBABWE - BEEF CUTS

Wildlife ranching may be best for meat production where vegetation is sparse, but beef production is sustainable in the more fertile areas of Zimbabwe. Beef cutting in Zimbabwe is documented, following the pattern of the Cold Storage Commission of Rhodesia (Figure 299).

The basic pattern seems to be a simplification of that used in London, but with enough alteration to make it unique. Cutting the wing end of the loin as a wingrib is a typical feature of London cutting, suggesting the original Rhodesian hindquarter retained three ribs. The rounded cut including the humerus was not named in the source, but probably is the clod.

(Sources: ADSRI, 1981).

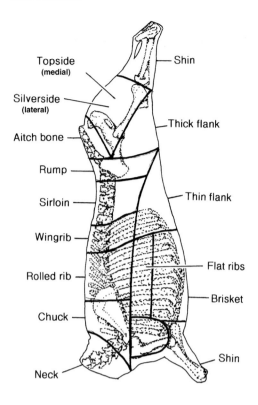

Figure 299. Zimbabwe - beef cuts.

Z-LINE

Also sometimes called the Z-disk, the Z-line forms a partition across the → myofibril and anchors the thin → myofilaments. The distance from one Z-line to the next defines the sarcomere. The Z-line is just about visible by light microscopy, where it appears as a simple thin line across the myofibril. Electron microscopy reveals a complex structure which changes as the myofibril contracts.

The Z-line appears as an electron-dense zig-zag line about 40 to 50 nm in width when the muscle sample is sectioned longitudinally. Z-line thickness is rather variable and Z-lines tend to be thicker in myofibres with a slow contraction speed relative to myofibres with a fast contraction speed. The electron density of the Z-line may increase in contracted sarcomeres so that its structure is more difficult to discern. In longitudinal sections, when the structure of the Z-line can be clearly seen, its appearance may change depending on the plane of sectioning in the depth of the sample block. Sometimes the thin myofilaments from the two sarcomeres adjacent to the Z-line meet the V-shaped filaments of the Z-line in an alternating pattern, but sometimes an extra set of thin myofilaments is directly lined up from one sarcomere to the next.

The appearance of the Z-line in transverse sections reveals a different perspective of its structure to that seen in longitudinal sections. In transverse sections through the outer part of the A-band, thin myofilaments are arranged in a hexagonal lattice.

However, as transverse sections are taken serially towards the Z-line, the hexagonal pattern is replaced by a square pattern. When this square pattern is followed farther into the Z-line it merges into a more complex pattern, as if oblique filaments had been woven among the longitudinally disposed thin myofilaments. The structure that underlies this complexity may be a tetragonal array of axial filaments interconnected by small-diameter Z-filaments. The axial filaments are arranged so that an axial filament from one sarcomere is connected to four axial filaments in an adjacent sarcomere. The backbone of the axial filaments is composed of actin. The filaments of the Z-line that anchor the ends of the actin filaments are composed of α-actinin. Zeugmatin and CapZ are also involved in the attachment of thin filaments to the Z-line. Zeugmatin appears to be part of the titin molecule and to be involved in Z-line formation.

The filamentous component of the Z-line has been identified as α-actinin and the amorphous residue has been identified as the protein amorphin. ß-actinin is probably located at the ends of thin myofilaments, and may participate in their assembly. It accelerates polymerization of F-actin from G-actin, but inhibits the irregular assembly of F-actin. Gamma-actinin has the opposite action, and inhibits polymerization of G-actin.

(Sources: Hikida, 1978; Goldstein *et al.*, 1986, 1990; Yamaguchi *et al.*, 1986; Zimmer and Goldstein, 1987; Schroeter *et al.*, 1996)

ZUNGENSTÜCK

In → German beef cutting, the tongue-piece or Zungenstück is connected to the neck or Kamm and is composed of the axial skeleton and muscles of the forequarter after the removal of the shoulder or Schulter.

CITED REFERENCES AND OTHER SOURCES

Abdallah, B. 2003. Personal communication. University of Jordan, Faculty of Agriculture, Department of Nutrition and Food Technology, Amman 11942, Jordan.

ABIF. Associação Brasileira de Indústria Frigorífica, and Ministério da Agricultura. *Padronizaçã de Cortes de Carne Bovina.*

Abiola, S.S. 2003. Personal communication. University of Agriculture, College of Animal Science and Livestock Production, Department of Animal Production and Health, Abeokuta, Ogun State, Nigeria.

Abu-Tarbousch, H.M., M. Atia and A.M. Al-Johany. 1996. Nutritional quality of Dhub (lizard) meat (Uromastys aegyptius Blanford 1874) and characterization of its protein using electrophoretic techniques. *Ecology of Food and Nutrition*, **35**: 275-284.

ADSRI, 1981. *The Cuts of a Beef Carcass. Die Snitte van 'n beeskarkas.* Animal and Dairy Science Research Institute, Private Bag X2, Irene 1675, South Africa.

Agriculture Canada. 1974. Lexicon. English-French. Meat. Ottawa, Canada.

Ahrens, M.A. 1985. Atlantic mackerel. *Underwater World*, 17. Fisheries and Oceans, Ottawa.

Aidos, I., O. Lie and M. Espe. 1999. Collagen content in farmed Atlantic salmon (Salmo salar L.). *Journal of Agricultural and Food Chemistry*, **47**: 1440-1444.

Alexander, R. McN. 1969. The orientation of muscle fibres in the myomeres of fishes. *Journal of the Marine Biological Association, U.K.*, **49**: 263-290.

Alexic, S. 1999. Personal communication. Institute for Animal Husbandry, Department of Meat Science, Autoput 16, 11081 Beograd-Zemun, Serbia & Montenegro.

Alhadrami, G. 1999. Personal communication. Faculty of Agricultural Sciences, United Arab Emirates University, United Arab Emirates.

Allen, C.E., D.C. Beitz, D.A. Cramer and R.G. Kauffman. 1976. *Biology of Fat in Meat Animals.* College of Agricultural and Life Sciences, University of Wisconsin - Madison. North Central Regional Publication No. 234.

Almenar-Queralt, A., C.C. Gregorio and V.M. Fowler. 1999. Tropomodulin assembles early in myofibrillogenesis in chick skeletal muscle: evidence that thin filaments rearrange to form striated myofibrils. *Journal of Cell Science*, **112**: 1111-1123.

Alvarez, C., I. Couso and M. Tejada. 1999. Microstructure of suwari and kamaboko sardine surimi gels. *Journal of the Science of Food and Agriculture*, **79**: 839-844.

Amaria, R.P. 1982. *Proceedings of the International Squid Symposium.* August 9-12, 1981. New England Fisheries Development Foundation, Boston, MA. ISBN 0-89059-026-5.

AMLC, 1996. *The Complete Guide to Ordering Australian Beef.* Australian Meat and Live-stock Corporation, 805 Third Avenue, New York.

AMLC, 1991. *Fresh Australian Range Lamb Product Guide.* Australian Meat & Live-stock Corporation, North American Section, 750 Lexington Avenue, New York.

AMLC, 1979. *Handbook of Australian Meat*, 3rd edition, Section 2. Australian Meat and Live-stock Corporation.

Ampola, V.G. 1980. The quality of squid held in chilled seawater versus conventional shipboard handling. *Marine Fisheries Review*, **42** (7-8): 74-76.

Anadu, P.A., P.O. Elamah and J.F. Oates. 1988. The bushmeat trade in southwestern Nigeria: a case study. *Human Ecology*, **16**: 199-208.

Andreev, D.P. and W.A. Wassilev. 1986. Specialized contacts between sarcolemma and sarcoplasmic reticulum at the ends of muscle fibres in the diaphragm of the rat. *Cell and Tissue Research*, **243**: 415-420.

Anjum, I. 2003. Personal communication. Deputy District Livestock Officer (B.I.), City District Government, Lahore, Punjab, Pakistan.

Ansorena, D., M.P. Depena, I. Astiasaran and J. Bello. 1997. Colour evaluation of chorizo de Pamplona, a Spanish dry fermented sausage: comparison between the CIE L*a*b* and the Hunter Lab systems with illuminants D65 and C. *Meat Science*, **46**: 313-318.

Arnold, A.F. 1901. *The Sea-beach at Ebb-tide*. Century, New York.

Ashton, F.T., J. Weisel and F.A. Pepe. 1992. The myosin filament XIV backbone structure. *Biophysical Journal*, **61**: 1513-1528.

Atkinson, D.B. 1993. Roundnose grenadier. *Underwater World*, 7. Fisheries and Oceans, Ottawa.

Atwood, H.L. 1973. An attempt to account for the diversity of crustacean muscles. *American Zoologist*, **13**: 357-378.

Babol, J., E.J. Squires and E.A. Gullett. 1996. Investigation of factors responsible for the development of boar taint. *Food Research International*, **28**: 573-581.

Baek, H.H. and K.R. Cadwallader. 1997a. Character-impact aroma compounds of crustaceans. *Flavor and Lipid Chemistry of Seafoods*, **674**: 85-94.

Baek, H.H. and K.R. Cadwallader. 1997b. Aroma volatiles in cooked alligator meat. *Journal of Food Science*, **62**: 321-325.

Baeza, E., M.R. Salichon, G. Marche and H. Juin. 1998. Effect of sex on growth, technological and organoleptic characteristics of the Muscovy duck breast muscle. *British Poultry Science*, **39**: 398-403.

Bailey, R. and G.S. Jamieson. 1990. The Atlantic snow crab. *Underwater World*, 6. Fisheries and Oceans, Ottawa.

Bao, Z.Z., M. Lakonishok, S. Kaufman and A.F. Horowitz. 1993. "$_7\beta_1$ Integrin is a component of the myotendinous junction on skeletal muscle. *Journal of Cell Science*, **106**: 579-590.

Barden, J.A. and P. Mason. 1978. Muscle crossbridge stroke and activity revealed by optical diffraction. *Science*, **199**: 1212-1213.

Barnabas, S.A.S. and M. Süth, 1998. Personal communication. Department of Food Hygiene, University of Veterinary Science, H-14000, Budapest, Hungary.

Barrett, J.H. and C.M. Yonge. 1958. *Collins Pocket Guide to the Sea Shore*. Collins, London.

Bate Smith, E.C. 1939. Changes in elasticity of mammalian muscle udergoing rigor mortis. *Journal of Physiology*, **96**: 176-193.

Batrukova, M.A., A.M. Rubtsov and A.A. Boldyrev. 1993. Effect of carnosine on Ca^{2+}-release channels of skeletal muscle sarcoplasmic reticulum. *Biokhimiya*, **57**: 904-910.

Batty, J. 1979. *Domesticated Ducks and Geese*. Spur Publications, Saiga Publishing, Hindhead, Surrey.

Beamish, R.J. and A. Cass. 1990. Lingcod. *Underwater World*, **30**. Fisheries and Oceans, Ottawa.

Bendall, J.R. 1954. The swelling effect of polyphosphates on lean meat. *Journal of the Science of Food and Agriculture*, **5**: 468-475.

Bendall, J.R. 1967. The elastin content of various muscles of beef animals. *Journal of the Science of Food and Agriculture*, **18**: 553-558.

Bendall, J.R. 1973. Postmortem changes in muscle. In G.H. Bourne (editor), *The Structure and Function of Muscle*, Volume II, Part 2, Academic Press, New York. pp. 244-309.

Bendall, J.R. and H.J. Swatland. 1988. A review of the relationships of pH with physical aspects of pork quality. *Meat Science*, **24**: 85-126.

Bertin, L. 1967. Birds (Class Aves), In *Larousse Encyclopedia of Animal Life*. Paul Hamlyn, London. pp. 331-467.

Bigford, T.E. Synopsis of biological data on the rock crab, *Cancer irroratus* Say. U.S. Department of Commerce, National Oceanic and Atmospheric Administration Technical Report, National Marine Fisheries Service, Circular 426.

Biscontini, T.M.B., M. Shimokomaki, S.F. Oliveira and T.M.T. Zorn. 1996. An ultrastructural observation on charquis, salted and intermediate moisture meat products. *Meat Science*, **43**: 351-358.

Bishop, C.A. 1993. White hake. *Underwater World*, 40. Fisheries and Oceans, Ottawa.

Bittner, G.D. 1973. Trophic dependence of fiber diameter in a crustacean muscle. *Experimental Neurology*, **41**: 38-53.

Blange, T., V.A. Vanderheide, B.W. Treijtel and E.L. Debeer. 1997. The effect of actin filament compliance on the interpretation of the elastic properties of skeletal muscle fibres. *Journal of Muscle Research and Cell Motility*, **18**: 125-131.

Bompas, G.C. 1886. *Life of Frank Buckland*. Smith, Elder & Co., London.

Bone, Q. 1966. On the function of the two types of myotomal muscle fibre in elasmobranch fish. *Journal of the Marine Biological Association, U.K.*, **46**: 321-349.

Boon, D.D. 1975. Discoloration in processed crabmeat. A review. *Journal of Food Science*, **40**: 756-761.

Boone, L.P. and G.D. Bittner. 1974. Morphological and physiological measures of trophic dependence in a crustacean muscle. *Experimental Neurology*, **89**: 123-144.

Bothma, J. du P. 1989. *Game Ranch Management*. J.L. van Schaik, Pretoria, South Africa.

Bourdelle, E. 1920. *Anatomie Regionale des Animaux Domestique, Volume 3,* Bailliere et Fils, Paris.

Bowering, W.R. 1990. Witch flounder. *Underwater World*, 12. Fisheries and Oceans, Ottawa.

Bozkurt, H., and O. Erkmen. 2002. Effects of starter cultures and additives on the quality of Turkish style sausage (sucuk). *Meat Science*, **61**: 149-156.

Bressan, M.C., S.H.I. Oda, M.D.G. CardosaG.Z. Miguel,J.O. Vieira, P.B. Faria, and T.V. Savian. 2003. Fat acid profile of the capybara (*Hydrochaeris hydrochaeris* L. 1766) . *49ᵗʰ International Congress of Meat Science and Technology. Pp. 173-174.*

Breul, K. 1952. *Cassell's German and English Dictionary*. Cassell, London.

Briggs, F.N., J.L.Poland and R.J. Solaro. 1977. Relative capabilities of sarcoplasmic reticulum in fast and slow mammalian skeletal muscles. *Journal of Physiology*, **266**: 587-594.

Broad, S.R. and R.A. Luxmore. 1989. Crocodile ranching and the potential impacts of commercial wildlife production on conservation. In R. Valdez (editor), *First International Wildlife Ranching Symposium*. New Mexico State University, Las Cruces, New Mexico. pp. 243-255.

Brodowski, G. and D. Beutling. 1999. The characterization of the meat quality of fallow deer. *Fleischwirtschaft*, **79**: 94-98.

Browning, R.J. 1974. *Fisheries of the North Pacific*. Alaska Northwest Publishing Company, Anchorage.

Buckland, F. 1894. *Notes and Jottings from Animal Life*. Smith, Elder, & Co., London.

Bulas, K., L.L. Thomas and F.J. Whitfield. 1982. *The KoÑciuszko Foundation Dictionary. Volume II. Polish-English*. The KoÑciuszko Foundation, New York.

Bull, S. 1951. *Meat for the Table*. McGraw-Hill Book Company, New York.

Burger, J., R.A. Kennamer, I.L. Brisbin and M. Gochfeld. 1997. Metal levels in mourning doves from South Carolina: potential hazards to doves and hunters. *Environmental Research*, **75**: 173-186.

Burson, D.E. and M.C. Hunt. 1986. Proportion of collagen Types I and III in four bovine muscles differing in tenderness. *Journal of Food Science*, **51**: 51-53.

Burt, W.H. and R.P. Grossenheider. 1976. *A Field Guide to the Mammals. North America North of Mexico. 3rd edition.* The Peterson Field Guide Series, Houghton Mifflin Company, Boston.

Burton, M. 1962. *Systematic Dictionary of Mammals of the World.* Crowell, New York.

Butler, T.H. 1988. Dungeness crab. *Underwater World*, 37. Fisheries and Oceans, Ottawa.

Butler, T.H., J. Boutillier and A.A. Denbigh. 1989. Selected shrimps of British Columbia. *Underwater World*, 29E. Fisheries and Oceans, Ottawa.

Butterfield, R.M. and N.D.S. May. 1966. *Muscles of the Ox.* University of Queensland Press, St. Lucia, Brisbane.

Calle Escobar, R. 1984. *Animal Breeding and Production of American Camelids.* Ron Hennig - Patience, Lima, Peru.

Canadian Meat Council. 1980. *Food Service Meat Manual.* Islington, Ontario. 1st edition, pp. 31.

Canadian Meat Council. 1988. *Food Service Meat Manual.* Islington, Ontario. 2nd edition, ISBN 0-920391-20-6. pp. 43.

Canadian Pork Council. 1975. *Pork.* Ottawa. pp. 112.

Carlsson, E. and L-E. Thornell. 1987. Diversification of the myofibrillar M-band in rat skeletal muscle during postnatal development. *Cell and Tissue Research*, **248**: 169-180.

Casella, J.F., S.W. Craig, D.J. Maack and A.E. Brown. 1987. CapZ($_{36/32}$), a barbed end actin-capping protein, is a component of the Z-line of skeletal muscle. *Journal of Cell Biology*, 105: 371-379.

Chadwick, D.H. 1996. A place for parks in the new South Africa. *National Geographic*, **190** (1): 2-41.

Chamberlain, B.J. 1991. *The Random House Portuguese Dictionary.* Random House, New York.

Chan, K.M., E.A. Decker and W.J. Means. 1993. Extraction and activity of carnosine, a naturally occurring antioxidant in beef muscle. *Journal of Food Science*, **58**: 1-4.

Chapple, W.D. 1982. Muscle. In D.E. Bliss, H.L. Atwood, and D.C. Sandeman (editors), *The Biology of the Crustacea*, Volume 3., Academic Press, New York. pp. 151-184.

Cheeke, P.R., N.M. Patton and G.S. Templeton. 1982. *Rabbit Production.* Interstate Publishers, Danville, Illinois.

Chizzolini, R. 1997. Personal communication. Istituto di Scienza e Tecnologia Degli Alimenti, Università Degli Studi di Parma, Facoltà di Medicina Veterinaria, Via del Taglio, Cornocchio, 43100 Parma, Italy.

Chizzolini, R., G. Delbon, P. Rosa, E. Novelli, P. Baldini, G. Parolari and S. Barbuti. 1992. Hot and cold cutting techniques. Comparative evaluation of heavy pig carcasses in Italy. *Fleischwirtschaft*, **72** (11): 1550-1552.

Chowrashi, P.K. and F.A. Pepe. 1982. The Z-band: 85,000-dalton amorphin and alpha-actinin and their relation to structure. *Journal of Cell Biology*, **94**: 565-573.

Clay, D. and T. Hurlbut. 1988. Bluefin tuna. *Underwater World*, 55. Fisheries and Oceans, Ottawa.

Cloney, R.A. and E. Florey. 1968. Ultrastructure of cephalopod chromatophore organs. *Zeitschrift für Zellforschung*, **89**: 250-280.

Clutton-Brock, J. 1987. *A Natural History of Domesticated Mammals.* Cambridge University Press and British Museum (Natural History), London.

Cohen, C. 1975. The protein switch of muscle contraction. *Scientific American*, **233**: 36-45.

Collins, 1997. *Collins Spanish-English English-Spanish Dictionary.* Harper Collins, New York.

Costello, W.J. and F. Lang. 1979. Development of the dimorphic claw closer muscles of the lobster *Homarus americanus*. IV. Changes in functional morphology during growth. *Biological Bulletin*, **156**: 179-195.

Crosbie, R.H., J. Heighway, D.P. Venzke, J.C. Lee and K.P. Campbell. 1997. Sarcospan, the 25-kDa transmembrane component of the dystrophin-glycoprotein complex. *Journal of Biological Chemistry*, **272**: 31221-31224.

Crosbie, R.H., C.S. Lebakken, K.H. Holt, D.P. Venzke, V. Straub, J.C. Lee, R.M. Grady, J.S. Chamberlain, J.R. Sanes and K.P. Campbell. 1999. Membrane targeting and stabilization of sarcospan is mediated by the sarcoglycan subcomplex. *Journal of Cell Biology*, **145**: 153-165.

Cullen, M.J., J.J. Fulthorpe and J.B. Harris. 1992. The distribution of desmin and titin in normal and dystrophic muscle. *Acta Neuropathologica*, **83**: 158-169.

Cyril, H.W. 1999. Personal communication. Department of Animal Science, Faculty of Agriculture, University of Peradeniya, Peradeniya, Sri Lanka.

Dadswell, M.J. 1990. The American shad. *Underwater World*, 39. Fisheries and Oceans, Ottawa.

Danowski, B.A., K. Imanaka-Yoshida, J.M. Sanger and J.W. Sanger. 1992. Costameres are sites of force transmission to the substratum in adult cardiomyocytes. *Journal of Cell Biology*, **118**: 1411-1420.

Daoudi, A. and N. Benkerroum. 2003. Personal communication. Département HIDAOA and. Département des Sciences Alimentaire et Nutritionnelles, Institut Agronomique et Vétérinaire Hassan II, BP. 6202, Instituts, 10101-Rabat, Maroc.

Dassow, J.A. and Learson, R.J. 1976. The crab and lobster fisheries. In M.E. Stansby (editor), *Industrial Fishery Technology*. 2nd edition, Krieger, New York.

Davidson, A. 1979. *North Atlantic Seafood*. Macmillan, London.

Dawson, T.J. 1995. *Kangaroos. Biology of the Largest Marsupials*. Comstock Publishing, Cornell University Press, Ithaca, New York.

Debelle, L. and A.M. Tamburro. 1999. Elastin: molecular description and function. *International Journal of Biochemistry and Cell Biology*, **31**: 261-272.

De Boer, W.F. and D.S. Baquete. 1998. Natural resource use, crop damage and attitudes of rural people in the vicinity of the Maputo Elephant Reserve, Mozambique. *Environmental Conservation*, **25**: 208-218.

Delgado, H. 1988. *Beef cutting in Costa Rica. Procedure, muscular component of cuts and cooking methods*. Thesis, University of Guelph, Ontario, Canada.

de Oliveira Roça, R. 1999. Personal communication. Universidade Estadual Paulista, Laboratório de Tecnologia dos Produtos de Origem Animal, Fazenda Lageado, Caixa Postal 237, Botucato, Brazil.

Déterville, P. 1982. *Technologie de la Viande*. 4th edition. Éditions André Casteilla, Paris.

Devear, Z. *Fresh Beef Meat Dishes*. Israeli Ministry of Agriculture. (In Hebrew).

Dewar, A.B., W.E. Andersson and S. Varga. 1972. The shelf life of canned queen crab meat. *Fisheries and Oceans, Regional Inspection and Technology Laboratory, Halifax, Nova Scotia*. Technical Report, 13. Halifax, Nova Scotia.

DFO, 1993. *Fish Borne Illness*. Scotia-Fundy Region, Department of Fisheries and Oceans, Ottawa.

Dhanda, J.S., D.G. Taylor, P.J. Murray and J.E. McCosker. 1999. The influence of goat genotype on the production of Capretto and Chevon carcasses. 2. Meat quality. *Meat Science*, **52**: 363-367.

Diesbourg, L., H.J. Swatland and B.M. Millman. 1988. X-ray diffraction measurements of postmortem changes in the myofilament lattice of pork. *Journal of Animal Science*, **66**: 1048-1054.

Dorst, J. and P. Dandelot. 1970. *A Field Guide to the Larger Mammals of Africa*. Collins, London.

Drvodelič, M. and Ž. Bujas. 1989. *Croatian-English Dictionary*. 6th edition. Školska Knjiga, Zagreb.

Dubois, M-M. 1960. *Dictionnaire Moderne Français-Anglais*. Librairie Larousse, Paris.

Dumonteil, M., J. Bouet and A.F. Cognet. 1981. *Mieux Connaître pour Mieux Choisir, Préparer, La Viande*. Éditions Brunétoile, Neuilly, France. ISBN 2-903101-29-9.

Ebashi, S. and Y. Nonomura. 1973. Proteins of the Myofibril, In G.H. Bourne (editor), *The Structure and Function of Muscle*. 2nd edition. Academic Press, New York. pp. 285-362.

Edman, A-C., J.M. Squire and M. Sjöström. 1988. Fine structure of the A-band in cryo-sections. Diversity of M-band structure in chicken breast muscle. *Journal of Ultrastructural and Molecular Structure Research*, **100**: 1-12.

Edwards, E. 1979. *The Edible Crab and its Fisheries in British Waters*. Fishing News Books, Farnham, England.

Einarsson, S. 1945. *Icelandic*. Johns Hopkins University Press, Baltimore, Maryland.

Eisenberg, B. and R.S. Eisenberg. 1968. Transverse tubular system in glycerol-treated skeletal muscle. *Science*, **160**: 1243-1244.

Eitenmiller, R.R., J.H. Orr and W.W. Wallis. 1982. Histamine formation in fish: microbiological and biochemical conditions. In R.E. Martin, G.J. Flick, C.E. Hebard and D.R. Ward (editors), *Chemistry and Biochemistry of Marine Food Products*. AVI Publishing Company, Westport, Connecticut. pp. 39-50.

Elgasim, E.A. and M.A. Alkanhal. 1992. Proximate composition, amino acids and inorganic mineral content of Arabian camel meat. *Food Chemistry*, **45**: 1-4.

Elkhateib, T. 1997. Microbiological status of Egyptian salted meat (basterma) and fresh sausage. *Journal of Food Safety*, **17**: 141-150.

Ellenberger, W., H. Baum and H. Dittrich. 1949. In L.S. Brown (editor), *An Atlas of Animal Anatomy for Artists*. 2nd revised edition. Dover Publications, New York.

Elliott, G.F. and C.R. Worthington. 1994. How muscle may contract. *Biochimica et Biophysica Acta*, **1200**: 109-116.

Eltringham, S.K. 1984. *Wildlife Resources and Economic Development*. John Wiley, Chichester.

ESS-Food. Axelborg, Axeltorv 3, DK-1609 Copenhagen V, Denmark.

Ezquerra, J.M., F.L. Garcia-Carreno, G. Arteaga and N.F. Haard. 1999. Effect of feed diet on aminopeptidase activities from the hepatopancreas of white shrimp (Penaeus vannamei). *Journal of Food Biochemistry*, **23**: 59-74.

Fabbricante, T. and W.J. Sultan. 1978. *Practical Meat Cutting and Merchandising. Beef*. Volume 1. 2nd edition. AVI, Westport, Connecticut.

Fabbricante, T. and W.J. Sultan. 1975. *Practical Meat Cutting and Merchandising. Pork, Lamb, Veal*. Volume 2. AVI, Westport, Connecticut.

Fadda, S. 1999. Personal communication. Cerela Centro de Referncia para Lactobacilos, Chabuco 145, S.M. de Tucumán 4000, Argentina.

Fahrenbach, W.H. 1967. The fine structure of fast and slow crustacean muscles. *Journal of Cell Biology*, **35**: 69-79.

FAO, 1981. *Atlas of the Living Resources of the Seas*. FAO Fisheries Series, 15.

Favati, F. 1971. *Dictionary of Agriculture. Dizionario di Agricoltura*. Edagricole, Edizioni Agricole, Bologna.

Field, R.A., F.C. Smith and W.G. Hepworth. 1973a. The mule deer carcass. *Agricultural Experiment Station, University of Wyoming, Laramie*. Bulletin 589.

Field, R.A., F.C. Smith and W.G. Hepworth. 1973b. The elk carcass. *Agricultural Experiment Station, University of Wyoming, Laramie*. Bulletin 594.

Findlay, C.J. and D.W. Stanley. 1984. Texture-structure relationships in scallop. *Journal of Texture Studies*, **15**: 75-85.

Fitzgerald, D.R., L.D. Thompson, M.F. Miller and L.C. Hoover. 1999. Cooking temperature, bird type, and muscle origin effects on sensory properties of broiled emu steaks. *Journal of Food Science*, **64**: 167-170.

Fletcher, N. 1983. *Venison. The Monarch of the Table*. Nichola Fletcher, Reediehill, Auchtermuchty, Fife, KY14 7HS, Scotland.

Fornias, O.V. 1998. Ciudad de la Habana, Cuba. Personal communication.

Franzini-Armstrong, C. 1973. Membranous systems in muscle fibers. In G.H. Bourne (editor), *The Structure and Function of Muscle*, 2nd edition, Volume II, Part 2. Academic Press, New York. pp. 578-581.

Fritz, J.D., M.C. Mitchell. B.B. Marsh and M.L. Greaser. 1993. Titin content of beef in relation to tenderness. *Meat Science*, **33**: 41-50.

Fürst, D.O., U. Vinkemeier and K. Weber. 1992. Mammalian skeletal muscle C-protein: purification from bovine muscle, binding to titin and the characterization of a full-length human cDNA. *Journal of Cell Science*, **102**: 769-778.

Gade, D.W. 1976. Horsemeat as human food in France. *Ecology of Food and Nutrition*, **5**: 1-11.

Gads, G.E.C. 1992. *Gads Engelsk Ordbog*. Gads Forlag, København.

Gallo, C.St. 2003. Personal communication. Instituto de Ciencia y Tecnología de Carnes, Facultad de Ciencias Veterinarias, Universidad Austral de Chile, Casilla 567, Valdivia, Chile. Including a translation of Official Chilean Standard NCh 1596.Of99 of the Instituto Nacional de Normalizacion.

Gangal, S. and N.G. Magar. 1967. Canning and storage of crabmeat. *Food Technology*, **21**: 79-82A.

Garcia, I., V. Diez and J.M. Zumalacarregui. 1997. Changes in proteins during the ripening of Spanish dried beef 'Cecina'. *Meat Science*, **46**: 379-385.

Gerrard, F. 1947. *The Book of the Meat Trade*. Volumes I and II. Caxton, London.

Gerrard, F. and F.J. Mallion. 1977. *The Complete Book of Meat*. Virtue, London.

Getchell, J.S. and M.E. Highlands. 1957. Processing lobster and lobster meat for freezing and storage. *Maine Agricultural Experiment Station, Orono*. Bulletin 558. pp. 15.

Giddings, G.G. and L.H. Hill. 1976. A scanning electron microscopic study of effects of processing on crustacean muscle. *Journal of Food Science*, **41**: 455-457

Gillies, M.T. 1971. *Seafood Processing*. Noyes Data Corporation, Park Ridge, New Jersey.

Goldstein, M.A., L.H. Michael, J.P. Schroeter and R.L. Sass. 1986. The Z-band lattice in skeletal muscle before, during and after tetanic contraction. *Journal of Muscle Research and Cell Motility*, **7**: 527-536.

Goldstein, M.A., J.P. Schroeter and R.L. Sass. 1990. Two structural states of the vertebrate Z band. *Electron Microscopy Research*, 3: 227-248.

Goll, D.E., M.L. Boehm, G.H. Geesink and V.F. Thompson. 1997. What causes postmortem tenderization? *Proceedings of the Reciprocal Meat Conference*, **50**: 60-67.

Goll, D.E., V.F. Thompson, H. Li, W. Wei and J. Cong. 2003. The calpain system. *Physiologicial Reviews*, **83**: 731-801.

Gómez, G.G., C.Y.K. Wong and W.Y. Toma. 1992. Carcass yield of Hawaii pork, 2. Yield of lean cuts and relationship to local carcass grade using the Hawaiian cutting method. *University of Hawaii, College of Tropical Agriculture and Human Resources*. Research Extension Series 135.

Gondret, F., H. Juin, J. Mourot and M. Bonneau. 1998. Effect of age at slaughter on chemical traits and sensory quality of longissimus lumborum muscle in the rabbit. *Meat Science*, **48**: 181-187.

Govind, C.K., D.E. Meiss, J. She and E. Yap-Chung. 1978. Fiber composition of the distal accessory flexor muscle in several decapod crustaceans. *Journal of Morphology*, **157**: 151-160.

Griebenow, R.L., F.A. Martz and R.E. Morrow. 1997. Forage-based beef finishing systems: a review. *Journal of Production Agriculture*, **10**: 84-91.

Grigor, P.N., P.J. Goddard, C.A. Littlewood, P.D. Warriss and S.N. Brown. 1999. Effects of preslaughter handling on the behaviour, blood chemistry and carcases of farmed fallow deer. *Veterinary Record*, **144**: 223-227.

Grove, B.K., B. Holmbom and L-E. Thornell. 1987. Myomesin and M protein: differential expression in embryonic fibres during pectoral muscle development. *Differentiation*, **34**: 106-114.

Gutworth, M.S., B.L. Tinker, and R.J. Learson. 1981. Textural evaluation of squid (Illex illecebrosus) as affected by cooking time: sensory and instrumental analysis. In: R.P. Amaria (editor), *Proceedings of the International Squid Symposium*. New England Fisheries Development Foundation, Boston, Massachussets.

Hall, J.B. and M.C. Hunt. 1982. Collagen solubility of A-maturity bovine longissimus muscle as affected by nutritional regimen. *Journal of Animal Science*, **55**: 321-328.

Halvorson, D.B. 1972. Differences in naming muscles of the pelvic limb of chicken. *Poultry Science*, **51**: 727-738.

Hamm, R. 1982. Postmortem changes in muscle with regard to processing of hot-boned beef. *Food Technology*, 105-115.

Hardy, A. 1959. *The Open Sea: Its Natural History. Part II. Fish and Fisheries*. Collins, London.

Hart, J.F.L. 1982. Crabs and their Relatives of British Columbia. *British Columbia Provincial Museum, Victoria, BC. Handbook* 40.

Hasselbach, W. and H. Oetliker. 1983. Energetics and electrogenicity of the sarcoplasmic reticulum calcium pump. *Annual Review of Physiology*, **45**: 325-339.

Havas, E., T. Parviainen, J. Vuorela, J. Toivanen, T. Nikula, and V. Vihko. 1997. Lymph flow dynamics in exercising human skeletal muscle as detected by scintography. *Journal of Physiology*, **504**: 233-239.

Hawley, A.W.L. 1989. Bison farming in North America. In R.J. Hudson, K.R. Drew and L.M. Baskin (editors), *Wildlife Production Systems. Economic Utilisation of Wild Ungulates*. Cambridge University Press, Cambridge. pp. 346-361.

Henry, M.D. and K.P. Campbell. 1996. Dystroglycan: an extracellular matrix receptor linked to the cytoskeleton. *Current Opinion in Cell Biology*, **8**: 625-631.

Hikida, R.S. 1978. Z-line extraction: comparative effects in avian skeletal muscle fiber types. *Journal of Ultrastructure Research*, **65**: 266-278.

Hilmarsson, O.T. 1999. Personal communication. MATRA, Technological Institute of Iceland, Keldnaholt, IS 112, Reykjavik, Iceland.

Hincks, M.J. and D.W. Stanley. 1975. Colour measurement of the squid Illex illecebrosus and its relationship to quality and chromatophore ultrastructure. *Canadian Institute of Food Science and Technology Journal*, **18**: 233-241.

Hinkle, A., A. Goranson, C.A. Butters and L.S. Tobacman. 1999. Roles for the troponin tail domain in thin filament assembly and regulation - a deletional study of cardiac troponin T. *Journal of Biological Chemistry*, **274**: 7157-7164.

Hoffman, L. 1999. Personal communication. University of Stellenbosch, Faculty of Agricultural Sciences, Department of Animal Sciences, Private Bag X1, Matieland, South Africa.

Holmsen, A.A. and H. McAllister. 1974. Technological and economic aspects of red crab harvesting and processing. *University of Rhode Island, Marine Technology Report*, 28. pp. 35.

Holthuis, L.B. 1980. Shrimps and prawns of the World. *FAO Species Catalogue*, Volume 1. FAO Fisheries Synopsis, 125.

Hongsprabhas, P. 1999. Personal communication. Division of Food Science and Nutrition, Faculty of Science, Srinakharinwirot University, Sukhumvit 23, Bangkok 10110, Thailand.

Honig, C.R., R.J. Connett and T.E.J. Gayeski. 1992. O_2 transport and its interaction with metabolism; a systems view of aerobic capacity. *Medicine and Science in Sports and Exercise*, **24**: 47-53.

Honikel, K.O. 1996. Personal communication. Bunderstalt für Fleischforschung, Institut für Chemie und Physik, Kulmbach, Germany.

Horowits, R., K. Maruyama and R.J. Podolsky. 1989. Elastic behavior of connectin filaments during thick filament movement in activated skeletal muscle. *Journal of Cell Biology*, **109**: 1989-2169.

Howgate, P. 1979. Fish. In J.G. Vaughan (editor), *Food Microscopy*. Academic Press, New York.

Huxley, H.E. 1965. The mechanism of muscular contraction. *Scientific American*, **213**: 18-27.

IAIS. Icelandic Agricultural Information Service. *The Shopper's Guide to Icelandic Food*. Icelandic Agriculture, Hagatorg 1, 107 Reykjavik, Iceland.

Ilian, M.A., A. El-Din, A. Bekhit and R. Bickerstaffe. 2004a. Does the newly discovered calpain 10 play a role in meat tenderization during post-mortem storage? *Meat Science*, **66**: 317-327.

Ilian, M.A., A. El-Din, A. Bekhit and R. Bickerstaffe. 2004b. The relationship between meat tenderization, myofibril fragmentation and autolysis of calpain 3 during post-mortem aging. *Meat Science*, **66**: 387-397.

Iniguez, L.C., R. Alem, A. Wauer and J. Mueller. 1998. Fleece types, fiber characteristics and production system of an outstanding llama population from Southern Bolivia. *Small Ruminant Research*, **30**, 57-65.

Irie, M. 1999. Personal communication. Osaka Agricultural Research Center, Livestock Division, Shakudo, Habikino-Shi, Osaka 583, Japan.

ITC, 1983. *The World Market for Horsemeat*. International Trade Centre, UNCTAD, GATT, Geneva.

Janz, J.A.M. 1999. Characterization of bison muscle tissue and evaluation of the efficacy of postmortem carcass treatments designed to influence the quality of bison meat. *M.Sc. Thesis, University of Alberta, Edmonton, Canada*.

Japan Meat Grading Association, 1979. *Cut Beef and Pork Trade Standards*. Approved by the Ministry of Agriculture, Forestry and Fisheries, 1976. Livestock A No. 366. Revision, 1979, Livestock A No. 4183.

Jessop, B.M. 1993. American eel. *Underwater World*, 39. Fisheries and Oceans, Ottawa.

Jessop, B.M. 1990. Alewife. *Underwater World*, 57. Fisheries and Oceans, Ottawa.

Johnson, J.A., D.P. Green and R.E. Martin. 1998. Industry perspectives: the hard blue crab fishery - Atlantic and Gulf. *Journal of Shellfish Research*, **17**: 371-374.

Jones, J.W. 1967. Fishes (Classes Marsipobranchii, Selachii, Bradyodonti, Pisces).In, *Larousse Encyclopedia of Animal Life*. Paul Hamlyn, London. pp. 207-267.

Joo, S.T. 1999. Personal communication. Gyeongsang National University, Department of Animal Science, Meat Laboratory, 900 Kajwa-Dong, Chinju, Gyeongnam, 600-701, South Korea.

Jordan, D.S. and B.W. Evermann. 1934. *American Food and Game Fishes*. Doubleday, Doran & Co., Garden City, New Jersey.

Jori, F., M. Lopez-Bejar and P. Houben. 1998. The biology and use of the African brush-tailed porcupine (Atherurus africanus, Gray, 1842) as a food animal. A review. *Biodiversity and Conservation*, **7**: 1417-1426.

Kannus, P., L. Jozsa, T.A.H. Jarvinen, T.L.N. Jarvinen, M. Kvist, A. Natri and M. Jarvinen. 1998. Location and distribution of non-collagenous matrix proteins in musculoskeletal tissues of rat. *Histochemical Journal*, **30**: 799-810.

Kauffman, R.G., L.E. St. Clair, and R.J. Reber. 1963. Ovine Myology. *University of Illinois College of Agriculture, Agricultural Experiment Station*, Bulletin 698.

Kauffman, R.G. and L.E. St. Clair. 1965. Porcine Myology. *University of Illinois College of Agriculture, Agricultural Experiment Station*, Bulletin 715.

Kawamoto, R.M., J-P. Brunschwig and A.H. Caswell. 1988. Localization by immunoelectron microscopy of spanning protein of triad junction in terminal cisternae/triad vesicles. *Journal of Muscle Research and Cell Motility*, **9**: 334-343.

Ke, P.J., B. Smith-Lall, A. B. Dewar. 1981. Quality Improvement Investigations for Atlantic Queen Crab (Chionoectes opilio). *Technical Report 1002, Fisheries and Marine Service, Fisheries and Oceans Canada*, Halifax, Nova Scotia.

Kier, W.M. 1985. The musculature of squid arms and tentacles: ultrastructural evidence for functional differences. *Journal of Morphology*, **185**: 223-239.

Kimura, S. and K. Maruyama. 1989. Isolation of α-connectin, an elastic protein, from rabbit skeletal muscle. *Journal of Biochemistry*, **106**: 952-954.

Kirkpatrick, T.H. and P.J. Amos. 1985. The kangaroo industry. In H.J. Lavery (editor), *The Kangaroo Keepers*. University of Queensland Press, St. Lucia. pp. 75-100.

Koh, T.J. and W. Herzog. 1998. Excursion is important in regulating sarcomere number in the growing rabbit tibialis anterior. *Journal of Physiology*, London, **508**: 267-280.

Komprda, T., J. Neznalovà, S. Standara and S. Bover-Cid. 2001. Effect of starter culture and storage temperature on the content of biogenic amines in dry fermented sausage polican. *Meat Science*, **59**: 267-276.

Kuo, J.C. 1999. Personal communication. College of Agriculture, Tunghai University, 181, Section 3, Taichung-Kan Road, Taichung, Taiwan 407.

Kutas, R. 1984. *Great Sausage Recipes and Meat Curing*. Sausage Maker, Buffalo, New York.

Labeit, S., T. Gibson, A. Lakey, K. Leonard, M. Zeviani, P. Knight, J. Wardale and J. Trinick. 1991. Evidence that nebulin is a protein-ruler in muscle thin filaments. *Federation of European Biochemical Societies*, **282**: 313-316.

Labeit, S. and B. Kolmerer. 1995a. The complete primary structure of human nebulin and its correlation to muscle structure. *Journal of Molecular Biology*, **248**: 308-315.

Labeit, S. and B. Kolmerer. 1995b. Titins: giant proteins in charge of muscle ultrastructure and elasticity. *Science*, **270**: 293-296.

Laborde, F. 1999. Personal communication. Department of Animal and Poultry Science, Unviersity of Guelph, Guelph, Ontario N1G 2W1, Canada.

Lagler, K.F., J.E. Bardach, R.S. Miller, and D.R.M. Passino. 1977. *Ichthyology*. Wiley, New York.

Lahucky, R. 1999. Personal communication. Research Institute of Animal Production, 949 92 Nitra, Hlohovská 2, Slovakia.

Laing, I. and A. Psimopoulous. 1998. Hatchery cultivation of king scallop (Pecten maximus) spat with cultured and bloomed algal diets. *Aquaculture*, **169**: 55-68.

Lamb, A. and P. Edgell. 1986. *Coastal Fishes of the Pacific Northwest*. Harbour Publishing, Madeira Park, BC.

Land, D.G. and A. Hobson-Frohock. 1977. Flavour, taint and texture in poultry meat. In K.N. Boorman and B.J. Wilson (editors), *Growth and Poultry Meat Production*. Longman, Edinburgh. British Poultry Science Symposium 12. pp. 301-334.

Lawrie, R.A. 1985. *Meat Science*. 4th edition. Pergamon Press, Oxford.

LeBlanc, R.J. and T.A. Gill. 1984. Ammonia as an objective quality index in squid. *Canadian Institute of Food Science and Technology Journal*, **17**: 195-201.

Ledger, H.P. 1968. Body composition as a basis for a comparative study of some East African mammals. *Symposium of the Zoological Society of London*, **21**: 289-310.

Lee, C.M. 1985. Microstructure of meat emulsions in relation to fat stabilization. *Food Microstructure*, **4**: 63-72.

Lewis, C.T. 1987. *An Elementary Latin Dictionary*. Oxford University Press, Oxford.

Li, M., D.W. Dickson and A.J. Spiro. 1997. Abnormal expression of laminin beta (1) chain in skeletal muscle of adult-onset limb-girdle muscular dystrophy. *Archives of Neurology*, **54**: 1457-1461.

Light, N., A.E. Champion, C. Voyle and A.J. Bailey. 1985. The rôle of epimysial, perimysial and endomysial collagen in determining texture in six bovine muscles. *Meat Science*, **13**: 137-149.

Lightfoot, C. 1977. Eland (Taurotragus oryx) as a ranching animal complementary to cattle in Rhodesia. 3. Production and marketing. *Rhodesia Agricultural Journal*, **74**: 85-91.

Lihateollisuuden Tutkimuskeskus, 1989. *Liha-opas* (Meat Guide). 11th edition. ISBN 951-99498-6-0.

Lipiński, M. 1973. The place of squids in the biological and fishery structure of the world ocean. In M. Lipiński (editor), *Squid Symposium, Gdynia*. National Marine Fisheries Service, Washington. pp. 4-13.

Lissitsyn, A. 1999. Personal communication. The All - Russia Meat Research Institute, VNIIMP, 26 Talalikhin Str., 109316 Moscow, Russia.

LMC, 2003. Livestock & Meat Commission, Northern Ireland. Technical Manual. Lissue House, 31 Ballinderry Road, Lisburn, Northern Ireland, BT28 2SL.

Locker, R.H. 1987. The non-sliding filaments of the sarcomere. *Meat Science*, **20**: 217-236.

Locker, R.H. and D.J.C. Wild. 1984. The N-lines of skeletal muscle. *Journal of Ultrastructure Research*, **88**: 207-222.

Lourdes Pérez, M. 1999. Personal communication. Depertamento de Biotecnologia, Universidad Autónoma Metropolitana, Unidad Iztapalapa, Apartado Postal 55-535, Distrito Federal, Mexico.

Lundström, K., B. Malmfors, S. Stern, L. Rydhmer, L. Eliasson-Selling, A.B. Mortensen and H.P. Mortensen. 1994. Skatole levels in pigs selected for high lean tissue growth rate on different dietary protein levels. *Livestock Production Science* **38**: 125-132.

Luther, P. and J. Squire. 1978. Three-dimensional structure of the vertebrate muscle M-region. *Journal of Molecular Biology*, **125**: 313-324.

Lydekker, R. 1894. *The Royal Natural History*. Volume II, Section IV. Frederick Warne, London.

MacGillivary, P.S., E.J. Anderson, G.M. Wright and M.E. Demont. 1999. Structure and mechanics of the squid mantle. *Journal of Experimental Biology*, **202**: 683-695.

MacLennan, D.H. and M.S. Phillips. 1992. Malignant hyperthermia. *Science*, **256**: 789-794.

Mågård, M.Å., H.E.B. Berg, V. Tagesson, M.L.G. Järemo, L.L.H. Karlsson, L.J.E. Mathiasson, M. Bonneau and J.Hansen-Møller. 1995. Determination of androstenone in pig fat using supercritical fluid extraction and gas chromatography - mass spectrometry. *Journal of Agricultural and Food Chemistry*, **43**: 114-120.

Magay, T. and L. Kiss. 1995. *Hungarian-English Standard Dictionary*. Hippocrene Books, New York, and Akadémiai Kiadó, Budapest.

Maier, A., E. Leberer and D. Pette. 1986. Distribution of sarcoplasmic reticulum Ca-ATPase and of calsequestrin in rabbit and rat skeletal muscle fibres. *Histochemistry*, **86**: 63-69.

Mäkelä, E., G. Hottinger and P. Immonen. 1995. *New Flavours from Finland*. Otava Publishing, Helsinki.

Mani, R.S., O.S. Herasymowych and C.M. Kay. 1980. Physical, chemical and ultrastructural studies on muscle M-line proteins. *International Journal of Biochemistry*, **12**: 333-338.

Marbek Meat Company. Beef: the different terms used in Israel.(In Hebrew with English and French equivalent terms).

Markle, D.F. 1989. Red hake. *Underwater World*, 8. Fisheries and Oceans, Ottawa.

Marsh, B.B. and W.A. Carse. 1974. Meat tenderness and the sliding-filament hypothesis. *Journal of Food Technology*, **9**: 129-139.

Marshall, A.M. 1900. *The Frog: an Introduction to Anatomy, Histology and Embryology*. Macmillan, New York.

Maruyama, K. 1997. Connectin/titin, giant elestic protein of muscle. *FASEB Journal*, **11**: 341-345.

May, N.D.S. 1964. *The Anatomy of the Sheep*, University of Queensland Press, St. Lucia.

McAndrew, I. 1990. *Ian McAndrew on Poultry and Game*. Van Nostrand Reinhold, New York.

McGlade, J.M. 1993. Pollock. *Underwater World*, 35. Fisheries and Oceans, Ottawa.

McGregor, A., A.D. Blanchard, A.J. Rowe and D.R. Critchley. 1994. Identification of the vinculin-binding site in the cytoskeletal protein "-actinin. *Biochemical Journal*, **301**: 225-233.

McKone, W.D. and E.M. LeGrow. 1983. Thorny and smooth skates. *Underwater World*, 21E. Fisheries and Oceans, Ottawa.

McKone, W.D. and E.M. LeGrow. 1990. Redfish (ocean perch). *Underwater World*, 33. Fisheries and Oceans, Ottawa.

MDF, 1998. Price list. Mid Devon Fallow, Keyethern Farm, Hatherleigh, Okehampton, Devon EX20 3LG.

Melton, S.L. 1990. Effects of feeds on flavor of red meat: a review. *Journal of Animal Science*, **68**: 4421-4435.

Merriam-Webster, 1991. *Webster's Dictionary of Word Origins*. Smithmark Publishers, New York.

Mickelson, J.R., C.M. Knudson, C.F.H. Kennedy, D-I. Yang, L.A. Litterer, W.E. Rempel, K.P. Campbell and C.F. Louis. 1992. Structural and functional correlates of a mutation in the malignant hyperthermia-susceptible pig ryanodine receptor. *Federation of European Biochemical Societies*, **301**: 49-52.

Minnaar, M. 1998. *The Emu Farmer's Handbook*. Volume 2. Nyoni Publishing Company, Groveton, Texas.

Monin, G. 1997. Personal communication. Station Viande, INRA, Theix 63122, France.

Moore, A.B. and R.R. Eitenmiller. 1980. Shrimp quality: biological and technological relationships. *University of Georgia College of Agriculture Experiment Station Research Bulletin*, 253. p. 31.

Moore, R., J. Stone and H. Tattersall. 1983. *The Meat Buyers' Guide for Caterers*. International Thompson Publishing, London. ISBN 7198 2517 2

Morin, W.A. and E. McLaughlin. 1973. Glycogen in crustacean fast and slow muscles. *American Zoologist*, **13**: 435-445.

Motohiro, T. and N. Inoue. 1970. pH of canned crab meat. I. Stages in the molting cycle in relation to pH. *Food Technology*, **24**: 71-73.

Muir, P.D. 1989. Deer farming in New Zealand. In R. Valdez (editor), *First International Wildlife Ranching Symposium*. New Mexico State University, Las Cruces, New Mexico. pp. 105-113.

Nagtegaal, D. 1985. Rockfish. *Underwater World*, 51. Fisheries and Oceans, Ottawa.

NLMB, 1937. National Livestock and Meat Board. *Cashing in on Lamb*. Chicago, Illinois. pp. 47.

NLMB. 1973. National Livestock and Meat Board. *Uniform Retail Meat Identity Standards*. Chicago, Illinois.

Norma Cubana, 1985. *Carne deshuesada de bovino para consumo. Especificaciones de calidad*. NC 79-12:84. Comite Estatal de Normalizacion, La Habana, Cuba.

NRC, 1983. *Little-Known Asian Animals with a Promising Economic Future*. Office of International Affairs, National Research Council, National Academy Press, Washington, D.C.

NZMPB, 1985. *The New Zealand Meat Trade Guide*. New Zealand Meat Producers Board, Wellington, NZ. pp.43.

O'Boyle, R.N. 1985. The haddock. *Underwater World*, 48, Fisheries and Oceans, Ottawa.

O'Brien, P.J., H. Shen, C.R. Cory and X. Zhang. 1993. Use of a DNA-based test for the mutation associated with porcine stress syndrome (malignant hyperthermia) in 10,000 breeding swine. *Journal of the American Veterinary Association*, **203**: 842-851.

OEEC, 1961. *Meat Cuts in O.E.E.C. Member Countries*. European Productivity Agency, Organisation for European Economic Co-operation. Project No. 7/11-1B. 3, Rue André-Pascal, Paris.

Ogonowski, M.M. and F. Lang. 1979. Histochemical evidence for enzyme differences in crustacean fast and slow muscle. *Journal of Experimental Zoology*, **207**: 143-151.

Ohtsuki, I. 1999. Calcium ion regulation of muscle contraction: the regulatory role of troponin T. *Molecular and Cellular Biochemistry*, **190**: 33-38.

Onions, C.T. 1972. *The Shorter Oxford English Dictionary on Historical Principles*. Prepared by W. Little, H.W. Fowler and J. Coulson. 3rd Edition. Clarendon Press, Oxford.

Onyango, C.A., M. Izumimoto and P.M. Kutima. 1998. Comparison of some physical and chemical properties of selected game meats. *Meat Science*, **49**: 117-125.

Oplatka, A. 1997. Critical review of the swinging crossbridge theory and of the cardinal active role of water in muscle contraction. *Critical Reviews in Biochemistry and Molecular Biology*, **32**: 307-360.

Opplysningskontoret for kyøtt. Postboks 95 Refstad, 0513, Oslo.

Orcutt, M.W., T.R. Dutson, F.Y. Wu and S.B. Smith. 1986. The fine structure of the endomysium, perimysium and intermyofibrillar connections in muscle. *Food Microstructure*, **5**: 41-51.

Otwell, W.S. and G.G. Giddings. 1980. Scanning electron microscopy of squid, Loligo pealei: raw, cooked, and frozen mantle. *Marine Fisheries Review*, **42**: 67-73.

Paleari, M.A., S. Camisasca, G. Beretta, P. Renon, P. Corsico, G. Bertolo and G. Grivelli. 1998. Ostrich meat: physico-chemical characteristics and comparison with turkey and bovine meat. *Meat Science*, **48**: 205-210.

Pardo, J.V., J. D'Angelo Siliciano and S.W. Craig. 1983. A vinculin-containing cortical lattice in skeletal muscle: transverse lattice elements ("costametres") mark sites of attachment between myofibrils and sarcolemma. *Proceedings of the National Academy of Sciences of the USA*, 80: 1008-1012.

Park, J.W., T.M. Lin and J. Yongsawatdigul. 1997. New developments in manufacturing of surimi and surimi seafood. *Food Reviews International*, **13**: 577-610.

Parmacek, M.S. and J.M. Leiden. 1991. Structure, function and regulation of troponin C. *Circulation*, **84**: 991-1003.

Pasquini, C. 1982. *Atlas of Bovine Anatomy*. Sudz Publishing, Eureka, California.

Peachey, L.D. 1967. Membrane systems of crab fibers. *American Zoologist*, **7**: 505-513.

Pearson, A.M. and F.W. Tauber. 1984. *Processed Meats*. AVI Publishing, Westport, Connecticut.

Pedrão, M.R., F.A.G. Coró, F. Lassances, F.E.Y. Youssef and M. Shimokomaki. 2003. Evaluation of humpback muscle (*M. rhomboideus*) texture from Zebu breed (*Bos indicus*). *49th International Congress of Meat Science and Technology*, pp. 33-34.

Perry, S.F., C. Daxboeck, B. Emmett, P.W. Hochachka and R.W. Brill. 1985. Effects of exhausting exercise on acid-base regulation in skipjack tuna (Katsuwonus pelamis) blood. *Physiological Zoology*, **58**: 421-429.

Perry, S.V. 1999. Troponin I: inhibitor or facilitator. *Molecular and Cellular Biochemistry*, **190**: 9-32.

Petchey, A.M. 1991. Squab production: a possible diversification enterprise for egg producers. *Farm Building Progress*, **103**: 15-18.

Petti, V. and K. Petti. 1997. *Hippocrene Standard Dictionary. English-Swedish. Swedish-English*. Hippocrene Books, New York. Originally published by Norstedts Förlag.

Pexara, E.S., J. Metaxopoulos and E.H. Drosinos. 2002. Evaluation of shelf life of cured, cooked, sliced turkey fillets and cooked pork sausages – 'piroski' – stored under vacuum and modified atmospheres at +4 and +10°C. *Meat Science*, **62**: 33-43.

Piedra, L., J.R. Bolanos and J. Sanchez. 1997. Evaluation of captive Crocodylus acutus (Crocodilia: Crocodylidae) neonate growth. *Revista de Biologia Tropical*, **45**: 289-293.

Pitt, T.K. 1989. American plaice. *Underwater World*, 15E. Fisheries and Oceans, Ottawa.

Pitt, T.K. 1990. Winter flounder. *Underwater World*, 34. Fisheries and Oceans, Ottawa.

Pitt, T.K. 1993. Yellowtail flounder. *Underwater World*, 20E. Fisheries and Oceans, Ottawa.

Pivnićka, K. and Černý, K. 1987. In P. Bristow (editor), *The Illustrated Book of Fishes*. Octopus Books, London.

Pla, M., L. Guerrero, D. Guardia, M.A. Oliver and A. Blasco. 1998. Carcass characteristics and meat quality of rabbit lines selected for different objectives. I. Between lines comparison. *Meat Science*, **54**: 115-123.

Poligné, I., A. Collignan and G. Trystram. 2001. Characterization of traditional processing of pork meat into *boucané*. *Meat Science*, **59**: 377-389.

Pomeroy, R.W. and D.R. Williams. 1976. *Beef Carcass. Methods of Dressing, Measuring, Jointing and Tissue Separation*. European Association of Animal Production, Publication 18. Meat Research Institute, Langford, UK.

Þorkelsson, G. and Ó.Þ. Hilmarsson. 1994. *Íslenska Kjötbókin Handbók Fyrir Kjötkaupendur*. Íslenskur Landbúnapur.

Portsmouth, J.I. 1979. *Commercial Rabbit Meat Production*. Saiga Publishing, Hindhead, Surrey, England.

Pospiech, E. 1998. Personal communication. Agricultural University of Pozna½, Institute of Meat Technology, Wojska Polskiego **31**, 60-624 Poznań, Poland.

Price, M.G. 1987. Skelemins: cytoskeletal proteins located at the periphery of M-discs in mammalian striated muscle. *Journal of Cell Biology*, **104**: 1325-1336.

Pyornila, A.E.I., A.P. Putaala and R.K. Hissa. 1998. Fibre types in breast and leg muscles of hand-reared and wild grey partridge (Perdix perdix). *Canadian Journal of Zoology*, **76**: 236-242.

Ramos, E.M., L.A.M. Gomide, J.F.M. Parreiras, S.M. Lima and L.A. Peternelli. 2003. Effect of sex and live weight on bullfrog (*Rana catesbeiana*) meat composition. *49th International Congress of Meat Science and Technology*. Pp. 29-30.

Randall, J.E. 1996. *Caribbean Reef Fishes*. 3rd edition. T.F.H. Publications, Neptune, New Jersey.

Rao, M.V. and N.F.S. Gault. 1989. The influence of fibre-type composition and associated biochemical characteristics on the acid buffering capacities of several beef muscles. *Meat Science*, **26**: 5-18.

Rayment, I., W.R. Rypniewski, K. Schmidt-Bäse, R. Smith, D.R. Tomchick, M.M. Benning, D.A. Winkelmann, G. Wesenberg and H.M. Holden. 1993. Three-dimensional structure of myosin subfragment-1: a molecular motor. *Science*, **261**: 50-65.

Rebora, P. 1977. *Cassell's Italian Dictionary*. MacMillan, New York.

Reinken, G. 1997. Re-distribution, use and naming of fallow deer Cervus dama in Europe. *Zeitschrift für Jagdwissenschaft*, **43**: 197-206.

Renecker, L.A. 1989. Wapiti farming in Canada. In R. Valdez (editor), *First International Wildlife Ranching Symposium*. New Mexico State University, Las Cruces, New Mexico. pp. 47-62.

Reynolds, P.C. 1963. *The Complete Book of Meat*. Barrows, New York.

Rixson, D. *Beef Cutting*. 1970. A Step by Step Guide. Meat Trades Journal, London. Northwood Publications, Thomson Organisation. pp. 28.

Rixson, D. 2000. *The History of Meat Trading*. Nottingham University Press, Nottingham.

Robert, G. 1988. The sea scallop. *Underwater World*, 5. Fisheries and Oceans, Ottawa.

Roca, R.D., N. Veiga, R.B.D. Neto and R.C. Cervi. 1999. Sensorial characteristics of smoked meat of capybara. *Pesquisa Agropecuaria Brasileira*, **34**: 487-492.

Romans, J.R. and Ziegler, P.T. 1974. *The Meat We Eat*. 11th edition. Interstate, Danville, Illinois.

Rombauer, I.S. and M.R. Becker. 1975. *Joy of Cooking*. Bobbs-Merrill Company, Indianapolis.

Ross, R.F. 1927. The preparation of lobster paste. *Biological Board of Canada, Bulletin* 10. pp. 24.

Rossner, K.L. and R.G. Sherman. 1978. Invaginated membrane in crustacean tonic muscle fibers: estimates of membrane capacitance. *American Journal of Physiology*, **235**: C220-226.

Royce, W.F. 1972. *Introduction to the Fishery Sciences*. Academic Press, New York.

Rue, L.L. III. 1973. *Game Birds of North America.* Outdoor Life, Harper Row, New York.

Rue, L.L. III. 1987. *Complete Guide to Game Animals.* Outdoor Life Books.

Ryan, P.M. 1988. Trout in Canada's Atlantic provinces. *Underwater World*, 58. Fisheries and Oceans, Ottawa.

Sack, W.O. 1982. *Essentials of Pig Anatomy.* Veterinary Textbooks, Ithaca, NY.

Salazar, R. 2001. Personal communication. Agropesa, Industria Agropecuaria Ecuatoriana S.A., Quito, Ecuador.

Sales, J. and J.G. Dingle. 1998. Kangaroo: as an alternative meat source. *Food Australia*, **50**: 531-534.

Sales, J. and B. Oliver-Lyons. 1996. Ostrich meat: a review. *Food Australia*, **48**(11), 504-511.

Sales, J., J.L. Navarro, L. Bellis, A. Manero, M. Lizurume and M.B. Martella. 1997. Carcase and component yields of rheas. *British Poultry Science*, **38**: 378-380.

Sandford, J.C. 1986. *The Domestic Rabbit.* 4th edition. Collins, London.

Santos, E.M., C. González-Fernández, I. Jaime and J. Rovira. 2003. Physicochemical and sensory characterization of *Morcilla de Burgos*, a traditional Spanish blood sausage. *Meat Science*, **65**: 893-898.

Schiereck, P., E.L. de Beer, R.L.F. Grundeman, T. Manussen, N. Kylstra and W. Bras. 1992. Tetragonal deformation of the hexagonal myofilament matrix in single skinned skeletal muscle fibres owing to change in sarcomere length. *Journal of Muscle Research and Cell Motility*, **13**: 573-580.

Schmalbruch, H. 1979. The membrane systems in different fibre types of the triceps surae muscle of cat. *Cell and Tissue Research*, **204**: 187-200.

Schmitt, O., T. Degas, P. Perot, M-R. Langlois and B.L. Dumont. 1979. Etude morpho-anatomique du périmysium (méthodes de description et d'évaluation). *Annals de Biologie Animale, Biochimie et Biophysique*, **19**: 1-30.

Schön, L. 1958. Vergleichende Zusammenstellung der Teilstückbezeichnungen bei den Schlachttierzerlegungen in verschiedenen Ländern Europas. *Fleischwirtschaft*, **12**: 853-857.

Schroeter, J.P., J-P. Bretaudiere, R.L. Sass and M.A. Golstein. 1996. Three-dimensional structure of the Z band in a normal mammalian skeletal muscle. *Journal of Cell Biology*, **133**: 571-583.

Scott, J.S. 1993. Sand lance. *Underwater World*, 46. Fisheries and Oceans, Ottawa.

Scott, W.B. and S.N. Messieh. 1976. *Common Canadian Atlantic Fishes.* The Huntsman Marine Laboratory, St. Andrews, New Brunswick.

Seishi, Y., C. Renchinmyadag, D. Namsrai, D. Batmunkh and T. Tserenulam. 1999. Production and consumption of sheep, goat, cattle, horse and camel meat in the Mongolian Gobi nomadic area. *45th International Congress of Meat Science and Technology, Yokohama, Japan.* Supplement 1, P8.

Selverston, A. 1967. Structure and function of the transverse tubular system in crustacean muscle fibers. *American Zoologist*, **7**: 515-525.

SGM, 1999. Kangaroo cuts. Southern Game Meat, 22 Churchill Street, Auburn, New South Wales 2144, Australia. http://www.sgm.com.au

Simela, L., L.R. Ndlovu and L.M. Sibanda. 1999. Carcass characteristics of the marketed Matabele goat from south-western Zimbabwe. *Small Ruminant Research*, **32**: 173-179.

Simms, A.E. and M. Quin. 1973. *Fish and Shell-fish.* Virtue and Company Limited, London and Coulsdon.

Simões, J.A. 1999. Personal communication. Ministério da Agricultura, Instituto Nacional de Investigação Agrária, Estação Zootécnica Nacional, 2000 Santarém, Portugal.

Simonne, A.H., N.R. Green and D.I. Bransby. 1996. Consumer acceptability and beta-carotene content of beef as related to cattle finishing diets. *Journal of Food Science*, **61**: 1254.

Simoons, F.J. 1996. Dogflesh eating by humans in sub-Saharan Africa. *Ecology of Food and Nutrition*, **34**: 251-292.

Singh, R.P. and D.E. Brown. 1980. Development of a squid skinning and eviscerating system. *Marine Fisheries Review*, **42**: 77-84.

Sirkis, R. 1975. *Cooking with Love*. Zmora, Bitan and Modan Publishers, Tel Aviv. (In Hebrew).

Sisson, S. and Grossman, J.D. 1953. *The Anatomy of the Domestic Animals*, Philadelphia: W.B. Saunders.

SKI, Svensk Kött Information. Swedish Meat Research Institute.

Skjervold, P.O., S.O. Fjaera and P.B. Ostby. 1999. Rigor in Atlantic salmon as affected by crowding stress prior to chilling before slaughter. *Aquaculture*, **175**: 93-101.

Skrede, G. and T. Storebakken. 1986. Instrumental colour analysis of farmed and wild Atlantic salmon when raw, baked and smoked. *Aquaculture*, **53**: 279-286.

Smith, E. 1876. *Foods*. International Scientic Series: Appleton, New York.

Smith, S.H. and M.D. Judge. 1991. Relationship between pyridinoline concentration and thermal stability of bovine intramuscular collagen. *Journal of Animal Science*, **69**: 1989-1993.

Snell, H. 1996. Carcass traits of German Fawn, Boer, Cashmere and crossbred kids. *Fleischwirtschaft*, **76**: 1335-1339.

Sokolov, V.E. and N.L. Lebedeva. 1968. Commercial hunting in the Soviet Union. In R.J. Hudson, K.R. Drew and L.M. Baskin (editors), *Wildlife Production Systems. Economic Utilisation of Wild Ungulates*. Cambridge University Press, Cambridge. pp. 170-185.

Sookhareea, R., D.G. Taylor, G.McL. Dryden and K.B. Woodford. 2001. Primal joints and hind-leg cuts of entire and castrated Javan rusa (*Cervus timorensis russa*) stages. *Meat Science,***58**: 9-15.

Stanley, D.W. and A.K. Smith. 1984. Microstructure of squid muscle and its influence on texture. *Canadian Institute of Food Science and Technology Journal*, **17**: 209-213.

Stanley, D.W. and H.O. Hultin. 1982. Quality factors in cooked North Atlantic squid. *Canadian Institute of Food Science and Technology Journal*, **15**: 277-282.

Stanton, C. and N. Light. 1990. The effects of conditioning on meat collagen: part 3 - evidence for proteolytic damage to endomysial collagen after conditioning. *Meat Science*, **27**: 41-54.

Storebakken, T., P. Foss, K. Schiedt, E. Austreng, S. Liaaen-Jensen and U. Manz. 1987. Carotenoids in diets for salmonids. IV. Pigmentation of Atlantic salmon with astaxanthin, astaxanthin dipalmitate and canthaxanthin. *Aquaculture*, **65**: 279-292.

Suwattana, D. 1999. Personal communication, Department of Husbandry, Faculty of Veterinary Sciences, Chulalongkorn, Thailand.

SVO. *Opleiding voor de Vleesector*. Pompelaan 8, 3572 LR, Utrecht, Netherlands.

Swatland, H.J. 1994. *Structure and Development of Meat Animals*. Technomic Publishing, Lancaster, Pennsylvania.

Swift Canadian. 1973. *Cuts of Meat. How You Can Identify Them*. Swift Canadian, Etobicoke, Ontario. pp. 28.

Takekura, H., H. Shuman and C. Franzini-Armstrong. 1993. Differentiation of membrane systems during development of slow and fast skeletal muscle fibres. *Journal of Muscle Research and Cell Motility*, **14**: 633-645.

Tamura, T. 1961. Carp cultivation in Japan. In G. Borgstrom (editor), *Fish as Food*. Volume 1. Academic Press, New York. pp. 103-120.

Terra, N.N. 1999. Personal communication. Departamento de Technologia e Ciençia dos Alimentos, Universidade Federal de Sante Maria, Campus Universitario, 97119-900 Santa Maria, Brazil.

Thiemig, F. and P. Oelker. 1998. Method of determining the gaping effect in frozen meat products before final cooking. *Fleischwirtschaft*, **78**: 221-224.

Thomas, F.B. 1971. Current and proposed catfish processing and marketing. *Proceedings of the Reciprocal Meat Conference*, **24**: 203-214.

Thomas, G.H., E.C. Newbern, C.C. Korte, M.A. Bales, S.V. Musa, A.G. Clark and D.P. Kiehart. 1998. Intragenic duplication and divergence in the spectrin superfamily of proteins. *Molecular Biology and Evolution*, **14**: 1285-1295.

Thompson, H.V. and A.N. Worden. 1956. *The Rabbit*. Collins, London.

Tidball, J.G. and D.J. Law. 1991. Dystrophin is required for normal thin filament - membrane associations at myotendinous junctions. *American Journal of Pathology*, **138**: 17-21.

TMDF. Taiwan Meat Development Foundation. 4th Floor, 14 Wenchow St., Taipei, Taiwan, R.O.C.

Toorop, R-M. 1998. Wageningen, the Netherlands. Personnal communication.

Tornberg, E. 1996. Biophysical aspects of meat tenderness. *Meat Science*, **43**: S175-S191.

Trombitás, K., P.H.W.W. Baatsen, M.S.Z. Kellermayer and G.H. Pollack. 1991. Nature and origin of gap filaments in striated muscle. *Journal of Cell Science*, **100**: 809-814.

Tucker, H.Q., M.M. Voegeli, and G.H. Wellington. 1952. *A Cross Sectional Muscle Nomenclature of the Beef Carcass*. Michigan State College Press, East Lansing.

Tume, R.K. 1979. Iodination of calsequestrin in the sarcoplasmic reticulum of rabbit skeletal muscle: a re-examination. *Australian Journal of Biological Science*, **32**: 177-185.

Turnacioglu, K.K., B. Mittal, G.A. Dabiri, J.M. Snager and J.W. Sanger. 1997. Zeugmatin is part of the Z-band targeting region of titin. *Cell Structure and Function*, **22**: 73-82.

Ucko, P.J. and G.W. Dimbleby. 1969. *The Domestication and Exploitation of Plants and Animals*. Aldine-Atherton, Chicago.

Unger, B. and J. Wilson. *Nourishing Meat from the Manchester Meat Market*. City of Manchester Markets Department, Manchester, UK.

van Rooyen, I., H. Ebedes and J.G. du Toit. 1989. Hunting for meat. Meat processing. In J. du P. Bothma (editor), *Game Ranch Management*. J.L. van Schaik, Pretoria. pp. 376-386.

Vareltzis, K. 1998. Personal communication. Aristotle University of Thessaloniki, Faculty of Veterinary Medicine, Department of Food Hygiene and Technology of Animal Origin, Laboratory of Food Technology of Animal Origin, Thessaloniki 54006, Greece.

Varga, S., A.B. Dewar, and W.E. Anderson. 1971. Prevention of struvite in canned queen crab by hexametaphosphate. *Fisheries and Oceans, Regional Inspection and Technology Laboratory*, Halifax, Nova Scotia. Technical Report, 8.

Vesey-Fitzgerald, B. 1946. *British Game*. Collins, London.

Voisey, P.W. 1976. Engineering assessment and critique of instruments used for meat tenderness evaluation. *Journal of Texture Studies*, **7**: 11-48.

Volpelli, L.A., R. Valusso, M. Morgante, P. Pittia and E. Piasentier. 2003. Meat quality in male fallow deer (*Dama dama*): effects of age and supplementary feeding. *Meat Science*, **65**: 555-562.

Wade, E.M. and J.G. Fadel. 1997. Optimization of caviar and meat production from white sturgeon (Acipenser transmontanus). *Agricultural Systems*, **54**: 1-21.

Wakabayashi, K., Y. Sugimoto, H. Tanaka, Y. Ueno, Y. Takezawa and Y. Amemiya. 1994. X-ray diffraction evidence for the extensibility of actin and myosin filaments during contraction. *Biophysical Journal*, **67**: 2422-2435.

Wakayama, Y. and S. Shibuya. 1991. Antibody-decorated dystrophin molecule of murine skeletal myofiber as seen by freeze-etching electron microscopy. *Journal of Electron Microscopy*, **40**: 143-145.

Walliman, T., D.C. Turner and H.M. Eppenberger. 1975. Creatine kinase and M-line structure. *Proceeedings of the Federation of European Biochemical Societies*, **31**: 119-124.

Walsh, S.J. Spiny dogfish. *Underwater World*, 13. Fisheries and Oceans, Ottawa.

Wang, K. and R. Ramirez-Mitchell. 1983. A network of transverse and longitudinal intermediate filaments is associated with sarcomeres of adult vertebrate skeletal muscle. *Journal of Cell Biology*, **96**: 562-570.

Wang, K. and C.L. Williamson. 1980. Identification of an N2 line protein of striated muscle. *Proceedings of the National Academy of Sciences of the USA*, **77**: 3254-3258.

Weber, A., C.R. Pennise, G.G. Babcock and V.M. Fowler. 1994. Tropmodulin caps the pointed ends of actin filaments. *Journal of Cell Biology*, **127**: 1627-1635.

Weber, M.L. and J.M. Thompson. 1998. Seasonal patterns in food intake, live mass, and body composition of meture female fallow deer (Dama dama). *Canadian Journal of Zoology*, **76**: 1141-1152.

Weyermann, F. and V. Dzapo. 1997. Studies on the post-mortem pH in horses. *Fleischwirtschaft*, **77**: 1119-1121.

White, R.G., B.A. Tiplady and P. Groves. 1989. Quiviut production from muskoxen. In R.J. Hudson, K.R. Drew and L.M. Baskin (editors), *Wildlife Production Systems. Economic Utilisation of Wild Ungulates.* Cambridge University Press, Cambridge. pp. 387-400.

Wiklund, E. 1996. Pre-slaughter Handling of Reindeer (Rangifer tarandus tarandus L.). Effects on Meat Quality. *Doctoral Thesis, Swedish University of Agricultural Sciences, Uppsala.*

Wiklund, E. 1999. Personal communication.

Wilder, D.G. 1966. Canadian Atlantic crab resources. *Fisheries Research Board of Canada*, St. Andrews, NB. General Series Circular, 50.

Wilding, P., N. Hedges and P.J. Lillford. 1986. Salt-induced swelling of meat: the effect of storage time, pH, ion-type and concentration. *Meat Science*, **18**: 55-75.

Wilkinson, J.M. and B.A. Stark. 1987. *Commercial Goat Production.* BSP Professional Books, Oxford.

Willam, A. 1999. Personal communication. University of Agricultural Science Vienna, Department of Livestock Science, Animal Breeding Group, Gregor Mendel Strasse 33, A-1180, Vienna, Austria.

Wright, J., Q-Q. Huang and K. Wang. 1993. Nebulin is a full-length template of actin filaments in the skeletal muscle sarcomere: an immunoelectron microscopic study of its orientation and span with site-specific monoclonal antibodies. *Journal of Muscle Research and Cell Motility*, **14**: 476-483.

Wu, F.Y. and S.B. Smith. 1987. Ionic strength and myofibrillar protein solubilization. *Journal of Animal Science*, **65**: 597-608.

Yamaguchi, M., H. Kamisoyama, S. Nada, S. Yamano, M. Izumimoto, Y. Hirai, R.G. Cassens, H. Nasu, M. Muguruma and T. Fukazawa. 1986. Current concepts of muscle ultrastructure with emphasis on Z-line architecture. *Food Microstructure*, **5**: 197-205.

Yan, X. 1999. Personal communication. Basic Courses Department, Gansu Agricultural University, Lanzhou, Gansu 730070, Peoples Republic of China.

Yoshinaka, R., K. Sato, H. Anbe, M. Sato and Y. Shimizu. 1988. Distribution of collagen in body muscle of fishes with different swimming modes. *Comparative Biochemistry and Physiology*, **89B**: 147-151.

Young, O.A. and T.J. Braggins. 1993. Tenderness of ovine semimembranosus: is collagen concentration or solubility the critical factor? *Meat Science*, **35**: 213-222.

Zeder, M.A. and B. Hesse. 2000. The initial domestication of goats (Capra hircus) in the Zagros mountains 10,000 years ago. *Science*, **287:** 2254-2257.

Zeuner, F.E. 1963. *A History of Domesticated Animals.* Harper and Row, New York.

Zimmer, D.B. and M.A. Goldstein. 1987. Immunolocalization of alpha-actinin in adult chicken skeletal muscles. *Journal of Electron Microscopy Technique*, **6**: 357-366.

Zwanenburg, K. 1990. Atlantic halibut. *Underwater World*, 36. Fisheries and Oceans, Ottawa.